Proceedings of the 3rd RILEM Spring Convention and Conference (RSCC 2020)

RILEM BOOKSERIES

Volume 34

RILEM, The International Union of Laboratories and Experts in Construction Materials, Systems and Structures, founded in 1947, is a non-governmental scientific association whose goal is to contribute to progress in the construction sciences, techniques and industries, essentially by means of the communication it fosters between research and practice. RILEM's focus is on construction materials and their use in building and civil engineering structures, covering all phases of the building process from manufacture to use and recycling of materials. More information on RILEM and its previous publications can be found on www.RILEM.net.

Indexed in SCOPUS, Google Scholar and SpringerLink.

More information about this series at http://www.springer.com/series/8781

José Sena-Cruz · Luis Correia · Miguel Azenha
Editors

Proceedings of the 3rd RILEM Spring Convention and Conference (RSCC 2020)

Volume 3: Service Life Extension of Existing Structures

 Springer

Editors
José Sena-Cruz
ISISE, Civil Engineering
University of Minho
Guimarães, Portugal

Luis Correia
ISISE, Civil Engineering
University of Minho
Guimarães, Portugal

Miguel Azenha
ISISE, Civil Engineering
University of Minho
Guimarães, Portugal

ISSN 2211-0844   ISSN 2211-0852   (electronic)
RILEM Bookseries
ISBN 978-3-030-76467-8   ISBN 978-3-030-76465-4   (eBook)
https://doi.org/10.1007/978-3-030-76465-4

This Springer imprint is published by the registered company Springer Nature Switzerland AG
The registered company address is: Gewerbestrasse 11, 6330 Cham, Switzerland

Committee

9–14 March 2020—Guimarães, Portugal

Executive Chair
Eduardo B. Pereira, ISISE—University of Minho, Portugal

Vice-chair
Fábio Figueiredo

Honorary Chair
Joaquim A. O. Barros

Topic 3: Service Life Extension of Existing Structures

Topic Lead:

Nele de Belie, University of Gent, Belgium

Topic Scientific Committee:

Bahman Ghiassi, University of Nottingham, United Kingdom
Christoph Gehlen, Technische Universität München, Germany
Hans Beushausen, University of Cape Town, South Africa
Kei-Ichi Imamoto, Tokyo University of Science, Japan
Luping Tang, Chalmers University of Technology, Sweden
Markus Krüger, Technical University of Graz, Austria
Mustafa Sahmaran, Hacettepe University, Turkey
Ravi Patel, Paul Scherrer Institute PSI, Switzerland
Roberto Torrent, Materials Advanced Services Ltd, Argentina

Preface

This volume gathers the proceedings of the 3rd Rilem Spring Convention 2020 on the topic "Service life extension of existing structures," held in Guimarães, Portugal in March 2020. The papers cover the most recent scientific and technological developments in the understanding of the evolution and degradation of construction materials and structural systems. Analytical and numerical, as well as experimental, approaches aimed at characterizing, modeling, and predicting the evolution of the physical, chemical, and mechanical properties of construction materials and structural systems are addressed. Multiphysics models are also considered, as well as other strategies that contribute for an accurate characterization and prediction of the service life of existing and novel construction materials under normal or extreme environmental exposure or loading conditions. New strategies to promote the smart repairing or the recovery of material properties, as well as the service life extension, are also considered. The following subtopics are included in this volume: (i) Service life models and multiphysics approaches; (ii) Smart structures: innovative monitoring and intervention strategies; (iii) Management and optimized maintenance strategies; and (iv) Integrated rehabilitation and strengthening approaches.

The editors of the topic "Service life extension of existing structures" appreciate all the efforts made by those who submitted and presented the high-quality papers included in this volume, under very difficult and challenging conditions. Together with the other volumes from the RSC2020 conference, the present book will form a valuable base for discussion and suggestions for future development and research.

Guimarães, Portugal

José Sena-Cruz
Luis Correia
Miguel Azenha

Contents

RILEM Publications

The following list is presenting the global offer of RILEM Publications, sorted by series. Each publication is available in printed version and/or in online version.

RILEM Proceedings (PRO)

PRO 1: Durability of High Performance Concrete (ISBN: 2-912143-03-9; e-ISBN: 2-351580-12-5; e-ISBN: 2351580125); *Ed. H. Sommer*

PRO 2: Chloride Penetration into Concrete (ISBN: 2-912143-00-04; e-ISBN: 2912143454); *Eds. L.-O. Nilsson and J.-P. Ollivier*

PRO 3: Evaluation and Strengthening of Existing Masonry Structures (ISBN: 2-912143-02-0; e-ISBN: 2351580141); *Eds. L. Binda and C. Modena*

PRO 4: Concrete: From Material to Structure (ISBN: 2-912143-04-7; e-ISBN: 2351580206); *Eds. J.-P. Bournazel and Y. Malier*

PRO 5: The Role of Admixtures in High Performance Concrete (ISBN: 2-912143-05-5; e-ISBN: 2351580214); *Eds. J. G. Cabrera and R. Rivera-Villarreal*

PRO 6: High Performance Fiber Reinforced Cement Composites—HPFRCC 3 (ISBN: 2-912143-06-3; e-ISBN: 2351580222); *Eds. H. W. Reinhardt and A. E. Naaman*

PRO 7: 1st International RILEM Symposium on Self-Compacting Concrete (ISBN: 2-912143-09-8; e-ISBN: 2912143721); *Eds. Å. Skarendahl and Ö. Petersson*

PRO 8: International RILEM Symposium on Timber Engineering (ISBN: 2-912143-10-1; e-ISBN: 2351580230); *Ed. L. Boström*

PRO 9: 2nd International RILEM Symposium on Adhesion between Polymers and Concrete ISAP '99 (ISBN: 2-912143-11-X; e-ISBN: 2351580249); *Eds. Y. Ohama and M. Puterman*

PRO 10: 3rd International RILEM Symposium on Durability of Building and Construction Sealants (ISBN: 2-912143-13-6; e-ISBN: 2351580257); *Ed. A. T. Wolf*

PRO 11: 4th International RILEM Conference on Reflective Cracking in Pavements (ISBN: 2-912143-14-4; e-ISBN: 2351580265); *Eds. A. O. Abd El Halim, D. A. Taylor and El H. H. Mohamed*

PRO 12: International RILEM Workshop on Historic Mortars: Characteristics and Tests (ISBN: 2-912143-15-2; e-ISBN: 2351580273); *Eds. P. Bartos, C. Groot and J. J. Hughes*

PRO 13: 2nd International RILEM Symposium on Hydration and Setting (ISBN: 2-912143-16-0; e-ISBN: 2351580281); *Ed. A. Nonat*

PRO 14: Integrated Life-Cycle Design of Materials and Structures—ILCDES 2000 (ISBN: 951-758-408-3; e-ISBN: 235158029X); (ISSN: 0356-9403); *Ed. S. Sarja*

PRO 15: Fifth RILEM Symposium on Fibre-Reinforced Concretes (FRC)—BEFIB'2000 (ISBN: 2-912143-18-7; e-ISBN: 291214373X); *Eds. P. Rossi and G. Chanvillard*

PRO 16: Life Prediction and Management of Concrete Structures (ISBN: 2-912143-19-5; e-ISBN: 2351580303); *Ed. D. Naus*

PRO 17: Shrinkage of Concrete—Shrinkage 2000 (ISBN: 2-912143-20-9; e-ISBN: 2351580311); *Eds. V. Baroghel-Bouny and P.-C. Aïtcin*

PRO 18: Measurement and Interpretation of the On-Site Corrosion Rate (ISBN: 2-912143-21-7; e-ISBN: 235158032X); *Eds. C. Andrade, C. Alonso, J. Fullea, J. Polimon and J. Rodriguez*

PRO 19: Testing and Modelling the Chloride Ingress into Concrete (ISBN: 2-912143-22-5; e-ISBN: 2351580338); *Eds. C. Andrade and J. Kropp*

PRO 20: 1st International RILEM Workshop on Microbial Impacts on Building Materials (CD 02) (e-ISBN 978-2-35158-013-4); *Ed. M. Ribas Silva*

PRO 21: International RILEM Symposium on Connections between Steel and Concrete (ISBN: 2-912143-25-X; e-ISBN: 2351580346); *Ed. R. Eligehausen*

PRO 22: International RILEM Symposium on Joints in Timber Structures (ISBN: 2-912143-28-4; e-ISBN: 2351580354); *Eds. S. Aicher and H.-W. Reinhardt*

PRO 23: International RILEM Conference on Early Age Cracking in Cementitious Systems (ISBN: 2-912143-29-2; e-ISBN: 2351580362); *Eds. K. Kovler and A. Bentur*

PRO 24: 2nd International RILEM Workshop on Frost Resistance of Concrete (ISBN: 2-912143-30-6; e-ISBN: 2351580370); *Eds. M. J. Setzer, R. Auberg and H.-J. Keck*

PRO 25: International RILEM Workshop on Frost Damage in Concrete (ISBN: 2-912143-31-4; e-ISBN: 2351580389); *Eds. D. J. Janssen, M. J. Setzer and M. B. Snyder*

PRO 26: International RILEM Workshop on On-Site Control and Evaluation of Masonry Structures (ISBN: 2-912143-34-9; e-ISBN: 2351580141); *Eds. L. Binda and R. C. de Vekey*

PRO 27: International RILEM Symposium on Building Joint Sealants (CD03; e-ISBN: 235158015X); *Ed. A. T. Wolf*

PRO 28: 6th International RILEM Symposium on Performance Testing and Evaluation of Bituminous Materials—PTEBM'03 (ISBN: 2-912143-35-7; e-ISBN: 978-2-912143-77-8); *Ed. M. N. Partl*

PRO 29: 2nd International RILEM Workshop on Life Prediction and Ageing Management of Concrete Structures (ISBN: 2-912143-36-5; e-ISBN: 2912143780); *Ed. D. J. Naus*

PRO 30: 4th International RILEM Workshop on High Performance Fiber Reinforced Cement Composites—HPFRCC 4 (ISBN: 2-912143-37-3; e-ISBN: 2912143799); *Eds. A. E. Naaman and H. W. Reinhardt*

PRO 31: International RILEM Workshop on Test and Design Methods for Steel Fibre Reinforced Concrete: Background and Experiences (ISBN: 2-912143-38-1; e-ISBN: 2351580168); *Eds. B. Schnütgen and L. Vandewalle*

PRO 32: International Conference on Advances in Concrete and Structures 2 vol. (ISBN (set): 2-912143-41-1; e-ISBN: 2351580176); *Eds. Ying-shu Yuan, Surendra P. Shah and Heng-lin Lü*

PRO 33: 3rd International Symposium on Self-Compacting Concrete (ISBN: 2-912143-42-X; e-ISBN: 2912143713); *Eds. Ó. Wallevik and I. Níelsson*

PRO 34: International RILEM Conference on Microbial Impact on Building Materials (ISBN: 2-912143-43-8; e-ISBN: 2351580184); *Ed. M. Ribas Silva*

PRO 35: International RILEM TC 186-ISA on Internal Sulfate Attack and Delayed Ettringite Formation (ISBN: 2-912143-44-6; e-ISBN: 2912143802); *Eds. K. Scrivener and J. Skalny*

PRO 36: International RILEM Symposium on Concrete Science and Engineering —A Tribute to Arnon Bentur (ISBN: 2-912143-46-2; e-ISBN: 2912143586); *Eds. K. Kovler, J. Marchand, S. Mindess and J. Weiss*

PRO 37: 5th International RILEM Conference on Cracking in Pavements—Mitigation, Risk Assessment and Prevention (ISBN: 2-912143-47-0; e-ISBN: 2912143764); *Eds. C. Petit, I. Al-Qadi and A. Millien*

PRO 38: 3rd International RILEM Workshop on Testing and Modelling the Chloride Ingress into Concrete (ISBN: 2-912143-48-9; e-ISBN: 2912143578); *Eds. C. Andrade and J. Kropp*

PRO 39: 6th International RILEM Symposium on Fibre-Reinforced Concretes—BEFIB 2004 (ISBN: 2-912143-51-9; e-ISBN: 2912143748); *Eds. M. Di Prisco, R. Felicetti and G. A. Plizzari*

PRO 40: International RILEM Conference on the Use of Recycled Materials in Buildings and Structures (ISBN: 2-912143-52-7; e-ISBN: 2912143756); *Eds. E. Vázquez, Ch. F. Hendriks and G. M. T. Janssen*

PRO 41: RILEM International Symposium on Environment-Conscious Materials and Systems for Sustainable Development (ISBN: 2-912143-55-1; e-ISBN: 2912143640); *Eds. N. Kashino and Y. Ohama*

PRO 42: SCC'2005—China: 1st International Symposium on Design, Performance and Use of Self-Consolidating Concrete (ISBN: 2-912143-61-6; e-ISBN: 2912143624); *Eds. Zhiwu Yu, Caijun Shi, Kamal Henri Khayat and Youjun Xie*

PRO 43: International RILEM Workshop on Bonded Concrete Overlays (e-ISBN: 2-912143-83-7); *Eds. J. L. Granju and J. Silfwerbrand*

PRO 44: 2nd International RILEM Workshop on Microbial Impacts on Building Materials (CD11) (e-ISBN: 2-912143-84-5); *Ed. M. Ribas Silva*

PRO 45: 2nd International Symposium on Nanotechnology in Construction, Bilbao (ISBN: 2-912143-87-X; e-ISBN: 2912143888); *Eds. Peter J. M. Bartos, Yolanda de Miguel and Antonio Porro*

PRO 46: Concrete Life'06—International RILEM-JCI Seminar on Concrete Durability and Service Life Planning: Curing, Crack Control, Performance in Harsh Environments (ISBN: 2-912143-89-6; e-ISBN: 291214390X); *Ed. K. Kovler*

PRO 47: International RILEM Workshop on Performance Based Evaluation and Indicators for Concrete Durability (ISBN: 978-2-912143-95-2; e-ISBN: 9782912143969); *Eds. V. Baroghel-Bouny, C. Andrade, R. Torrent and K. Scrivener*

PRO 48: 1st International RILEM Symposium on Advances in Concrete through Science and Engineering (e-ISBN: 2-912143-92-6); *Eds. J. Weiss, K. Kovler, J. Marchand, and S. Mindess*

PRO 49: International RILEM Workshop on High Performance Fiber Reinforced Cementitious Composites in Structural Applications (ISBN: 2-912143-93-4; e-ISBN: 2912143942); *Eds. G. Fischer and V. C. Li*

PRO 50: 1st International RILEM Symposium on Textile Reinforced Concrete (ISBN: 2-912143-97-7; e-ISBN: 2351580087); *Eds. Josef Hegger, Wolfgang Brameshuber and Norbert Will*

PRO 51: 2nd International Symposium on Advances in Concrete through Science and Engineering (ISBN: 2-35158-003-6; e-ISBN: 2-35158-002-8); *Eds. J. Marchand, B. Bissonnette, R. Gagné, M. Jolin and F. Paradis*

PRO 52: Volume Changes of Hardening Concrete: Testing and Mitigation (ISBN: 2-35158-004-4; e-ISBN: 2-35158-005-2); *Eds. O. M. Jensen, P. Lura and K. Kovler*

PRO 53: High Performance Fiber Reinforced Cement Composites—HPFRCC5 (ISBN: 978-2-35158-046-2; e-ISBN: 978-2-35158-089-9); *Eds. H. W. Reinhardt and A. E. Naaman*

PRO 54: 5th International RILEM Symposium on Self-Compacting Concrete (ISBN: 978-2-35158-047-9; e-ISBN: 978-2-35158-088-2); *Eds. G. De Schutter and V. Boel*

PRO 55: International RILEM Symposium Photocatalysis, Environment and Construction Materials (ISBN: 978-2-35158-056-1; e-ISBN: 978-2-35158-057-8); *Eds. P. Baglioni and L. Cassar*

PRO 56: International RILEM Workshop on Integral Service Life Modelling of Concrete Structures (ISBN 978-2-35158-058-5; e-ISBN: 978-2-35158-090-5); *Eds. R. M. Ferreira, J. Gulikers and C. Andrade*

PRO 57: RILEM Workshop on Performance of cement-based materials in aggressive aqueous environments (e-ISBN: 978-2-35158-059-2); *Ed. N. De Belie*

PRO 58: International RILEM Symposium on Concrete Modelling—CONMOD'08 (ISBN: 978-2-35158-060-8; e-ISBN: 978-2-35158-076-9); *Eds. E. Schlangen and G. De Schutter*

PRO 59: International RILEM Conference on On Site Assessment of Concrete, Masonry and Timber Structures—SACoMaTiS 2008 (ISBN set: 978-2-35158-061-5; e-ISBN: 978-2-35158-075-2); *Eds. L. Binda, M. di Prisco and R. Felicetti*

PRO 60: Seventh RILEM International Symposium on Fibre Reinforced Concrete: Design and Applications—BEFIB 2008 (ISBN: 978-2-35158-064-6; e-ISBN: 978-2-35158-086-8); *Ed. R. Gettu*

PRO 61: 1st International Conference on Microstructure Related Durability of Cementitious Composites 2 vol., (ISBN: 978-2-35158-065-3; e-ISBN: 978-2-35158-084-4); *Eds. W. Sun, K. van Breugel, C. Miao, G. Ye and H. Chen*

PRO 62: NSF/ RILEM Workshop: In-situ Evaluation of Historic Wood and Masonry Structures (e-ISBN: 978-2-35158-068-4); *Eds. B. Kasal, R. Anthony and M. Drdácký*

PRO 63: Concrete in Aggressive Aqueous Environments: Performance, Testing and Modelling, 2 vol., (ISBN: 978-2-35158-071-4; e-ISBN: 978-2-35158-082-0); *Eds. M. G. Alexander and A. Bertron*

PRO 64: Long Term Performance of Cementitious Barriers and Reinforced Concrete in Nuclear Power Plants and Waste Management—NUCPERF 2009 (ISBN: 978-2-35158-072-1; e-ISBN: 978-2-35158-087-5); *Eds. V. L'Hostis, R. Gens and C. Gallé*

PRO 65: Design Performance and Use of Self-consolidating Concrete—SCC'2009 (ISBN: 978-2-35158-073-8; e-ISBN: 978-2-35158-093-6); *Eds. C. Shi, Z. Yu, K. H. Khayat and P. Yan*

PRO 66: 2nd International RILEM Workshop on Concrete Durability and Service Life Planning—ConcreteLife'09 (ISBN: 978-2-35158-074-5; ISBN: 978-2-35158-074-5); *Ed. K. Kovler*

PRO 67: Repairs Mortars for Historic Masonry (e-ISBN: 978-2-35158-083-7); *Ed. C. Groot*

PRO 68: Proceedings of the 3rd International RILEM Symposium on 'Rheology of Cement Suspensions such as Fresh Concrete (ISBN 978-2-35158-091-2; e-ISBN: 978-2-35158-092-9); *Eds. O. H. Wallevik, S. Kubens and S. Oesterheld*

PRO 69: 3rd International PhD Student Workshop on 'Modelling the Durability of Reinforced Concrete (ISBN: 978-2-35158-095-0); *Eds. R. M. Ferreira, J. Gulikers and C. Andrade*

PRO 70: 2nd International Conference on 'Service Life Design for Infrastructure' (ISBN set: 978-2-35158-096-7, e-ISBN: 978-2-35158-097-4); *Eds. K. van Breugel, G. Ye and Y. Yuan*

PRO 71: Advances in Civil Engineering Materials—The 50-year Teaching Anniversary of Prof. Sun Wei' (ISBN: 978-2-35158-098-1; e-ISBN: 978-2-35158-099-8); *Eds. C. Miao, G. Ye and H. Chen*

PRO 72: First International Conference on 'Advances in Chemically-Activated Materials—CAM'2010' (2010), 264 pp., ISBN: 978-2-35158-101-8; e-ISBN: 978-2-35158-115-5; *Eds. Caijun Shi and Xiaodong Shen*

PRO 73: 2nd International Conference on 'Waste Engineering and Management—ICWEM 2010' (2010), 894 pp., ISBN: 978-2-35158-102-5; e-ISBN: 978-2-35158-103-2, *Eds. J. Zh. Xiao, Y. Zhang, M. S. Cheung and R. Chu*

PRO 74: International RILEM Conference on 'Use of Superabsorsorbent Polymers and Other New Addditives in Concrete' (2010) 374 pp., ISBN: 978-2-35158-104-9; e-ISBN: 978-2-35158-105-6; *Eds. O.M. Jensen, M.T. Hasholt, and S. Laustsen*

PRO 75: International Conference on 'Material Science—2nd ICTRC—Textile Reinforced Concrete—Theme 1' (2010) 436 pp., ISBN: 978-2-35158-106-3; e-ISBN: 978-2-35158-107-0; *Ed. W. Brameshuber*

PRO 76: International Conference on 'Material Science—HetMat—Modelling of Heterogeneous Materials—Theme 2' (2010) 255 pp., ISBN: 978-2-35158-108-7; e-ISBN: 978-2-35158-109-4; *Ed. W. Brameshuber*

PRO 77: International Conference on 'Material Science—AdIPoC—Additions Improving Properties of Concrete—Theme 3' (2010) 459 pp., ISBN: 978-2-35158-110-0; e-ISBN: 978-2-35158-111-7; *Ed. W. Brameshuber*

PRO 78: 2nd Historic Mortars Conference and RILEM TC 203-RHM Final Workshop—HMC2010 (2010) 1416 pp., e-ISBN: 978-2-35158-112-4; *Eds. J. Válek, C. Groot and J. J. Hughes*

PRO 79: International RILEM Conference on Advances in Construction Materials Through Science and Engineering (2011) 213 pp., ISBN: 978-2-35158-116-2, e-ISBN: 978-2-35158-117-9; *Eds. Christopher Leung and K.T. Wan*

PRO 80: 2nd International RILEM Conference on Concrete Spalling due to Fire Exposure (2011) 453 pp., ISBN: 978-2-35158-118-6; e-ISBN: 978-2-35158-119-3; *Eds. E.A.B. Koenders and F. Dehn*

PRO 81: 2nd International RILEM Conference on Strain Hardening Cementitious Composites (SHCC2-Rio) (2011) 451 pp., ISBN: 978-2-35158-120-9; e-ISBN: 978-2-35158-121-6; *Eds. R.D. Toledo Filho, F.A. Silva, E.A.B. Koenders and E.M. R. Fairbairn*

PRO 82: 2nd International RILEM Conference on Progress of Recycling in the Built Environment (2011) 507 pp., e-ISBN: 978-2-35158-122-3; *Eds. V.M. John, E. Vazquez, S.C. Angulo and C. Ulsen*

PRO 83: 2nd International Conference on Microstructural-related Durability of Cementitious Composites (2012) 250 pp., ISBN: 978-2-35158-129-2; e-ISBN: 978-2-35158-123-0; *Eds. G. Ye, K. van Breugel, W. Sun and C. Miao*

PRO 84: CONSEC13—Seventh International Conference on Concrete under Severe Conditions—Environment and Loading (2013) 1930 pp., ISBN: 978-2-35158-124-7; e-ISBN: 978-2- 35158-134-6; *Eds. Z.J. Li, W. Sun, C.W. Miao, K. Sakai, O.E. Gjorv and N. Banthia*

PRO 85: RILEM-JCI International Workshop on Crack Control of Mass Concrete and Related issues concerning Early-Age of Concrete Structures—ConCrack 3—Control of Cracking in Concrete Structures 3 (2012) 237 pp., ISBN: 978-2-35158-125-4; e-ISBN: 978-2-35158-126-1; *Eds. F. Toutlemonde and J.-M. Torrenti*

PRO 86: International Symposium on Life Cycle Assessment and Construction (2012) 414 pp., ISBN: 978-2-35158-127-8, e-ISBN: 978-2-35158-128-5; *Eds. A. Ventura and C. de la Roche*

PRO 87: UHPFRC 2013—RILEM-fib-AFGC International Symposium on Ultra-High Performance Fibre-Reinforced Concrete (2013), ISBN: 978-2-35158-130-8, e-ISBN: 978-2-35158-131-5; *Eds. F. Toutlemonde*

PRO 88: 8th RILEM International Symposium on Fibre Reinforced Concrete (2012) 344 pp., ISBN: 978-2-35158-132-2; e-ISBN: 978-2-35158-133-9; *Eds. Joaquim A.O. Barros*

PRO 89: RILEM International workshop on performance-based specification and control of concrete durability (2014) 678 pp., ISBN: 978-2-35158-135-3; e-ISBN: 978-2-35158-136-0; *Eds. D. Bjegović, H. Beushausen and M. Serdar*

PRO 90: 7th RILEM International Conference on Self-Compacting Concrete and of the 1st RILEM International Conference on Rheology and Processing of Construction Materials (2013) 396 pp., ISBN: 978-2-35158-137-7; e-ISBN: 978-2-35158-138-4; *Eds. Nicolas Roussel and Hela Bessaies-Bey*

PRO 91: CONMOD 2014—RILEM International Symposium on Concrete Modelling (2014), ISBN: 978-2-35158-139-1; e-ISBN: 978-2-35158-140-7; *Eds. Kefei Li, Peiyu Yan and Rongwei Yang*

PRO 92: CAM 2014—2nd International Conference on advances in chemically-activated materials (2014) 392 pp., ISBN: 978-2-35158-141-4; e-ISBN: 978-2-35158-142-1; *Eds. Caijun Shi and Xiadong Shen*

PRO 93: SCC 2014—3rd International Symposium on Design, Performance and Use of Self-Consolidating Concrete (2014) 438 pp., ISBN: 978-2-35158-143-8; e-ISBN: 978-2-35158-144-5; *Eds. Caijun Shi, Zhihua Ou and Kamal H. Khayat*

PRO 94 (online version): HPFRCC-7—7th RILEM conference on High performance fiber reinforced cement composites (2015), e-ISBN: 978-2-35158-146-9; *Eds. H.W. Reinhardt, G.J. Parra-Montesinos and H. Garrecht*

PRO 95: International RILEM Conference on Application of superabsorbent polymers and other new admixtures in concrete construction (2014), ISBN: 978-2-35158-147-6; e-ISBN: 978-2-35158-148-3; *Eds. Viktor Mechtcherine and Christof Schroefl*

PRO 96 (online version): XIII DBMC: XIII International Conference on Durability of Building Materials and Components (2015), e-ISBN: 978-2-35158-149-0; *Eds. M. Quattrone and V.M. John*

PRO 97: SHCC3—3rd International RILEM Conference on Strain Hardening Cementitious Composites (2014), ISBN: 978-2-35158-150-6; e-ISBN: 978-2-35158-151-3; *Eds. E. Schlangen, M.G. Sierra Beltran, M. Lukovic and G. Ye*

PRO 98: FERRO-11—11th International Symposium on Ferrocement and 3rd ICTRC—International Conference on Textile Reinforced Concrete (2015), ISBN: 978-2-35158-152-0; e-ISBN: 978-2-35158-153-7; *Ed. W. Brameshuber*

PRO 99 (online version): ICBBM 2015—1st International Conference on Bio-Based Building Materials (2015), e-ISBN: 978-2-35158-154-4; *Eds. S. Amziane and M. Sonebi*

PRO 100: SCC16—RILEM Self-Consolidating Concrete Conference (2016), ISBN: 978-2-35158-156-8; e-ISBN: 978-2-35158-157-5; *Ed. Kamal H. Kayat*

PRO 101 (online version): III Progress of Recycling in the Built Environment (2015), e-ISBN: 978-2-35158-158-2; *Eds I. Martins, C. Ulsen and S. C. Angulo*

PRO 102 (online version): RILEM Conference on Microorganisms-Cementitious Materials Interactions (2016), e-ISBN: 978-2-35158-160-5; *Eds. Alexandra Bertron, Henk Jonkers and Virginie Wiktor*

PRO 103 (online version): ACESC'16—Advances in Civil Engineering and Sustainable Construction (2016), e-ISBN: 978-2-35158-161-2; *Eds. T.Ch. Madhavi, G. Prabhakar, Santhosh Ram and P.M. Rameshwaran*

PRO 104 (online version): SSCS'2015—Numerical Modeling—Strategies for Sustainable Concrete Structures (2015), e-ISBN: 978-2-35158-162-9

PRO 105: 1st International Conference on UHPC Materials and Structures (2016), ISBN: 978-2-35158-164-3; e-ISBN: 978-2-35158-165-0

PRO 106: AFGC-ACI-fib-RILEM International Conference on Ultra-High-Performance Fibre-Reinforced Concrete—UHPFRC 2017 (2017), ISBN: 978-2-35158-166-7; e-ISBN: 978-2-35158-167-4; *Eds. François Toutlemonde and Jacques Resplendino*

PRO 107 (online version): XIV DBMC—14th International Conference on Durability of Building Materials and Components (2017), e-ISBN: 978-2-35158-159-9; *Eds. Geert De Schutter, Nele De Belie, Arnold Janssens and Nathan Van Den Bossche*

PRO 108: MSSCE 2016—Innovation of Teaching in Materials and Structures (2016), ISBN: 978-2-35158-178-0; e-ISBN: 978-2-35158-179-7; *Ed. Per Goltermann*

PRO 109 (2 volumes): MSSCE 2016—Service Life of Cement-Based Materials and Structures (2016), ISBN Vol. 1: 978-2-35158-170-4; Vol. 2: 978-2-35158-171-4; Set Vol. 1&2: 978-2-35158-172-8; e-ISBN : 978-2-35158-173-5; *Eds. Miguel Azenha, Ivan Gabrijel, Dirk Schlicke, Terje Kanstad and Ole Mejlhede Jensen*

PRO 110: MSSCE 2016—Historical Masonry (2016), ISBN: 978-2-35158-178-0; e-ISBN: 978-2-35158-179-7; *Eds. Inge Rörig-Dalgaard and Ioannis Ioannou*

PRO 111: MSSCE 2016—Electrochemistry in Civil Engineering (2016); ISBN: 978-2-35158-176-6; e-ISBN: 978-2-35158-177-3; *Ed. Lisbeth M. Ottosen*

PRO 112: MSSCE 2016—Moisture in Materials and Structures (2016), ISBN: 978-2-35158-178-0; e-ISBN: 978-2-35158-179-7; *Eds. Kurt Kielsgaard Hansen, Carsten Rode and Lars-Olof Nilsson*

PRO 113: MSSCE 2016—Concrete with Supplementary Cementitious Materials (2016), ISBN: 978-2-35158-178-0; e-ISBN: 978-2-35158-179-7; *Eds. Ole Mejlhede Jensen, Konstantin Kovler and Nele De Belie*

PRO 114: MSSCE 2016—Frost Action in Concrete (2016), ISBN: 978-2-35158-182-7; e-ISBN: 978-2-35158-183-4; *Eds. Marianne Tange Hasholt, Katja Fridh and R. Doug Hooton*

PRO 115: MSSCE 2016—Fresh Concrete (2016), ISBN: 978-2-35158-184-1; e-ISBN: 978-2-35158-185-8; *Eds. Lars N. Thrane, Claus Pade, Oldrich Svec and Nicolas Roussel*

PRO 116: BEFIB 2016—9th RILEM International Symposium on Fiber Reinforced Concrete (2016), ISBN: 978-2-35158-187-2; e-ISBN: 978-2-35158-186-5; *Eds. N. Banthia, M. di Prisco and S. Soleimani-Dashtaki*

PRO 117: 3rd International RILEM Conference on Microstructure Related Durability of Cementitious Composites (2016), ISBN: 978-2-35158-188-9; e-ISBN: 978-2-35158-189-6; *Eds. Changwen Miao, Wei Sun, Jiaping Liu, Huisu Chen, Guang Ye and Klaas van Breugel*

PRO 118 (4 volumes): International Conference on Advances in Construction Materials and Systems (2017), ISBN Set: 978-2-35158-190-2; Vol. 1: 978-2-35158-193-3; Vol. 2: 978-2-35158-194-0; Vol. 3: ISBN:978-2-35158-195-7; Vol. 4: ISBN:978-2-35158-196-4; e-ISBN: 978-2-35158-191-9; *Ed. Manu Santhanam*

PRO 119 (online version): ICBBM 2017—Second International RILEM Conference on Bio-based Building Materials, (2017), e-ISBN: 978-2-35158-192-6; *Ed. Sofiane Amziane*

PRO 120 (2 volumes): EAC-02—2nd International RILEM/COST Conference on Early Age Cracking and Serviceability in Cement-based Materials and Structures, (2017), Vol. 1: 978-2-35158-199-5, Vol. 2: 978-2-35158-200-8, Set: 978-2-35158-197-1, e-ISBN: 978-2-35158-198-8; *Eds. Stéphanie Staquet and Dimitrios Aggelis*

PRO 121 (2 volumes): SynerCrete18: Interdisciplinary Approaches for Cementbased Materials and Structural Concrete: Synergizing Expertise and Bridging Scales of Space and Time, (2018), Set: 978-2-35158-202-2, Vol.1: 978-2-35158-211-4, Vol.2: 978-2-35158-212-1, e-ISBN: 978-2-35158-203-9; *Eds. Miguel Azenha, Dirk Schlicke, Farid Benboudjema, Agnieszka Knoppik*

PRO 122: SCC'2018 China—Fourth International Symposium on Design, Performance and Use of Self-Consolidating Concrete, (2018), ISBN:

978-2-35158-204-6, e-ISBN: 978-2-35158-205-3; *Eds. C. Shi, Z. Zhang, K. H. Khayat*

PRO 123: Final Conference of RILEM TC 253-MCI: Microorganisms-Cementitious Materials Interactions (2018), Set: 978-2-35158-207-7, Vol.1: 978-2-35158-209-1, Vol.2: 978-2-35158-210-7, e-ISBN: 978-2-35158-206-0; *Ed. Alexandra Bertron*

PRO 124 (online version): Fourth International Conference Progress of Recycling in the Built Environment (2018), e-ISBN: 978-2-35158-208-4; *Eds. Isabel M. Martins, Carina Ulsen, Yury Villagran*

PRO 125 (online version): SLD4—4th International Conference on Service Life Design for Infrastructures (2018), e-ISBN: 978-2-35158-213-8; *Eds. Guang Ye, Yong Yuan, Claudia Romero Rodriguez, Hongzhi Zhang, Branko Savija*

PRO 126: Workshop on Concrete Modelling and Material Behaviour in honor of Professor Klaas van Breugel (2018), ISBN: 978-2-35158-214-5, e-ISBN: 978-2-35158-215-2; *Ed. Guang Ye*

PRO 127 (online version): CONMOD2018—Symposium on Concrete Modelling (2018), e-ISBN: 978-2-35158-216-9; *Eds. Erik Schlangen, Geert de Schutter, Branko Savija, Hongzhi Zhang, Claudia Romero Rodriguez*

PRO 128: SMSS2019—International Conference on Sustainable Materials, Systems and Structures (2019), ISBN: 978-2-35158-217-6, e-ISBN: 978-2-35158-218-3

PRO 129: 2nd International Conference on UHPC Materials and Structures (UHPC2018-China), ISBN: 978-2-35158-219-0, e-ISBN: 978-2-35158-220-6

PRO 130: 5th Historic Mortars Conference (2019), ISBN: 978-2-35158-221-3, e-ISBN: 978-2-35158-222-0; *Eds. José Ignacio Álvarez, José María Fernández, Íñigo Navarro, Adrián Durán, Rafael Sirera*

PRO 131 (online version): 3rd International Conference on Bio-Based Building Materials (ICBBM2019), e-ISBN: 978-2-35158-229-9; *Eds. Mohammed Sonebi, Sofiane Amziane, Jonathan Page*

PRO 132: IRWRMC'18—International RILEM Workshop on Rheological Measurements of Cement-based Materials (2018), ISBN: 978-2-35158-230-5, e-ISBN: 978-2-35158-231-2; *Eds. Chafika Djelal, Yannick Vanhove*

PRO 133 (online version): CO2STO2019—International Workshop CO2 Storage in Concrete (2019), e-ISBN: 978-2-35158-232-9; *Eds. Assia Djerbi, Othman Omikrine-Metalssi, Teddy Fen-Chong*

PRO 134: 3rd ACF/HNU International Conference on UHPC Materials and Structures—UHPC'2020, ISBN: 978-2-35158-233-6, e-ISBN: 978-2-35158-234-3; *Eds. Caijun Shi and Jiaping Liu*

RILEM Reports (REP)

Report 19: Considerations for Use in Managing the Aging of Nuclear Power Plant Concrete Structures (ISBN: 2-912143-07-1); *Ed. D. J. Naus*

Report 20: Engineering and Transport Properties of the Interfacial Transition Zone in Cementitious Composites (ISBN: 2-912143-08-X); *Eds. M. G. Alexander, G. Arliguie, G. Ballivy, A. Bentur and J. Marchand*

Report 21: Durability of Building Sealants (ISBN: 2-912143-12-8); *Ed. A. T. Wolf*

Report 22: Sustainable Raw Materials—Construction and Demolition Waste (ISBN: 2-912143-17-9); *Eds. C. F. Hendriks and H. S. Pietersen*

Report 23: Self-Compacting Concrete state-of-the-art report (ISBN: 2-912143-23-3); *Eds. Å. Skarendahl and Ö. Petersson*

Report 24: Workability and Rheology of Fresh Concrete: Compendium of Tests (ISBN: 2-912143-32-2); *Eds. P. J. M. Bartos, M. Sonebi and A. K. Tamimi*

Report 25: Early Age Cracking in Cementitious Systems (ISBN: 2-912143-33-0); *Ed. A. Bentur*

Report 26: Towards Sustainable Roofing (Joint Committee CIB/RILEM) (CD 07) (e-ISBN 978-2-912143-65-5); *Eds. Thomas W. Hutchinson and Keith Roberts*

Report 27: Condition Assessment of Roofs (Joint Committee CIB/RILEM) (CD 08) (e-ISBN 978-2-912143-66-2); *Ed. CIB W 83/RILEM TC166-RMS*

Report 28: Final report of RILEM TC 167-COM 'Characterisation of Old Mortars with Respect to Their Repair (ISBN: 978-2-912143-56-3); *Eds. C. Groot, G. Ashall and J. Hughes*

Report 29: Pavement Performance Prediction and Evaluation (PPPE): Interlaboratory Tests (e-ISBN: 2-912143-68-3); *Eds. M. Partl and H. Piber*

Report 30: Final Report of RILEM TC 198-URM 'Use of Recycled Materials' (ISBN: 2-912143-82-9; e-ISBN: 2-912143-69-1); *Eds. Ch. F. Hendriks, G. M. T. Janssen and E. Vázquez*

Report 31: Final Report of RILEM TC 185-ATC 'Advanced testing of cement-based materials during setting and hardening' (ISBN: 2-912143-81-0; e-ISBN: 2-912143-70-5); *Eds. H. W. Reinhardt and C. U. Grosse*

Report 32: Probabilistic Assessment of Existing Structures. A JCSS publication (ISBN 2-912143-24-1); *Ed. D. Diamantidis*

Report 33: State-of-the-Art Report of RILEM Technical Committee TC 184-IFE 'Industrial Floors' (ISBN 2-35158-006-0); *Ed. P. Seidler*

Report 34: Report of RILEM Technical Committee TC 147-FMB 'Fracture mechanics applications to anchorage and bond' Tension of Reinforced Concrete Prisms—Round Robin Analysis and Tests on Bond (e-ISBN 2-912143-91-8); *Eds. L. Elfgren and K. Noghabai*

Report 35: Final Report of RILEM Technical Committee TC 188-CSC 'Casting of Self Compacting Concrete' (ISBN 2-35158-001-X; e-ISBN: 2-912143-98-5); *Eds. Å. Skarendahl and P. Billberg*

Report 36: State-of-the-Art Report of RILEM Technical Committee TC 201-TRC 'Textile Reinforced Concrete' (ISBN 2-912143-99-3); *Ed. W. Brameshuber*

Report 37: State-of-the-Art Report of RILEM Technical Committee TC 192-ECM 'Environment-conscious construction materials and systems' (ISBN: 978-2-35158-053-0); *Eds. N. Kashino, D. Van Gemert and K. Imamoto*

Report 38: State-of-the-Art Report of RILEM Technical Committee TC 205-DSC 'Durability of Self-Compacting Concrete' (ISBN: 978-2-35158-048-6); *Eds. G. De Schutter and K. Audenaert*

Report 39: Final Report of RILEM Technical Committee TC 187-SOC 'Experimental determination of the stress-crack opening curve for concrete in tension' (ISBN 978-2-35158-049-3); *Ed. J. Planas*

Report 40: State-of-the-Art Report of RILEM Technical Committee TC 189-NEC 'Non-Destructive Evaluation of the Penetrability and Thickness of the Concrete Cover' (ISBN 978-2-35158-054-7); *Eds. R. Torrent and L. Fernández Luco*

Report 41: State-of-the-Art Report of RILEM Technical Committee TC 196-ICC 'Internal Curing of Concrete' (ISBN 978-2-35158-009-7); *Eds. K. Kovler and O. M. Jensen*

Report 42: 'Acoustic Emission and Related Non-destructive Evaluation Techniques for Crack Detection and Damage Evaluation in Concrete'—Final Report of RILEM Technical Committee 212-ACD (e-ISBN: 978-2-35158-100-1); *Ed. M. Ohtsu*

Report 45: Repair Mortars for Historic Masonry—State-of-the-Art Report of RILEM Technical Committee TC 203-RHM (e-ISBN: 978-2-35158-163-6); *Eds. Paul Maurenbrecher and Caspar Groot*

Report 46: Surface delamination of concrete industrial ffioors and other durability related aspects guide—Report of RILEM Technical Committee TC 268-SIF (e-ISBN: 978-2-35158-201-5); *Ed. Valerie Pollet*

Experimental Analysis of the Bending Behavior of Structural Metal Joints Based on the Use of Girder Clamps to Service Life Extension of Existing Structures

Manuel Cabaleiro, Cristina González-Gaya, and Fernando Gonzalez

Abstract The objective of this work is mainly focused on the structures that support the facilities and machinery present in industrial plants. Service life extension of existing structures is one of the current objectives of the industry. The lay-out of the production has to constantly adapt to changes in production, variations in demand and new products. These changes of model or demand, usually imply important changes in the facilities and machinery and therefore in the structures that support them. Currently, these support structures are destroyed at each lay-out change, mainly because an easy reconfiguration of the support structures is not possible. To achieve service life extension of existing structures it is necessary to use easily reconfigurable, reusable and removable structures. Currently there are already solutions of structures that are removable and reconfigurable but mainly in aluminum beams. Nowadays, there are no definitive solutions for steel structures manufactured with standard shapes (standard steel beams). The use of clamp joints for steel profiles would be a very interesting solution and would allow the manufacture of completely removable and reconfigurable structures. This paper proposes and uses a methodology for the experimental analysis of the bending behavior of this type of joints depending on the length of the profile flange.

Keywords Bolt · Joint · Steel beam · Reconfigurable · Removable

M. Cabaleiro (✉) · C. González-Gaya
Department of Construction and Manufacturing Engineering, School of Industrial Engineering,
UNED, 28040 Madrid, Spain
e-mail: mcabaleir15@alumno.uned.es

M. Cabaleiro · F. Gonzalez
Department of Materials Engineering, Applied Mechanics and Construction, School of Industrial
Engineering, University of Vigo, C.P. 36208 Vigo, Spain

1

1 Introduction

Service life extension of existing structures is one of the current objectives of the industry and it is enclosed within the objectives of industry 4.0. The objective of this work is mainly focused on the structures that support the facilities and machinery present in industrial plants. The lay-out of the production has to constantly adapt to changes in production, variations in demand and new products. These changes are usually, for example in the automotive industry the change of car model is done on average every 8 years. These changes of model or demand, usually imply important changes in the facilities and machinery and therefore in the structures that support them. Currently, these support structures are destroyed at each lay-out change, mainly because an easy reconfiguration of the support structures is not possible. To achieve service life extension of existing structures it is necessary to use easily reconfigurable, reusable and removable structures [1, 2].

In the market there are structural systems with aluminum profiles that allow structures that are completely removable and reconfigurable, but these types of structures are designed for small constructions, not being economically viable for structures that have to support large loads [2]. In the case of steel structures manufactured with standard profiles, there is not yet a definitive solution that is widely used. In this line, a promising solution would be the steel structures made by joints whit girder clamps [1–3]. This type of union with girder clamps would allow the manufacture of structures totally removable and reconfigurable, so that they can be reused every time there is a lay-out change.

The girder clamps joints are mainly formed by girder clamps, which by means of a bolt and a nut allow the joint of the elements that form the connection. The girder clamps work by the principle of the lever [3] which mean that when it is applied a preload (P) on the bolt, this preload produces two reactions, one at point b (Fb) and another at point a (Fa). It is the Fa force that keeps the elements of the union rigidly united (Fig. 1). The value of the force Fa responds to the principle of the lever so that its value is always proportional to the value of the preload and the levers a and b, therefore Fa $=$ P $*$ b $/$ (a $+$ b).

Fig. 1 Girder clamps joints and lever operating principle

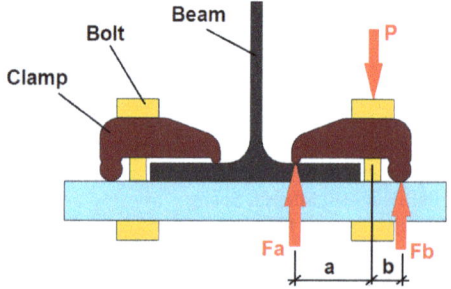

Fig. 2 Joints with I-cross-section profiles using clamps

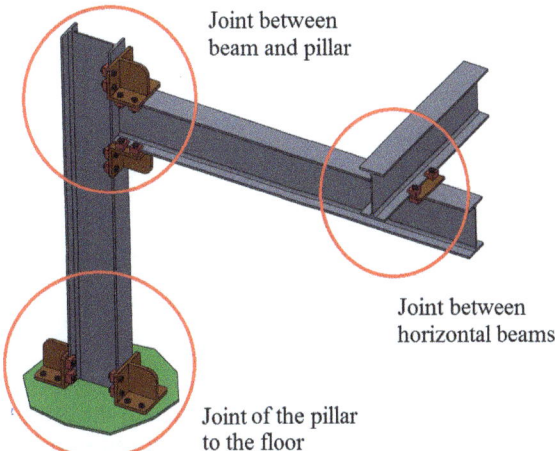

Joint between beam and pillar

Joint between horizontal beams

Joint of the pillar to the floor

There are currently several manufacturers of clamps on the market [4, 5], which have clamps with different measures of levers a and b as well as different bolts sizes, according to the strength of union that it is need to do.

By using clamps, different types of joints can be obtained between I-shaped profiles [2, 3]. Among the different types of joints can be mentioned: the union between horizontal beams, the union between beams and pillars or the union of the pillars to the floor (see Fig. 2).

For classic bolted joints there are extensive regulations that include analytical formulations [6, 7] that allow easy design and calculation, also on classic bolted joints there are a large number of studies and test [8–12]. However, in the case of clamps unions, there are very few researches.

Some previous work has already been carried out on the behaviour of the clamps joints, such as the application of an analytical model for the resistant calculation based on the T-Stub methodology used by the Eurocode for the study of the classic bolted joints [3], but adapted for the analysis of structural metal joints made using clamps. Other work [13] has also studied the behaviour of clamps joints according the shear stress. The shear stress is made in the clamps joints exclusively by the effect of friction between the pieces due to the preload of the bolts. Another work directly related with the aim of this paper is the Cabaleiro et al. [2] where the analysis of the effective length of the beam flange was carried out by FEM simulation, performing an analysis of the stress distribution.

But despite the works mentioned above, there is currently a significant lack of studies and test about this type of clamps joints, as well as the fact that they are not included in any regulations. These reasons make that it is difficult to use this type of joint for the construction of removable and reconfigurable structures. Therefore it is very important that the scientific community carry out more research and tests on the behavior of this type of joints. Among the parameters to study of the clamp union is which is the minimum length that the profile must have so that the transmission of

Fig. 3 Example of
minimum profile length
needed to make a 90° clamp
joint

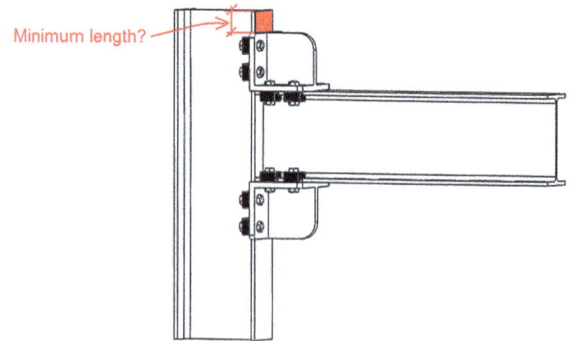

the bending is the same as if the length of the profile were infinite. This parameter is an important value because for the manufacturing process of the structures it is necessary to know which is the minimum length that must be left at the end of the beam or pillar to not reduce the strength of the union (see Fig. 3).

Although some simulation work has been carried out on the effective length, these were focused on the analysis of the stress distribution and not towards the analysis of the deformation of the joint. In addition, the data calculated by simulation at the work Cabaleiro et al. [2] could not be tested in the laboratory. Measuring in the laboratory when the length that has been cut to the profile if it affects the strength and deformation of the joint is a complex task.

The objective of this work is to carry out a first approach and application of a methodology that allows us to find out in a practical way in the laboratory when the length that is cut from the profile really affects its deformation and resistance. This methodology will also allow analyzing the value of the minimum profile length depending on the length of the clamp.

2 Proposed Methodology

In order to analyze from which profile length the joint begins to lose stiffness (and therefore the ability to transmit bending moment), the following experimental methodology is proposed (see Figs. 4 and 5): a clamp joint formed by a pillar and a beam is done (the pillar is made rigidly anchored at the ends and with a length and moment of inertia enough to make its deformation by bending negligible).

This deformation (Δi) must be measured for each test. Progressively and on both sides of the upper clamps, the length (L) of the profile flange that contributes to the strength of the joint will be shortened. At the end of the cantilever beam, the fixed load (P) will be applied for each different profile flange length (Li) and the achieved deformation (Δi) for each different profile flange length will be measured. At the moment when there is an increase in the measured deformation ($\Delta i > \Delta_0$), this will mean that the profile length Li is the minimum necessary for the joint. This

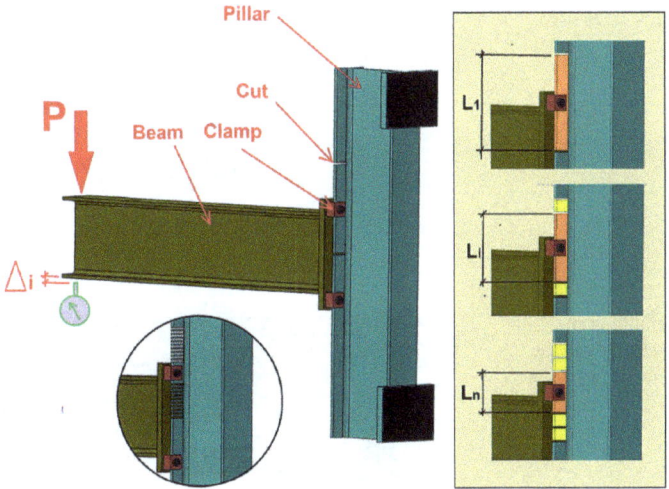

Fig. 4 Assembly of the profiles and methodology used

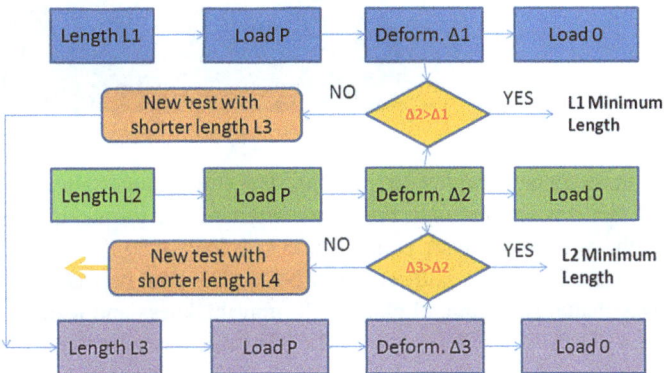

Fig. 5 Scheme of the proposed methodology

methodology will be repeated for various sizes of clamps, in order to analyze how the size of the clamp affects the effective length of the profile in the joint.

3 Materials

Three sizes of clamps with a width of 40 mm and a length of 30 mm (short clamp), 55 mm (medium clamp) and 67 mm (long clamp) were used to perform these tests (see Fig. 6). The bolts used are size M16 and a grade 12.9 with a yield strength of 10.8 N/mm^2. Bold preload is controlled by applying a torque with a torque wrench

Fig. 6 Clamps sizes used

Fig. 7 Assembly of the
beams to be carried out the
tests in the laboratory press

and also by strain gauges located on the bolts. The strain gauges allow to verify the
bolt preload that it is applied with the torque wrench. The torque applied with the
torque wrench has been 220 N.m, equivalent to a preload of 59 KN. The profile used
was an IPE220 for both pillar and cantilever beam. The joint between the beam and
the pillar was made using an end plate of 20 mm thickness and using a total of 4
clamps (Fig. 4).The steel used was grade S235. The measurement of the deformation
of the end cantilever beam was performed using digital comparators with precision
± 0.01 mm. High precision comparators ± 0.001 were also used at different points of
the joint to verify that there were no plastic deformations (Figs. 7and 8). A grinder
machine was used to cut the profile flange and gradually reduce the length (Li) of
the profile.

4   Results

The results found for each clamp size were as follows (Fig. 9). In the case of the long
clamp and with a 10 KN load, the initial deformation was 3.68 mm. When the length
of the pillar flange (Li) is shortened, the deformation ($\Delta 0$) remains constant to Li =
100 mm. From this value when the length of the profile flange (Li) is reduced, the
deformation (Δi) increases. In the case of the medium clamp and with a load of 10

Fig. 8 a Beam end plate (of cantilever beam) attached by means of clamps to the pillar. **b** Cuts made in the pillar flange to research the profile effective length by means of the tests

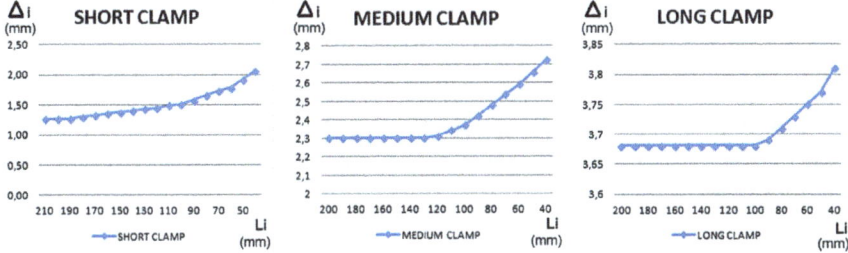

Fig. 9 Deformation (Δi) in the beam cantilever end of the beam for each clamp size and according to the length (Li) of the pillar flange

KN, the initial deformation was 2.3 mm. When the length of the pillar flange (Li) is shortened, the deformation ($\Delta 0$) remains constant to $Li = 130$ mm. From this value when the length of the profile flange (Li) is reduced, the deformation (Δi) increases. In the case of the short clamp and with a 5 KN load, the initial deformation was 1.26 mm. The deformation value (Δi) began to increase as the length of the profile (Li) decreases 190 mm. When the deformation begins to increase, it is verified that when the load was removed, the deformation of the cantilever beam returned to 0. With this verification it could be ensured that there were no plastic deformations in the joint. In this way it can be ensured that the deformation produced is in the steel elastic area. Due to the need to always work in the elastic area, in the case of the short clamp it was necessary to use a load of 5 KN and not 10 KN to always work in an elastic area.

The Fig. 10 shows the increase in deformation in the joint for each clamp size. In this graph it can be seen that when the length (Li) of the profile decreases, at the moment when the length (Li) of the profile begins to contribute completely to the joint, then there is a variation in the deformation of the beam end. If we consider that the length begins to be really all effective just before the deformation begins to vary, then we would have that for the large clamp the minimum length would be 100 mm,

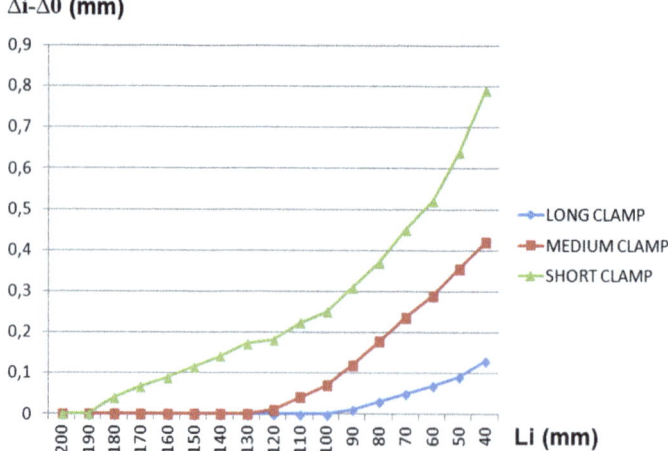

Fig. 10 Increased deformation (Δi–$\Delta 0$) in the beam cantilever end of the beam for each clamp size and according to the length (Li) of the pillar flange

for the medium clamp is 130 mm and for the short clamp is 190 mm. As expected, to greater clamp, the profile length necessary to contribute to the stiffness of the joint is much smaller.

5   Conclusions

The methodology proposed in this work to determine experimentally from what minimum profile length is necessary for the total transmission of the bending moment in the clamp union has been verified as adequate. Through the tests carried out, it has been found that the use of shorter clamps implies that a longer profile length is necessary to be able to transmit the whole bending moment in the joint. The data obtained in this work and the proposed methodology serves as a starting point for obtaining experimental data that allow us to calibrate 3D model for finite element analysis. This research means an important advance in the manufacturing processes of structures by girder clamps joints. The use of clamps as a joining method would allow the construction of completely removable and reconfigurable structures, the only requirement being the cut of the profiles. The use of these structures made with clamps is an adequate solution for service life extension of existing structures.

References

1. Cabaleiro, M., Caamaño, J.C., Riveiro, B., Conde B.: Analysis of the behavior of an innovative removable joint using flanges in connections of structural steel square tubes. In: The 13th International Conference on Computational Structures Technology (CST2018) 4–6 of September 2018 Sitges, Barcelona, Spain (2018)
2. Cabaleiro, M., González, C., Conde, B.: Analyzing the effective length of I cross-section beams in connections with girder clamps for totally removable, reusable and reconfigurable structures. In: 8th MESIC Manufacturing Society Engineering International Conference, 19–21 June 2019, Madrid, Spain, p. 16 (2019).
3. Cabaleiro, M., Riveiro, B., Conde, B., Caamaño, J.C.: Analytical T-stub model for the analysis of clamps in structural metal joints. J. Constr. Steel Res. **130**, 138–147 (2017)
4. Cabaleiro, M., Conde, B., Caamaño,J.C., Riveiro, B., Gonzalez, L.: Analysis of the shear behavior of structural metal joints based on the use of girder clamps. In: 1st Iberic Conference on Theoretical and Experimental Mechanics and Materials/11th National Congress on Experimental Mechanics. Porto/Portugal 4–7 November 2018, pp. 705–710. ISBN: 978-989-20-8771-9 (2018)
5. Lindapter.: www.lindapter.com. Lindapter International 2011. Accessed 18 Mar 2018
6. LNA Solutions.: www.lnasolutions.com. LNA Solutions, Steel Connection Solutions. Accessed 18 Mar 2018
7. Eurocode 3.: Design of Steel StructuresPart 1–8: Design of Joints (EN 1993-1-8: 2003), European Committee for Standardization, Brussels (2003)
8. Eurocode 3.: Design of Steel StructuresPart 1–1: General Rules and Rules for Building (EN 1993-1-1:2005), European Committee for Standardization, Brussels (2005)
9. Arguelles-Alvarez, R. Arguelles-Bustillo, J.M., Arguelles-Bustillo, R., Arriaga-Martitegui, F., Atienza-Reales, J. R.: Estructuras de acero:Uniones y sistemas Estructurales, Bellisco Ediciones, Spain, Madrid, (Spanish) (2007)
10. Faella, C., Piluso, V., Rizzano, G.: Structural Steel Semirigid Connections. CRC Press LLC, Boca Ratón, FL (2000)
11. Reinosa, J.M., Loureiro, A., Gutierrez, R., Lopez, M.: Analytical frame approach for the axial stiffness prediction of preloaded T-stubs. J. Constr. Steel Res. **90**, 156–163 (2013)
12. Zhu, X., Wang, P., Liu, M., Tuoya, W., Hu, S.: Behaviors of one-side bolted T-stub through thread holes under tension strengthened with backing plate. J. Constr. Steel Res. **134**, 53–65 (2017)
13. Abidelah, A., Bouchaïr, D.E.: Kerdal, Influence of the flexural rigidity of the bolt on the behavior of the T-stub steel connection. Eng. Struct. **81**, 181–194 (2017)

Corrosion of Carbonated Structures. Real Cases of Structures in Spain

Nuria Rebolledo, Julio E. Torres, and Javier Sánchez

Abstract One of the main causes of deterioration of structures is the corrosion of the reinforcement, and within this one possible source of deterioration is the carbonation of the concrete cover. There are numerous studies about carbonation of concrete but the great majority have been carried out with samples taken in the laboratory and their behaviour in real structures is not known. This work presents the evaluation of three types of structures during several months in which the conditions to which the structure is exposed and the corrosion speed have been measured by means of in-situ measurements.

Keywords Durability · Concrete · Carbonation · Corrosion

1 Introduction

Carbonation is one of the main problems affecting the durability of reinforced concrete structures, since it reduces the pH and triggers corrosion of the reinforcements [1–4]. There are numerous studies on the corrosion behaviour of hard weapons embedded in mortars and carbonated concretes in laboratory tests, see for example the review by Stefaroni, Angs and Elsener where various tests are reported [3]. However, there are few studies on the behaviour of structures that suffer from this process [5].

There are numerous studies on the mechanism and chemical reactions that take place during the carbonation of concrete [1, 6–8]. In the case of carbonation, corrosion of the reinforcements is induced due to a lowering of the pH [9], which causes the instability of the passive layer of the steel and therefore the corrosion of the reinforcements [10, 11].

The carbonation process is influenced by numerous variables such as relative humidity, type of binder, etc. Thus, there are several models that, based on the variables mentioned above, allow predicting the carbonation rate of a concrete [1, 3, 5, 9, 12].

N. Rebolledo · J. E. Torres · J. Sánchez (✉)
Instituto de Ciencias de la Construcción "Eduardo Torroja", (IETcc-CSIC), Madrid, Spain
e-mail: javier.sanchez@csic.es

© The Author(s), under exclusive license to Springer Nature Switzerland AG 2022
J. Sena-Cruz et al. (eds.), *Proceedings of the 3rd RILEM Spring Convention and Conference (RSCC 2020)*, RILEM Bookseries 34,
https://doi.org/10.1007/978-3-030-76465-4_2

11

Table 1 Mean values of carbonation and CO_2 concentration

	No risk of corrosion	Exposure class for risk of corrosion induced by carbonation			
	X0	XC1	XC2	XC3	XC4
Maximum w/c	–	0.65	0.60	0.55	0.50
Minimum strength class	C12/15	C20/25	C25/30	C30/37	C30/37
Minimum cement content (kg/m^3)	–	260	280	280	300

As referenced above, mainly three climatic parameters influence carbonation: relative humidity, temperature, and atmospheric carbon dioxide concentration. It has been found that one of the environmental parameters that mainly affect the carbonation rate is the CO_2 concentration in the air [13, 14].

European concrete standard EN 206-1 showed requirements for specifying concrete durability. Table 1 gives the requirement for the four XC expositions.

The present work shows the study carried out on structures that only suffer corrosion by carbonation, without the influence of other aggressive agents or environments, such as chloride ions.

2 Methodology

2.1 Description of the Structures

The location, description and environment of the structures under study are shown below. The structures are located around the cities of Tarragona and Oviedo in Spain.

The structure A, located in the province of Tarragona, consists of circular reinforced or prestressed concrete precast pipes. The sections of the galleries that can be visited are supported by concrete feet or resting on sand/gravel cradles. These elements are protected from the rain but in an industrial environment with a high concentration of CO_2 in certain places in the galleries.

The structure B is a railway bridge over the Nalón River located in Oviedo. The structure is formed by a series of spans of IP beams, with spans between 12.60 and 14 m and a height of 6–8 m.

In Fig. 1, it is possible to see one of the areas selected for in-situ corrosion measurements during the study. During this phase, reinforcement connections were made to for to be monitored (see Fig. 2).

Fig. 1 Location of Oviedo and Tarragona on the map of Spain

Fig. 2 Selected areas during the repair phase

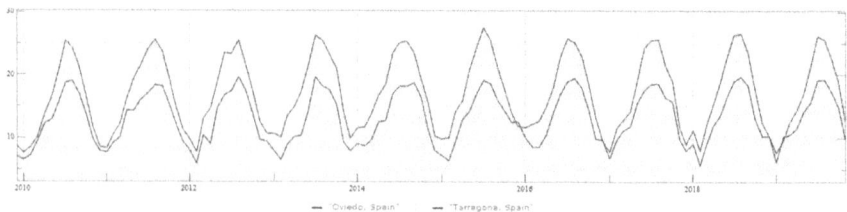

Fig. 3 Temperature (monthly means, °C) in Oviedo and Tarragona between 2009 and 2019

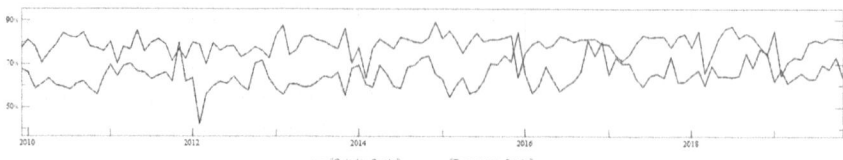

Fig. 4 Relative humidity in Oviedo and Tarragona between 2009 and 2019

2.2 Climate to Which Structures are Exposed

The following figures show the monthly means of temperature and relative humidity for the towns of Oviedo and Tarragona.

The temperature in Oviedo varies with monthly means between 5 and 25 °C, with extreme temperatures that occasionally reach minimums of −3 °C and maximums of 37 °C. The average relative humidity for this period of time is 79% (see Figs. 3 and 4).

The temperature in Tarragona varies with monthly means between 6 and 27 °C, with extreme temperatures that occasionally reach minimums of −7 °C and maximums of 40 °C. The average relative humidity for this period of time is 65%.

2.3 Corrosion Rate Measurements

The electrochemical measurements have been carried out with the Gecor portable corrosion rate meter which allows to measure the resistivity, corrosion potential and corrosion rate based on the measurement of the R_p with a guard ring [15–18].

3   Results and Discussions

3.1   Structure A (Tarragona)

The following table shows the percentage values of the coating that is carbonated and the average CO_2 concentration in the air during 1 month of measurement. In certain locations, such as A05, A06 or A10, high concentrations of CO_2 in the air are reached. The concentration of chlorides has been measured in all cases to check that it is below the threshold for depassivation (see Tables 2 and 3).

The following figures show a detail of one of the pipes where it can be seen that the reinforcements are corroded and the verification of the existence of carbonation (see Figs. 5 and 6).

The following figure shows the probability histogram of the corrosion rate of the data measured in structure A. The data have been fitted to a log-normal distribution obtaining adjustment parameters of $[-1.572, 0.639]$. Taking into account this fit, the value corresponding to 50% of probability is that of $0.208 \, \mu A/cm^2$. The values corresponding to percentiles 0.05, 0.10, 0.90 and 0.95 are respectively 0.073, 0.092, 0.471 and $0.590 \, \mu A/cm^2$ (see Fig. 7).

Table 2 Mean values of carbonation and CO_2 concentration

Location	% of concrete cover carbonated	CO_2 (ppm)	Exposure class (EN 206-1)
A01	100% 100%	418	XC4
A02	-100% -93% -63	512	XC3
A03	-30%	613	XC3
A04	0%	1100	XC3
A05	35–$50\%^{*}$ 100%	3200	XC3
A06	$0\%^{*}$	300	XC3
A07	-60% -97%	411	XC3
A08	100% 100% 100% 100%	484	XC4
A9	0%	510	XC3
A10	0%	827	XC3
A11	$0\%^{*}$	581	XC3

* Measured of high concentrations of CO_2 in the air

Table 3 In situ corrosion parameters

Location	Ecorr versus Cu/CuSO$_4$ (mV)	ρ (Ω m)	Icorr (μA/cm^2)
A01	$-2,255$	1133,4	0,282
	$-76,38$	1634,25	0,588
A02	$-193,52$	164,06	0,244
	$-88,76$	199,95	0,268
	$-123,01$	138,10	0,169
A03	$-38,925$	628,80	0,116
A04	$-223,43$	89,33	0,100
A05	23,240	1407,2	0,128
	$-208,35$	779,00	0,370
A06	$-$	94,17	$-$
A07	113,35	1058,65	0,102
	$-80,575$	691,73	0,244
	59,505	373,17	0,260
	$-$	$-$	$-$
A08	$-156,340$	1189,5	0,324
	114,617	627,00	0,372
	$-153,450$	468,80	0,562
	$-120,085$	593,33	0,430
	158,603	932,83	0,368
A9	$-220,84$	729,00	0,081
A10	$-53,601$	1788,60	0,101
A11	$-289,87$	386,40	0,107

Fig. 5 Corrosion of the reinforcements in one of the pipes

Fig. 6 Picture of the corrosion rebars in the structure A in a carbonation zone

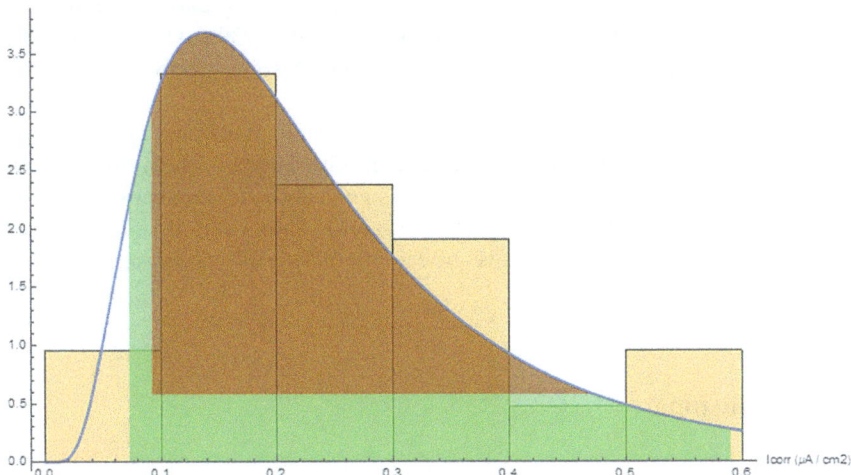

Fig. 7 Corrosion rate. Probability density function and its fit to lognormal distribution derived from a normal distribution with mean 0.208 μA/cm^2 and standard deviation 0.639 μA/cm^2

Table 4 In situ corrosion parameters. Oviedo structure

Identificación del tramo	Orientación	Antes de la reparación (Octubre del 2014)		
		Icorr (μA/cm^2)	Ecorr versus Cu/CuSO$_4$ (mV)	ρ 4 puntas (Ω.m)
Vano 6	Norte	0,592	$-267,84$	2250,0
	Sur	0,182	$-236,90$	1091,7
Vano 7	Norte	0,163	$-199,24$	1005,8
	Sur	0,333	$-207,52$	1179,2
Vano 10	Norte	0,215	$-246,66$	1700,0
	Sur	0,261	$-160,83$	1342,0
Vano 11	Norte	0,371	$-247,97$	2461,3
	Sur	0,182	$-182,99$	2247,5
Vano 12	Norte	0,188	$-170,22$	2452,8
	Sur	0,250	$-128,21$	2196,0
Vano 13	Norte	0,169	$-116,39$	2144,0
	Sur	0,197	$-219,00$	2421,2

3.2 Structure B (Oviedo)

In this case, the structure is subjected to the same conditions of carbonation. The concrete covers studied show 100% carbonation in all the cases (see Table 4).

The following figure shows the probability histogram of the corrosion rate of the data measured in structure B. The data have been fitted to a log-normal distribution obtaining adjustment parameters of $[-1.431, 0.371]$. Taking into account this fit, the value corresponding to 50% of probability is that of 0.239 μA/cm^2. The values corresponding to percentiles 0.05, 0.10, 0.90 and 0.95 are respectively 0.13, 0.15, 0.38 and 0.44 μA/cm^2 (see Fig. 8).

The following figure shows the cumulative distribution function for the corrosion rate of both structures. The corrosion rate and the distribution for both structures are very close. We cannot conclude that there is a significant difference between the two distributions to consider them different. The average values vary between 0.21 and 0.24 μA/cm^2, while for a cumulative probability of 90% all values are below 0.47 μA/cm^2 (see Fig. 9).

4   Conclusions

From the results obtained on two structures that only suffer from carbonation corrosion, it can be concluded that, regardless of the type of structure, an average corrosion rate of 0.21–0.24 μA/cm^2, while for a cumulative probability of 90% all values are below 0.47 μA/cm^2.

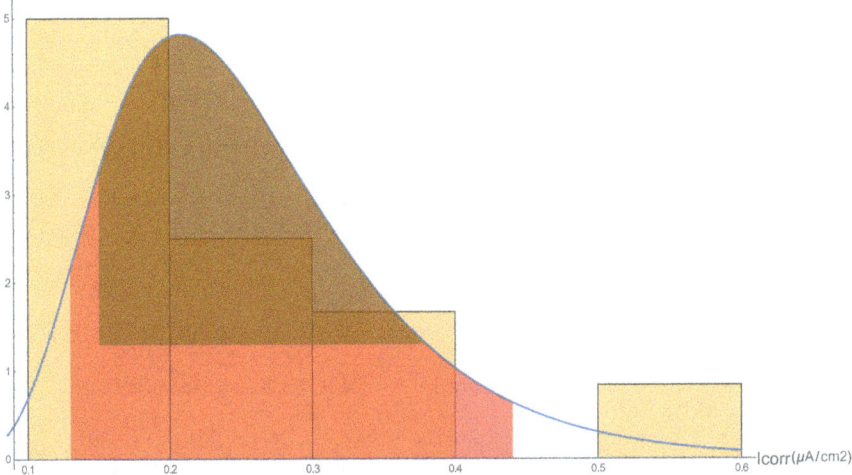

Fig. 8 Corrosion rate. Probability density function and its fit to lognormal distribution derived from a normal distribution with mean 0.239 μA/cm^2 and standard deviation 0.371 μA/cm^2

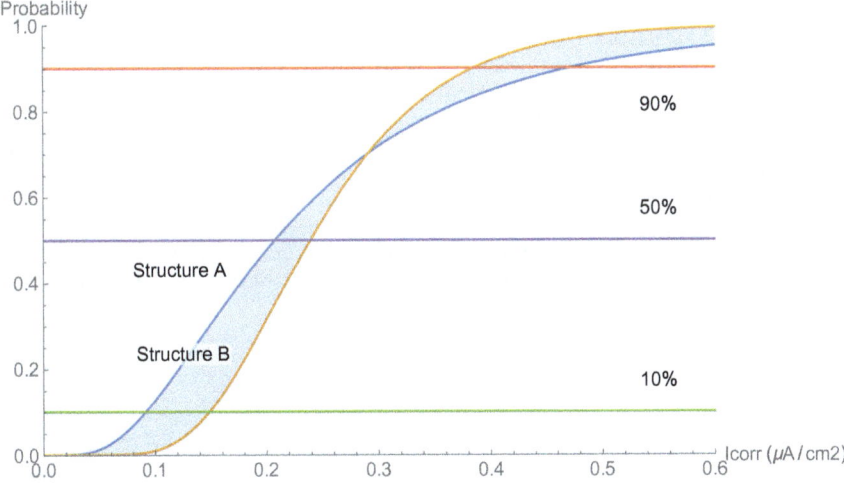

Fig. 9 Comparison between both distributions

References

1. Galan, I., et al.: Carbonation profiles in cement paste analyzed by neutron diffraction. J. Phys. Conf. Ser. **340**(1), (2012)
2. Andrade, C., Sarria, J., Alonso, C.: Relative humidity in the interior of concrete exposed to natural and artificial weathering. Cem. Concr. Res. **29**(8), 1249–1259 (1999)
3. Stefanoni, M., Angst, U., Elsener, B.: Corrosion rate of carbon steel in carbonated concrete—a critical review. Cem. Concr. Res. **103**, 35–48 (2018)

4. Otieno, M., Ikotun, J., Ballim, Y.: Experimental investigations on the influence of cover depth and concrete quality on time to cover cracking due to carbonation-induced corrosion of steel in RC structures in an urban, inland environment. Constr. Build. Mater. **198**, 172–181 (2019)
5. Benítez, P., et al.: Carbonated structures in Paraguay: Durability strategies for maintenance planning. Procedia Struct. Integr. **11**, 60–67 (2018)
6. Xi, F., et al.: Substantial global carbon uptake by cement carbonation. Nat. Geosci. **9**(12), 880–883 (2016)
7. Parrott, L.J.: Variations of water absorption rate and porosity with depth from an exposed concrete surface: Effects of exposure conditions and cement type. Cem. Concr. Res. **22**(6), 1077–1088 (1992)
8. Parrott, L.J., Killoh, D.C.: Carbonation in a 36 year old, in-situ concrete. Cem. Concr. Res. **19**(4), 649–656 (1989)
9. Silva, A., Neves, R., de Brito, J.: Statistical modelling of carbonation in reinforced concrete. Cem. Concr. Compos. **50**, 73–81 (2014)
10. Maurice, V., Marcus, P.: Progress in corrosion science at atomic and nanometric scales. Prog. Mater. Sci. **95**, 132–171 (2018)
11. Pourbaix, M.: Thermodynamics and corrosion. Corros. Sci. **30**(10), 963–988 (1990)
12. Galan, I., et al.: Neutron diffraction for studying the influence of the relative humidity on the carbonation process of cement pastes. J. Phys. Conf. Ser. **325**(1), (2011)
13. Saetta, A.V., Vitaliani, R.V.: Experimental investigation and numerical modeling of carbonation process in reinforced concrete structures: Part II. Practical applications. Cem. Concr. Res. **35**(5), 958–967 (2005)
14. Saetta, A.V., Vitaliani, R.V.: Experimental investigation and numerical modeling of carbonation process in reinforced concrete structures: Part I: Theoretical formulation. Cem. Concr. Res. **34**(4), 571–579 (2004)
15. Andrade, C., et al.: On-site corrosion rate measurements: 3D simulation and representative values. Mater. Corros. **63**(12), 1154–1164 (2012)
16. Feliu, S., Gonzalez, J.A., Andrade, C.: Electrochemical methods for on-site determinations of corrosion rates of rebars. In: Berke, N.S., et al. (eds.) Techniques to Assess the Corrosion Activity of Steel Reinforced Concrete Structures, pp. 107–118 (1996)
17. Feliu, S., et al.: Possibilities of the guard ring for electrical signal confinement in the polarization measurements of reinforcements. Corrosion **46**(12), 1015–1020 (1990)
18. Vennesland, Ø., Raupach, M., Andrade, C.: Recommendation of Rilem TC 154-EMC: "Electrochemical techniques for measuring corrosion in concrete"—measurements with embedded probes. Mater. Struct. **40**(8), 745–758 (2007)

Bond Behaviour of NSM FRP Strengthening Systems on Concrete Elements Under Sustained Load

J. Gómez, C. Barris, M. Baena, and R. Perera

Abstract During the last decades, Near Surface Mounted (NSM) Fibre Reinforced Polymer (FRP) strengthening system has been widely accepted as an efficient methodology in structural rehabilitation. The bond behaviour of the FRP-concrete joint has resulted to be governing the overall performance of the strengthening system. Moreover, due to the susceptibility of the materials used in the NSM strengthening system in front of high temperatures and sustained loads, their performance under service conditions can be affected. Although the short-term bond behaviour has been largely studied in the literature, there is still a lack of experimental data to better understanding the long-term performance. The paper aims to experimentally study the evolution of bond damage of the concrete-FRP bonded joint of an NSM FRP strengthened concrete element, and how it affects to the slip between the strengthening system and the concrete For this purpose, a total of eight NSM pull-out specimens with two different bonded lengths (150 and 225 mm) and two different groove thicknesses (7.5 and 10 mm) have been loaded with two different levels of sustained load (15 and 30% of the ultimate load) at room conditions up to 40 days. Specimens were previously characterized under a short-term test. Slip at the loaded end was measured to capture the evolution of slip during time. Results showed that the slip caused by creep increased with the sustained load level, whilst it decreased with the bonded length.

Keywords FRP · Concrete · NSM · Bond · Sustained load · Ambient conditions

J. Gómez (✉) · C. Barris · M. Baena
University of Girona, Girona, Spain
e-mail: javier.gomez@udg.edu

R. Perera
Technical University of Madrid, Madrid, Spain

© The Author(s), under exclusive license to Springer Nature Switzerland AG 2022
J. Sena-Cruz et al. (eds.), *Proceedings of the 3rd RILEM Spring Convention and Conference (RSCC 2020)*, RILEM Bookseries 34,
https://doi.org/10.1007/978-3-030-76465-4_3

1   Introduction

In the last decades, Fibre Reinforced Polymer (FRP) materials have been widely used as strengthening materials for concrete structures. Externally Bonded Reinforcement (EBR) and Near Surface Mounted (NSM) are the most used strengthening techniques using FRP materials. While in the EBR technique the FRP is bonded to the tensile face of the concrete surface, in the NSM system the FRP is placed into a groove previously cut in the concrete cover and bonded with an adhesive, typically an epoxy resin.

The short-term bond behaviour of these strengthening systems has been widely studied by many authors, and response has been standardized by guidelines and standards. However, due to the nature of the materials involved in the strengthening system (FRP, adhesive and concrete) their mechanical properties can be affected by sustained load and harmful environmental conditions, affecting the bond between the strengthening system and the concrete element. Nevertheless, only little research has been focussed on their bond behaviour under service conditions, leading to a lack of knowledge in their long-term bond behaviour. In that sense, during the last years, some experimental tests on long-term deformations caused by sustained loading and environmental conditions have been carried out to study the bond behaviour of EBR [8, 13] and NSM [2, 5] reinforcement. These works showed that the long-term behaviour of the FRP-concrete joint may be affected by creep deformations caused by sustained load and environmental conditions.

Moreover, some analytical models were proposed in [2, 5, 8, 13] to predict the bond performance of the FRP-strengthening systems under sustained loading. In most of these works, the authors adjusted different closed-form equations to their experimental results for taking into account the creep behaviour of the FRP-concrete joint. In particular, in [5] the authors used an approach presented in [3], but introducing the creep effects into a bilinear bond-slip model, and obtained good agreement between experimental and analytical results. However, in that work, the failure of the instantaneous test was attained by rupture of the laminate and the set-up completely confined the top surface of the concrete element.

This research aims to experimentally study the slip of FRP-strengthened concrete elements under sustained service load at room temperature conditions. An experimental programme on eight pull-out tests submitted to sustained loading is presented. The effect of the groove thickness, bonded length and sustained load level on the global bond behaviour is studied.

2   Experimental Programme

This paper presents the preliminary results of an extensive experimental programme aiming at studying bond behaviour of NSM FRP-concrete interface under sustained loading conditions.

In this work, eight reinforced concrete blocks have been strengthened with Carbon-FRP (CFRP) strips of 3.0 mm × 10 mm using the NSM technique and have been tested under sustained loading during 40 days at room temperature conditions. Two bonded lengths (L_b), 150 and 225 mm, two groove thicknesses (t_g), 7.5 and 10 mm, and two levels of sustained load (P_s), 15 and 30% of the instantaneous maximum load (P_u) have been used.

Previously, for each NSM CFRP-concrete element combination, four specimens were tested under monotonic pull-out testing, in order to obtain their average instantaneous bond response, giving a total of 16 tested specimens.

2.1 Materials

CFRP Sika CarboDur S strips of 3 mm × 10 mm were used for the NSM reinforcement. The CFRP mechanical properties were characterised according to ISO 527-5 standard [11], obtaining an elastic modulus of 169.3 GPa (CoV 7.7%) and a tensile strength of 3206 MPa (CoV 2.1%).

Sikadur 30 bi-component epoxy resin was used as adhesive. Its mechanical properties were characterized according to ISO 527-2 standard [10], resulting in an elastic modulus of 10.7 GPa (CoV 4.4%) and a tensile strength of 27.5 MPa (CoV 11.2%).

All concrete specimens were cast from the same batch. Concrete mechanical properties were assessed according the ASTM C469/C469M-10 [1] and UNE 12390-3 [15] standards. A compressive strength of 33.8 MPa (CoV 4.4%), an elastic modulus of 33.1 GPa (CoV 3.3%) and a tensile strength of 5.7 MPa (CoV 4.7%) were obtained.

2.2 Short-Term Direct Pull-Out Tests

The NSM CFRP-concrete elements were tested to obtain their short-term load-slip response under a monotonic loading test. The concrete block dimensions were 200 mm × 200 mm × 370 mm for the short-bonded lengths $(L_b = 150$ mm) and 200 mm × 200 mm × 420 mm for the long-bonded lengths $(L_b = 150$ mm). Four different NSM CFRP-concrete elements combining two groove thicknesses (7.5 and 10 mm) and two bonded lengths (150 and 225 mm) were tested. Figure 1 shows the set-up used for the monotonic testing. The set-up used in this work did not confine the concrete block as in [5, 6]; instead, a 60 mm wide steel reaction element partially confined the specimen [14], as can be seen in Fig. 1.

The load was applied with a servo-hydraulic testing machine at a 0.2 mm/min displacement controlled rate.

The instrumentation consisted in one LVDT at the loaded end to measure the relative slip between the FRP and concrete. Besides, a Digital Image Correlation (DIC) system was installed in order to capture the displacement and strain field of the front surface during the monotonic test, as it is shown in Fig. 2. 2D and 3D systems

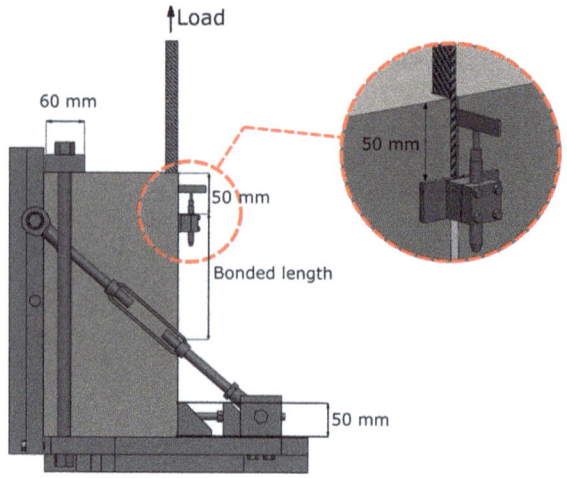

Fig. 1 Short-term test set-up

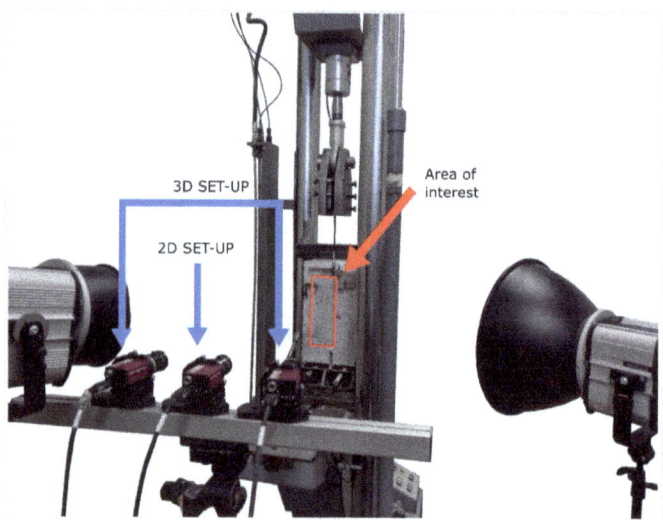

Fig. 2 2D and 3D set-ups of the Digital Image Correlation system and Area Of Interest (AOI)

were installed. The 2D system consists on one camera located perpendicular to the surface that captures the displacement of the studied surface in one plane. On the other hand, a 3D configuration with an angle between the camera optics was installed. This inclination allows the cameras to record out-of-plane movement; thus, displacements in the three directions can be measured.

To measure the field of displacements of the surface, a speckle pattern was needed to be tracked, therefore, the surfaces of the concrete blocks were white-painted, and

Table 1 Parameters of the DIC test and correlation

Parameters of the cameras		
Distance between the camera and the surface	840 mm	
Distance between cameras (3D system)	125 mm	
Diaphragm aperture	f/5.6	
Closing of the diaphragm	1000 μsec.	
Image acquisition frequency	1 image/sec.	
Correlation parameters		
	2D	3D
Subset	21 pixels	29 pixels
Stepsize	5 pixels	7 pixels
Correlation criterion	Zero-Normalized Cross-Correlation (ZNCC)	Zero-Normalized Cross-Correlation (ZNCC)

afterwards, a pattern of black speckles was applied. The displacement and strain fields during time were obtained using a correlation software. The correlation details (subset size, stepsize and the correlation criteria) are shown in Table 1 for the 2D and 3D systems, together with the DIC configuration parameters.

2.3 Long-Term Direct Pull-Out Tests

To study the long-term performance of the NSM-FRP bonded joint, eight NSM CFRP-concrete elements were tested under sustained loading, at 20 °C and 55% RH during 40 days. Sustained load levels of 15 and 30% of the experimental short-term maximum load were chosen to study loading levels similar to actual service conditions. Besides, two different groove thicknesses (7.5 and 10 mm) and two different bonded lengths (150 and 225 mm) were used.

The set-up used for the long-term tests is shown in Fig. 3. The instantaneous monotonic set-up was adapted to obtain a lever arm with a multiplying factor of 8.3.

Similarly to the instantaneous test, one LVDT was placed at the loaded end to capture the slip between the strengthening system and the concrete element. Additionally, one strain gauge was placed at the loaded end to monitor the FRP strain during the loading process and during the 40 days of testing.

Fig. 3 Long-term set-up

3 Analysis and Discussion of Results

3.1 Pull-Out Short-Term Tests Results

From the instantaneous tests, the maximum load and the failure mode of each configuration of bonded joint was obtained. Table 2 lists the NSM CFRP-concrete elements
tested in the pull-out short-term experimental programme. Each combination of specimens was named as ST-Lb-G, where ST stands for Short-term, Lb stands for the
bonded length (150 and 225 mm), and G stands for the width of the groove (7.5 and
10 mm).

The average load-slip curves obtained from instantaneous direct pull-out tests are
shown in Fig. 4. It can be seen that all the NSM specimens followed the same initial

Table 2 List of monotonic pull-out tests

Nomenclature	Bonded length [mm]	Groove thickness [mm]	Maximum load [kN]	Failure mode
ST-150-10	150	10	42.1	F-A
ST-150-7.5	150	7.5	45.5	C
ST-225-10	225	10	53.3	F-A
ST-225-7.5	225	7.5	58.8	C

Note F-A = FRP-adhesive interface failure, C = cohesive concrete failure

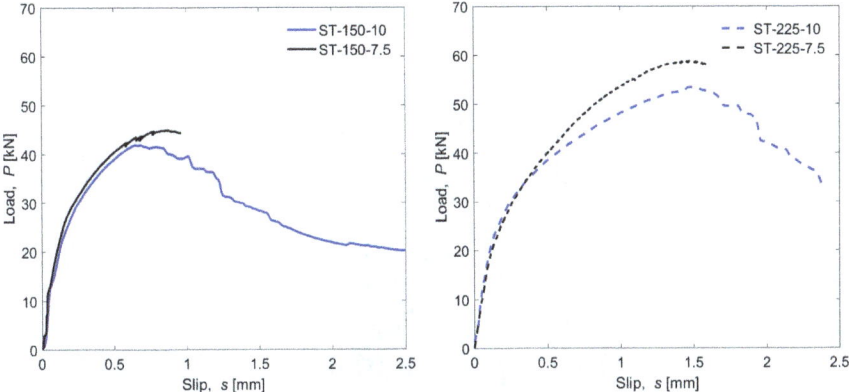

Fig. 4 Load-slip curve obtained from the monotonic test

trend in terms of stiffness. However, higher failure load was obtained in 7.5 mm grooved specimens than in the 10 mm grooved ones. Moreover, the failure modes observed were different for each groove thickness. For 10 mm grooved specimens, an adhesive failure in the FRP-epoxy interface was obtained, while for 7.5 mm grooved specimens, the failure was cohesive and it was attained in the concrete cover along inclined planes near the groove. This difference could be attributed to the different distribution of shear stresses along the groove in the FRP-concrete joint. Wider adhesive layers redistribute better the bond shear stress [5, 9, 12], therefore, as the groove thickness decreases, higher stresses were transmitted to the concrete.

On the other hand, 10 mm grooved specimens showed a clear friction stage once the maximum load was attained (Fig. 4). A descending branch could be observed in which the load smoothly decreased with the increase of slip. This stage could be caused because of friction between the FRP and the adhesive surfaces. On the contrary, failure in 7.5 mm grooved specimens was attained on the concrete surface once the maximum load was achieved, and a sudden drop of the load was observed and the specimen suddenly failed.

To better understand the stress transfer between FRP, adhesive and concrete, the strain field of the front surface of specimens was obtained by DIC. Figure 5a shows the area of interest of the specimen, indicated with dashed blue lines, which includes the groove and the right part of the concrete surface. Figure 5b, c show the strain field of the 10 mm and 7.5 mm grooved specimens, respectively, for specimens with a bonded length of 225 mm at the moment before the maximum load is attained. The red lines in Fig. 5a indicate the position of the groove in the pull-out surface, while in Fig. 5b, c the red lines mark the position of the groove in the strain field.

It can be observed that for 7.5 mm grooved specimen, the maximum strain in the concrete surface at the maximum load was approximately 4.500 $\mu\varepsilon$. On the other hand, in 10 mm grooved specimen the maximum strain was around 2.000 $\mu\varepsilon$. This means that for 10 mm grooved specimens, the maximum compressive capacity of the concrete was still not attained, whilst for 7.5 mm grooved specimen, the value of the

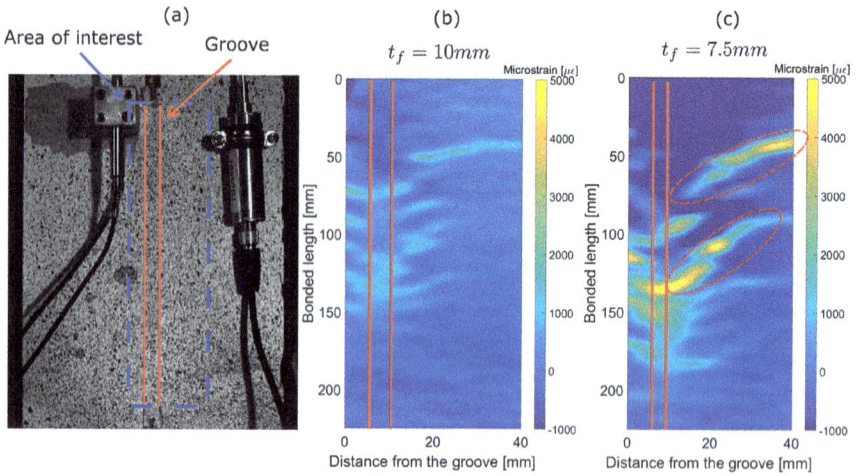

Fig. 5 **a** Area of interest, **b** strain field for 10 mm grooved specimens and **c** strain field for 7.5 mm grooved specimens

maximum strain was already higher than the usually accepted maximum compressive concrete strain [4]. Additionally, in Fig. 5c the typical compressive struts can be somehow appreciated, meaning that narrower grooves transmitted higher gradient of stresses to the concrete than wider grooves.

3.2   Pull-Out Long-Term Tests Results

In the long-term experimental programme, each specimen was named using the nomenclature defined in the short-term campaign, but this time the first parameter is LT, standing for long-term, and a forth parameter regarding the percentage of maximum load (P_u) under sustained load is added. Table 3 lists the specimens under long-term conditions. For each specimen, the initial slip (s_0) was obtained from the average curve of the instantaneous monotonic test.

Figure 6 shows the total slip of the NSM CFRP-concrete elements obtained during 40 days. Continuous and dashed lines refer to 10 mm and 7.5 mm groove thickness, respectively. Even though the percentage of sustained load level was the same for the different groove thicknesses, the applied load was higher for 7.5 mm grooved specimens, thus, the initial slip for these specimens was higher than for 10 mm grooved specimens.

The curves of NSM CFRP-concrete slip with time show an initial zone with high slope, and a second one after approximately 10 days of sustained loading where the increase of slip is much lower. These two stages were also observed in [5–7].

Moreover, it could be seen that specimens with 7.5 mm groove thickness (dashed curves in Fig. 6) attained higher slips than specimens with 10 mm, except for the

Table 3 List of specimens under sustained load

Nomenclature	Bonded length [mm]	Groove thickness [mm]	Applied load [kN]
LT-150-10-15	150	10	6.39
LT-150-7.5-15	150	7.5	6.54
LT-150-10-30	150	10	12.86
LT-150-7.5-30	150	7.5	14.14
LT-225-10-15	225	10	8.19
LT-225-7.5-15	225	7.5	8.94
LT-225-10-30	225	10	15.24
LT-225-7.5-30	225	7.5	17.49

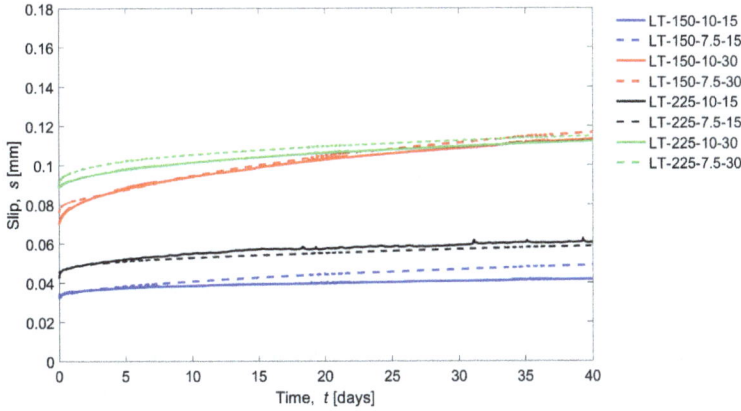

Fig. 6 Slip in the loaded end during time

LT-225-7.5-30 specimen, which obtained very similar slips in comparison with LT-225-10-30 specimen. The reason could be that for 10 mm grooved specimens, the redistribution of bond shear stress along the bonded length was better, due to the increase of the adhesive layer thickness [9, 12].

For specimens loaded at 30% of P_u, at t = 30 days approximately, the slip obtained in NSM specimens with 150 mm (red lines in Fig. 6) surpassed the slip obtained in $L_b = 225$ mm specimens (green lines), although the initial slip of 150 mm specimens was lower than that of 225 mm ones. This could be caused by the fact that specimens with short-bonded lengths may develop a higher average bond shear stress along the given FRP length and, consequently, higher damage was caused to the bonded joint.

On the other hand, for low sustained load values (15% of P_u), NSM CFRP-concrete elements with 150 and 225 mm were equally damaged, meaning that at this level of sustained load the increase of slip of the short and long bonded lengths were very similar.

In [5] it was observed that for sustained load levels up to 40% of P_u the behaviour of the bonded joint is in a linear viscoelastic stage, thus, the total slip should be proportional to the load applied. The total slip obtained in specimens with $L_b = 225$ mm was proportional to the load applied. When the sustained load was 15% of P_u (black lines) the final slip was 0.061 mm and 0.058 mm, for the 10 mm and 7.5 mm grooved specimens, respectively, and when the load applied was 30% of P_u (green lines), the slip was 0.112 mm and 0.115 mm, respectively. On the other hand, the slip obtained in specimens with $L_b = 150$ mm under the 15% and 30% of P_u were not proportional. LT-150-10-15 and LT-150-7.5-15 obtained 0.042 mm and 0.049 mm, respectively, and LT-150-10-30 and LT-150-7.5-0 obtained 0.113 mm and 0.117 mm, respectively.

4   Conclusions

This paper presents the preliminary results of an experimental programme aiming at studying the effect of sustained loading and room temperature conditions on the bond behaviour of NSM CFRP-concrete specimens. Eight pull-out tests under sustained loading during 40 days and 16 pull-out tests under instantaneous loading were carried out in the experimental programme.

From the pull-out monotonic tests, it can be concluded that:

- Higher failure loads were obtained for 7.5 mm grooved specimens than for 10 mm grooved ones.
- For 7.5 mm grooved specimens, a cohesive failure in concrete was attained, while for 10 mm grooved specimens the failure was located in the FRP-adhesive interface.
- In specimens with 10 mm groove, a friction stage after reaching the maximum load was observed. This phenomenon was not observed in 7.5 mm groove specimens, where failure was attained at the maximum load.
- From the Digital Image Correlation results, it was observed that 7.5 mm grooved bonded joints transmitted higher strains to the concrete surface.

From the pull-out long-term tests can be stated that:

- For equal bonded lengths, as the sustained load level increased, the slip in the loaded end increased as well.
- High sustained load levels (30% P_u) caused higher increment of slip in short bonded lengths ($L_b = 150$ mm) than in long-bonded lengths ($L_b = 225$ mm), caused by the fact that short-bonded length specimens might be under higher average bond shear stress.
- For a sustained load level of 15% P_u, similar increment of slip were obtained in short and long bonded lengths. For low sustained load levels, increasing the bonded length and the sustained load (in terms of absolute values) did not affect on the slip in the loaded end.

- Slightly higher slips were obtained in 7.5 mm grooved specimens. 10 mm grooved specimens seems to redistribute better the bond shear stresses along the bonded length.

Acknowledgements The authors acknowledge the support provided by the Spanish Government (MINECO), Project Ref. BIA2017-84975-C2-2-P. The first author acknowledges the University of Girona for conceding the IFUdG grant (IFUdG2018/28). The authors also wish to acknowledge the support of Sika Group for supplying the strips and the epoxy resin used in this study.

References

1. ASTM C469/C469M-10: Standard Test Method for Static Modulus of Elasticity and Poisson's Ratio of Concrete in Compression, West Conshohocken, PA (2010)
2. Borchert, K., Zilch, K.: Bond behaviour of NSM FRP strips in service. Struct. Concr. **9**(3), 127–142 (2008)
3. Comité Euro-International du Betón: Serviceability Models—behaviour and modelling in serviceability limit states including repeated and sustained load sustained loads. CEB, Bulletin d'information No. 235, Lausanne, Switzerland (1997)
4. Comité Europeen de Normalisation: Eurocode 2: Design of concrete structures: Part 1-1: 437 General rules and rules for buildings (2004)
5. Emara, M., Barris, C., Baena, M., Torres, L., Barros, J.: Bond behavior of NSM CFRP laminates in concrete under sustained loading. Constr. Build. Mater. **177**, 237–246 (2018)
6. Emara, M., Torres, L., Baena, M., Barris, C., Cahís, X.: Bond response of NSM CFRP strips in concrete under sustained loading and different temperature and humidity conditions. Compos. Struct. **192**, 1–7 (2018)
7. Emara, M., Torres, L., Baena, M., Barris, C., Moawad, M.: Effect of sustained loading and environmental conditions on the creep behavior of an epoxy adhesive for concrete structures strengthened with CFRP laminates. Compos. Part B Eng. **129**, 88–96 (2017)
8. Ferrier, E., Michel, L., Jurkiewiez, B., Hamelin, P.: Creep behavior of adhesives used for external FRP strengthening of RC structures. Constr. Build. Mater. **25**(2), 461–467 (2011)
9. Hassan, T., Rizkalla, S.: Investigation of bond in concrete structures strengthened with near surface mounted carbon fiber reinforced polymer strips. J. Compos. Constr. **7**(3), 248–257 (2003)
10. ISO 527-2:2012: Determination of tensile properties-Part 2: Test conditions for moulding and extrusion plastics, Geneva, Switzerland (2012)
11. ISO 527-5:2009: Determination of tensile properties-Part 5: Test conditions for unidirectional fibre-reinforced plastic composites, Geneva, Switzerland (2009)
12. Jeong, Y., Lee, J., Kim, W.S.: Modeling and measurement of sustained loading and temperature-dependent deformation of carbon fiber-reinforced polymer bonded to concrete. Materials **8**(2), 435–450 (2015)
13. Mazzotti, C., Savoia, M.: Stress redistribution along the interface between concrete and FRP subject to long-term loading. Adv. Struct. Eng. **12**(5), 651–662 (2009)
14. Mazzotti, C., Savoia, M., Ferracuti, B.: A new single-shear set-up for stable debonding of FRP-concrete joints. Constr. Build. Mater. **23**(4), 1529–1537 (2009)
15. UNE 12390-3: Testing hardened concrete. Part 3: Compressive strength of test specimens. AENOR, Madrid (2003)

Assessment of Different Coastal Defence Structures to Promote Wave Energy Dissipation and Sediments Retention

B. F. V. Vieira, J. L. S. Pinho, and J. A. O. Barros

Abstract Coastal areas are an apprized environment by society that will continue to expand rapidly. Traditional coastal protection structures are commonly deployed to protect coastal areas endangered by natural extreme weather events. However, due to their limited efficiency and very high costs, more efficient and sustainable strategies to deal with coastal erosion are imperative. This research work focuses on the assessment of engineering solutions to mitigate and delay coastal erosion. Three different structure geometries (triangular prism shape, single detached breakwater and group of two detached breakwaters) are analysed on a realistic bathymetry, using a combination of numerical models (SWAN and XBeach) to study the influence of those structures on the coastal hydro- and morphodynamics. SWAN was used for hydrodynamics and XBeach for hydrodynamics and morphodynamics assessments. In addition, a comparison between SWAN and XBeach hydrodynamics results was also performed. Structures considered in this study have regular shaped geometries, and are characterized in terms of their efficiency regarding wave height and wave energy dissipation considering different wave regimes and performance in terms of long-term beach morphodynamic impact (sediments accumulation and erosion). The analysis is concentrated in two scenarios, one for low and the other for highly energetic hydrodynamics (the most challenging to coastal zones defence). The obtained results allowed classifying their performance in terms of the impact on wave energy and wave height dissipation, and sediment erosion/deposition patterns.

Keywords Hydrodynamics · Morphodynamics · Numerical modelling · Coastal structures

B. F. V. Vieira (✉) · J. L. S. Pinho · J. A. O. Barros
Department of Civil Engineering, School of Engineering, University of Minho, Braga 4710-057, Portugal

© The Author(s), under exclusive license to Springer Nature Switzerland AG 2022
J. Sena-Cruz et al. (eds.), *Proceedings of the 3rd RILEM Spring Convention and Conference (RSCC 2020)*, RILEM Bookseries 34,
https://doi.org/10.1007/978-3-030-76465-4_4

33

1 Introduction

Coastal areas are a much-appreciated environment by society, and support a large amount of economic and leisure activities [1]. Due to coastal ecosystems vulnerability to natural and anthropogenic hazards, a range of challenges, including coastal erosion and rapid urbanization are contributing to their environmental degradation [2–4].

Projections presented by the Intergovernmental Panel on Climate Change (IPCC) indicate that global climate change may rise sea level, and, in some areas, increase the frequency and severity of storms [5]. Coastal protection on zones prone to shoreline retreat due to high tide/wave energy action and high sediment transport deficit may involve different solutions to control coastal erosion. However, traditional hard structures are not necessarily the most adequate solution since they can generate adverse effects such as: aggravation of erosion downdrift; disturbance of sediment supply and beach reduction; and adverse visual impacts [6, 7]. Coastal defence structures should contribute to the dissipation of wave energy before reaching the beach, minimizing, this way, erosion. It is important to highlight that the implementation of coastal structures may locally reduce risks of exposure to sea action, but does not eliminate them, and these should not be used as an excuse to allow building in areas of risk.

Behind the search for more efficient and sustainable strategies to deal with coastal retreat, this study focuses on a comparison between the performance of two traditional coastal protection solutions (single detached breakwater and group of two detached breakwaters) and a different structure shape on a particular coastal stretch. In order to analyse the hydro- (wave height and wave energy dissipation) and morphodynamics (sediments accumulation and erosion areas) of the structures and beach interactions, two computer programs were used: SWAN [8] for hydrodynamics and XBeach [9] for hydrodynamics and morphodynamics. In addition, a comparison between SWAN and XBeach hydrodynamics results was also performed, using three SWAN models for each one of the assessed structures and three XBeach models for the same structures.

2 Methodology

2.1 SWAN and XBeach numerical models

SWAN (Simulation WAves Nearshore) can be used as stand-alone application, but it is also included in the Delft3D 4 Suite [10]. SWAN [8] is a spectral wave model for obtaining realistic estimates of wave propagation in coastal areas, lakes and estuaries. The model is based on the wave action balance equation, and simulates the wave propagation from deep waters to the transition zone considering the physical processes of refraction, diffraction, shoaling, currents interaction, wave growth by wind action, wave breaking under the influence of excess slope, power dissipation due to bottom friction, blocking and reflection by opposing currents and transmission through obstacles. The spread of wave propagation in stationary or non-stationary modes, in the geographical and spectral spaces is performed using implicit numerical schemes. The data required for the implementation of SWAN are bathymetry of the model area and wave conditions at the open boundaries. Among the several results obtained by SWAN, these are the ones that stand out: significant wave heights (H_s), peak and average time periods (T), peak and average directions, directional dispersion, and level of water anywhere in the computational domain [11]. The application of SWAN at ocean scales is not recommended from an efficiency point of view. SWAN does not calculate wave-induced currents, and is not applicable to shallow waters (it is only valid to deep and transitional waters).

XBeach [9] numerical model is used for the computation of 2D-horizontal nearshore hydrodynamics due to wave propagation, including surf-beat (long period waves), average flow, and wave-induced currents in combination with non-cohesive sediment transports, overwash (wave uprush over a natural or artificial coastal barrier), scour around hard structures, and morphological changes of the nearshore beaches and dunes during storm events. XBeach concurrently solves the time-dependent short wave action balance, the roller energy equations, the nonlinear shallow water equations (NSWE) of mass and momentum, the sediment transport and bed change equations. As boundary conditions, XBeach requires tidal levels, deeper-water (outside the surf zone) wave conditions and bathymetry [12]. Users are allowed to choose which mode options to implement: stationary wave mode, surf-beat mode (instationary), and non-hydrostatic mode (wave-resolving: the most computationally expensive mode, because it requires higher spatial resolution and associated smaller time steps) [13]. Further information on these modes can be consulted in [14]. The model accounts for feedback between the evolving bathymetry and the hydrodynamics at each time step.

2.2 Simulated Scenarios

For this study, three different geometries for coastal protection structures were analysed. For better understanding the performance of a different shaped structure (Fig. 1a), the impact of this structure on a coastal zone was compared with a typical detached breakwater (Fig. 1b) and a group of detached breakwaters (Fig. 1c). The case study developed by [15] is taken as a reference for the modelling of the different shaped structure that resembles a triangular prism, where the structure performance was conducted in terms of 'surfability' and coastal protection. Reference [15] analysed the performance of two triangular prism structure geometries differing in their opening angles (45° and 66°) for two different incident wave conditions (frequent wave: $H_s = 1.5$ m, T $= 9$ s; and storm wave: $H_s = 4$ m, T $= 15$ s) and concluded that both geometries contribute to sedimentation. For the current study, results analysis will focus on coastal protection purposes and it will consider the structure geometry of 45° opening angle due to its wider shadow zone benefits for coastal protection. As geometrical considerations, all structures share the same length (250 m), except the

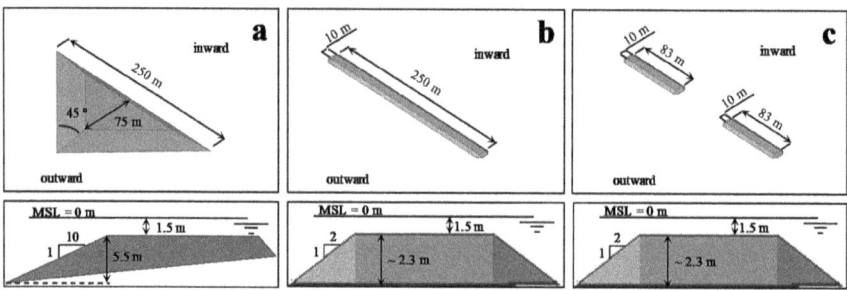

Fig. 1 Geometrical shapes considered in this study (upper panel: plan view; lower panel: cross-section): **a** triangular prism structure; **b** regular detached breakwater; **c** group of detached breakwaters

two detached breakwaters in Fig. 1c that presents a third of the length considered in the other structures (83 m). The detached breakwaters were designed to have a crest width of 10 m and side-slopes of 1:2, which are representative of regular structures of this type as presented in [16]. The triangular prism structure followed [15] design with a crest width of 75 m and side slopes of 1:10. All structures have their crests submerged at -1.5 m relatively to mean sea level and are located at the same distance from the shoreline (440 m).

In order to study the structure's influence on significant wave heights attenuation and sediments accumulation, a realistic bathymetry was used for the model simulations, based on [15]'s study. Regarding model conditions, the computational domain is 1670 m × 1870 m (crosshore × longshore) with a node spacing of dx = dy = 5.0 m for both hydro- and morphodynamics analysis. The total simulation time was two hours for the hydrodynamics analysis (using SWAN and XBeach) and one day for morphodynamics (XBeach) with a morphological acceleration factor to speed up the morphological time scale relative to the hydrodynamic timescale (morfac) of 50.

The study by [17] on the simulation time for morphodynamics has concluded that the necessary time (in days) to study the morphological development process of a significant salient due to a detached breakwater placed 500 m away from the shoreline was seventy-five days with a morfac equal to 100 (approximately twenty years). Even though the simulation time considered in this study is significantly shorter than the [17]'s (justified by the need to reduce calculation time), it is adequate to study the sediments transport tendency to erosion or accretion.

For hydrodynamics (using SWAN and XBeach), the frequent wave and the storm wave conditions considered in [15] were adopted in this study, with waves incoming from the West direction (270°), while the tide level considered was 0 m (mean sea level). For morphodynamics analysis (using XBeach), only the frequent wave scenario was analysed, in order to give insights on the response to a mean wave climate. Regarding seabed composition, the sediments dimensions considered were 200 μm for D_{50} and 300 μm for D_{90}, being D_{50} and D_{90} common metrics used to describe particle size distributions. In this case, D_{50} means that 50% of the sample has a size of 200 μm or smaller and D_{90} that 90% has a size of 300 μm or smaller. Boundary conditions for SWAN model (hydrodynamics) were defined for North, West and South boundaries (frequent and storm wave conditions), while for XBeach model (hydrodynamics and morphodynamics) absorbing-generating (weakly-reflective) boundary in 2D (abs_2d) for front and back boundaries, and wall boundary condition (simple no flux boundary condition) for left and right boundaries were defined. In XBeach model the left and right designations correspond to North and South, while front and back to West and East, respectively.

For hydrodynamics (SWAN and XBeach), the wave type considered was a JONSWAP spectrum, whereas for morphodynamics (XBeach) the stationary mode was selected. The consideration of stationary mode is justified by the need to reduce model calculation time for morphodynamics. The numerical model results obtained by the SWAN model are the significant wave heights, while XBeach model estimates the sediments accumulation and erosion near shoreline as well as the bed level. XBeach presents H_{rms} values (root mean square wave heights), which require a conversion to H_s (significant wave heights), in order to be compared to SWAN hydrodynamics results. The wave energy dissipation also requires separate calculation.

3   Results and Discussion

3.1   Hydrodynamics

Numerical simulations for the analysis of significant wave heights dissipation for two different scenarios (frequent wave and storm wave conditions) were performed for each structure: triangular prism structure (Fig. 2a1 and a2), detached breakwater (Fig. 2b1 and b2) and group of detached breakwaters (Fig. 2c1 and c2). Comparatively to a situation without structure, an analysis of the influence of each structure on the significant wave heights on a storm wave condition was also performed (Fig. 2a3, b3 and c3). For a more legible analysis, the presented results are centred in a limited window around the structures. The contour lines are also depicted in all presented results.

From the results presented in Fig. 2, wave shoaling (increase of the wave height) in every structure shape is evident due to a decrease of the depth. This phenomenon is visible at the apex of the triangular prism structure for both wave conditions (represented with number 1 in Fig. 2a1 and a2); at the North extremity of the detached breakwater and group of detached breakwaters for the storm wave condition (represented with number 2 in Fig. 2b2 and c2); a small wave shoaling along the detached breakwater for both wave conditions (represented with number 3 in Fig. 2b1 and b2), and also a small wave shoaling along the group of detached breakwaters for both wave conditions (also represented with number 3 in Fig. 2c1 and 2c2). A more intense variation of bottom elevation due to the presence of the detached breakwaters near the North extremities, relatively to the South extremities, may explain the wave

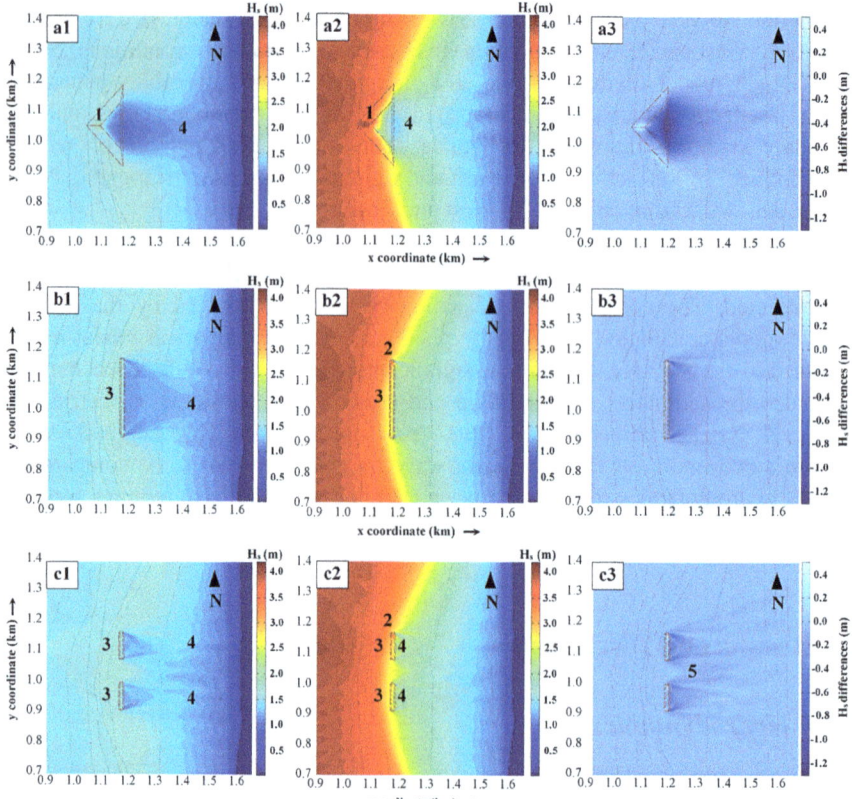

Fig. 2 Significant wave heights dissipation for frequent wave condition: triangular prism structure (**a1**); detached breakwater (**b1**), and group of detached breakwaters (**c1**); and storm wave condition: triangular prism structure (**a2**), detached breakwater (**b2**), and group of detached breakwaters (**c2**). Significant wave heights difference with and without any structure on a storm wave condition: triangular prism structure (**a3**), detached breakwater (**b3**), and group of detached breakwaters (**c3**)

shoaling at this particular area (2). The most evident wave shoaling is at the apex of the triangular prism structure (1), while on the other mentioned cases (2 and 3) the differences are more subtle. Regarding significant wave heights decrease, it is clear the effect for every structure shape under both wave conditions. This dissipation is more significant for the storm wave condition. For the triangular prism structure, a progressive increase on significant wave heights is visible from the structure inward (protected) extremity to position 1.45 km for frequent wave condition (represented with number 4 in Fig. 2a1) and 1.28 km for storm wave condition (also represented with number 4 in Fig. 2a2). After the position 1.28 km for storm wave condition, the significant wave heights progressively decrease towards shoreline. These phenomena are explained by the shoaling and breaking due to depth decrease. For the detached breakwater this phenomenon is not visible for the storm wave condition, but it is present for the frequent wave where an increase of significant wave height is perceptible from the structure inward extremity to position 1.42 km (also represented with number 4 in Fig. 2b1). From this position towards shoreline, the significant wave heights decrease gradually.

Finally, for the group of detached breakwaters a small increase on the significant wave height is visible from the structure inward extremity to position 1.21 km for the storm wave condition (also represented with number 4 in Fig. 2c2), and also an increase of significant wave height from the structure inward extremity to position 1.42 km for the frequent wave condition (also represented with number 4 in Fig. 2c1). After those positions, the significant wave heights decrease towards shoreline. Results for the North and South detached breakwater from the group of detached breakwaters are similar.

Relatively to the influence of a structure on significant wave heights, it is clear that all structures contribute to a decrease and that the triangular prism structure reduces significant wave heights at a larger scale (Fig. 2a3) than the detached breakwater and group of detached breakwaters. On the group of detached breakwaters, even though there is a gap between the structures, a small significant wave height reduction at the inward side is observable in Fig. 2c3 (represented with number 5). Near the shoreline, although the three cases (Fig. 2a3, b3 and c3) do not present any significant differences, it can be concluded that the triangular prism structure presents a more significant and larger shadow zone than the other two solutions. This effect can bring protection benefits if the structure is positioned closer to shoreline.

The XBeach numerical results for root mean square wave heights (H_{rms}) were converted to significant wave heights (H_s) using Eq. (1) [18].

$$H_s = \sqrt{2} \times H_{rms} \qquad (1)$$

The plots depicted in Fig. 3 present the similarities and differences between the significant wave heights results for SWAN and XBeach. The plots show the results for a cross-section at each structure. Results for the South detached breakwater from the group of detached breakwaters are not presented in this study due to results similarity to the North structure. The indicated values in the plots were selected for four positions: before, immediately before, immediately after and after the structures. The vertical lines plotted represent relevant seabed slope changes.

As mentioned before, the wave shoaling due to a sudden depth change immediately before the structure is visible for all cases with both numerical models. From the analysis of Fig. 3, it is clear that both SWAN and XBeach simulate this phenomenon for all structures and that for the all storm wave conditions, SWAN simulates a small wave shoaling due to the slope change at positions 625 and 1125 m, and a wave shoaling for all frequent wave condition at position 1125 m. The XBeach model only represents a significant wave shoaling for the frequent and storm wave condition for the triangular prism structure at position 1125 m.

Equations (2), (3) and (4) were, respectively, applied in order to proceed to: an overall comparison between SWAN and XBeach hydrodynamics results (Δ); an overall percentage of significant wave height results reduction before and after the

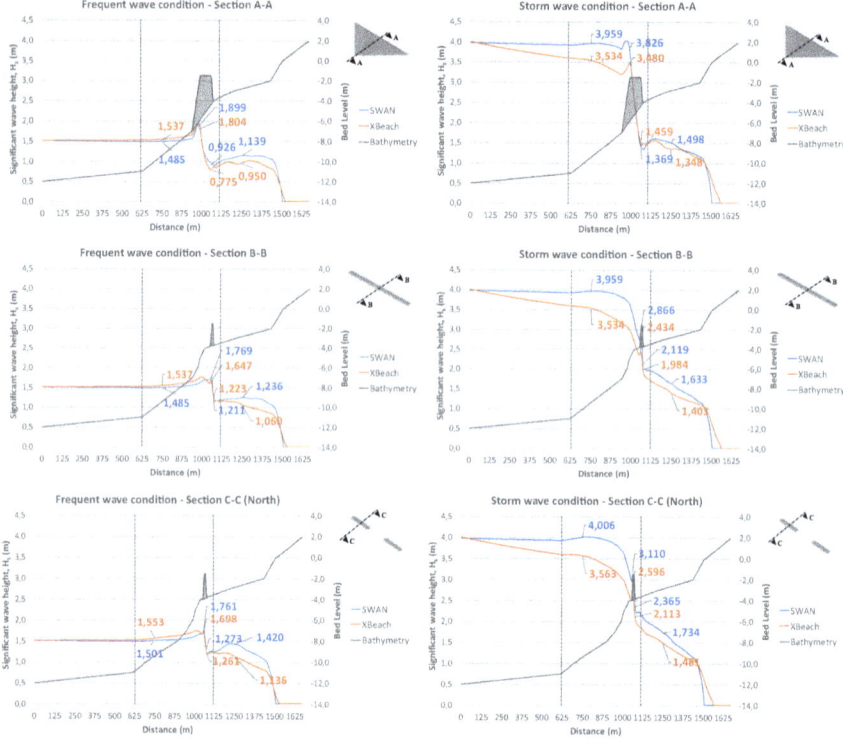

Fig. 3 Comparison between SWAN and XBeach significant wave height results for frequent (H_s = 1.5 m, T = 9 s) and storm (H_s = 4 m, T = 15 s) conditions for: triangular prism structure (Section A-A), detached breakwater (Section B-B) and North detached breakwater from the group of detached breakwaters (Section C-C)

structures for both wave conditions (α); and their wave energy reduction results (β). Eq. (5) represents the wave energy equation (E) used for this study.

$$\Delta_{SWAN-XBeach} = H_{sSWAN} - H_{sXBeach} \qquad (2)$$

$$\alpha = \left(H_{sBefore} - H_{sAfter}\right)/H_{sBefore} \times 100 \qquad (3)$$

$$\beta = \left(E_{Before} - E_{After}\right)/E_{Before} \times 100 \qquad (4)$$

$$E = \frac{1}{8}\left(\rho g H^2\right) \qquad (5)$$

where $\Delta_{SWAN\text{-}XBeach}$ is the comparison between significant wave height results between both models for a specific location (m); $H_{s\,SWAN}$ and $H_{s\,XBeach}$ are the significant wave height computed by, respectively, SWAN and XBeach model for a specific location (m); α is the numerical model H_s reduction for a specific location (%); $H_{s\,Before}$ and $H_{s\,After}$ are the significant wave height, respectively, before and after, a structure (m); β is the numerical model E reduction for a specific location (%); E_{Before} and E_{After} are the wave energy, respectively, before and after, a structure (J/m^2); E is the wave energy per unit area (J/m^2); ρ is the water density (kg/m^3); g is the gravity acceleration (m/s^2); and H is the wave height (m). Considering $\rho_{seawater} = 1025$ kg/m^3 and $g = 9.81$ m/s^2 the wave energy was computed for all scenarios.

From Fig. 3 and using Eq. (2), it is clear that for the frequent wave condition, the SWAN results before the structures (triangular prism; detached breakwater; and group of detached breakwaters) calculates smaller significant wave heights than the XBeach model (-0.05; -0.05; -0.05), and greater significant wave heights after the structures (0.19, 0.18, 0.28). For the storm wave condition, SWAN model computes greater significant wave heights than the XBeach model before (0.42; 0.42; 0.44) and after (0.15; 0.23; 0.25) the structures.

Regarding the difference between significant wave heights before and after the structures, using Eq. (3), XBeach presents greater reduction values than the SWAN model. Similar reduction values for significant wave height (SWAN, XBeach) can be seen for the storm wave conditions [(62.16, 62.85%); (58.75, 60.30%); (56.71, 58.43%)], while for the frequent wave condition significant differences are evident [(23.30, 38.21%); (16.77, 31.00%); (5.40, 26.84%)] (triangular prism; detached breakwater; and group of detached breakwaters). For both scenarios, the triangular prism structure has the best performance, due to higher reduction values, whereas the group of detached breakwater is the least effective.

Overall, for extreme wave conditions, results for significant wave height reductions for both SWAN and XBeach models are similar, which indicates that, even though the significant wave heights calculated are different, the performance for each structure is comparable. Since wave energy is proportional to wave heights (Eq. (5)), the same conclusion for wave energy can be taken (Eq. (4)).

3.2 Morphodynamics

One of the suggestions proposed by [15] was to study morphodynamics around the structure to enable a deeper understanding on sedimentation and erosion areas. In order to develop that study, XBeach morphodynamic models for each one of the structures were performed. Also, a study developed by [17] where the XBeach model was used to analyse salient and tombolo formations for different detached breakwater conditions was considered for testing the results quality and ensure the adequate models conditions.

Numerical simulations for the analysis of cumulative sedimentation and erosion for the frequent wave condition were conducted for each shape: triangular prism

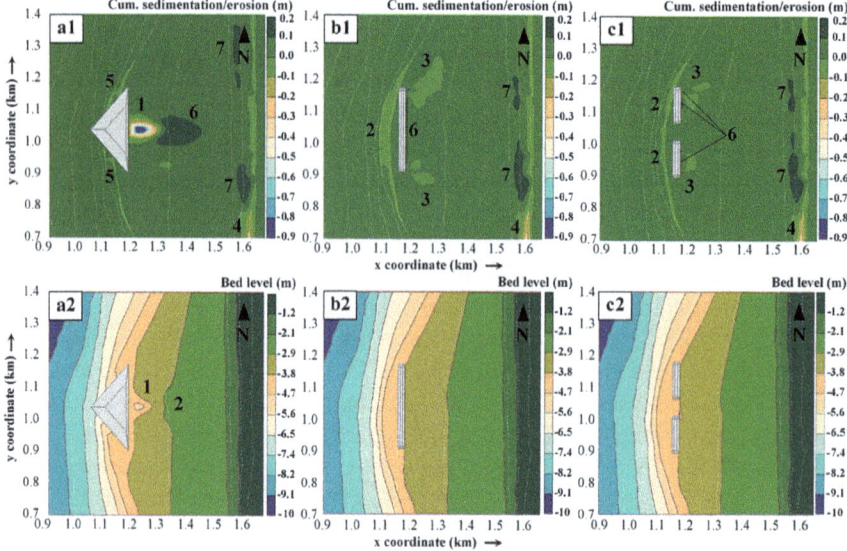

Fig. 4 Cumulative sedimentation and erosion for frequent wave condition: triangular prism structure (**a1**); detached breakwater (**b1**), group of detached breakwaters (**c1**). Bed level for frequent wave condition: triangular prism structure (**a2**); detached breakwater (**b2**), group of detached breakwaters (**c2**)

structure (Fig. 4a1), detached breakwater (Fig. 4b1) and group of detached breakwaters (Fig. 4c1). Fig. 4a2, b2 and c2 show the bed level evolution at the end of the XBeach simulation for each structure shape. The study was performed for one day with a morfac of 50, which insights results for fifty days. The simulations were taken for a frequent wave condition and the results presented are cropped for a more legible analysis. From the results depicted in Fig. 4a1, b1, and c1, it is evident the most significant erosion areas immediately after the triangular prism structure (1) (over −0.9 m); outwards the detached breakwater and group of detached breakwaters (2) due to waves reflection (−0.1 m); at the North and South extremities of the detached breakwater and group of detached breakwaters (3) due to waves diffraction (−0.1 m); and downdrift near the shoreline of all the three structures (4) (−0.3 m). Along the shoreline, a very small erosion with the same magnitude (−0.1 m) for the three scenarios is also evident. Outwards the triangular prism structure (5), it is also noticeable a slight erosion due to waves reflection (−0.1 m). Regarding the erosion on the detached breakwater and group of detached breakwaters (marked with numbers 2 and 3), it is noticeable a more intense phenomenon on the detached breakwater scenario. These erosion areas near the structures may put at risk the structures stability due to scouring. Near the shoreline, the downdrift erosion on the detached breakwater and group of detached breakwaters scenarios (4) is slightly more intense than the erosion in the triangular prism structure scenario (4). In the group of detached breakwaters there is no erosion at the gap between the structures.

Regarding sedimentation areas, it is visible a significant sediments accretion of 0.2 m inwards the triangular prism structure (6) immediately after the erosion area, which may suggest that part of the sediments on the eroded area settled further ahead. Along the shoreline, in the triangular prism structure there is a significant sedimentation updrift and downdrift the structure (7) (0.2 m), while on the detached breakwater and group of detached breakwaters the corresponding sedimentation (7) is located at the structures protected region. Immediately after the detached breakwater and group of detached breakwaters (6) there is also a slight sediments accretion (0.2 m). From the results presented in Fig. 4a2, b2, and c2, it can be concluded that there were no significant changes in morphology in all scenarios, except for immediately after the triangular prism structure (Fig. 4a2) where the erosion (1) and sedimentation (2) phenomena altered the bed level.

4 Conclusions

This study aimed to assess the performance of a triangular prism structure and two traditional coastal protection solutions in dissipating wave energy and protecting the beach using numerical models implemented with SWAN and XBeach. In addition, a comparison between SWAN and XBeach hydrodynamics results was also performed. The triangular prism structure characteristics were based on a previous [15] research work, and all structures analysed were simulated for storm and frequent wave conditions. Regarding significant wave heights results, a substantial decrease between before and after every structure shape is clear for both wave scenarios (especially during storm wave conditions), which indicates that all structures have great influence on reducing significant wave heights and wave energy. Amongst all structures, for both scenarios, the triangular prism was the best for reducing significant wave heights at a larger scale and a larger shadow zone compared to the detached breakwater and group of detached breakwaters, which can bring protection benefits if the structure is positioned closer to shoreline. The least effective structure is the group of detached breakwaters.

Comparing both models, it can be concluded that, overall, SWAN numerical model tends to present greater significant wave heights results; and that XBeach presents greater percentages of significant wave heights reduction for frequent wave conditions, and similar values for storm wave conditions. This similarity indicates that, even though the significant wave heights calculated are different, the performance for each structure is comparable. The same conclusions can be taken for wave energy results.

Regarding morphodynamics, the obtained results present a favourable tendency to sediments accretion near the shoreline, and at the inward areas for the three structures, since the greatest values for sediments accumulation are located at these sites. The most significant sediments accretion at the shoreline is noticeable for the group of detached breakwaters, while the largest overall sediments accumulation is visible for the triangular prism structure. The accretion and erosion patterns along the shoreline for the three structures scenarios are similar for the fifty days insights. It is important to note that erosion areas near all structures jeopardise their stability due to local scouring. In the simulated numerical models, the widely known erosion effects near the detached breakwaters (simple and group) are evident due to waves reflection and diffraction phenomena. The obtained results suggest that the longer the detached breakwater, the more intense the erosion is near the structure. The largest overall erosion is located inwards the triangular prism structure.

In this study, contrary to [17]'s, patterns of salient formations are not created for these conditions, since there were no evidences of significant changes in bed level results. It is important to mention that the author's study highlighted an emerged structure, which ensures greater sediments retention, and a seventy-five days simulation time. For further studies, emerged structures and/or structures placed closer to shoreline should be better analysed. Also, a longer morphodynamics simulation time and the consideration of a JONSWAP spectrum for the XBeach morphodynamics analysis may present more realistic morphodynamics results.

Acknowledgements This work is supported by the Portuguese Foundation for Science and Technology (FCT) [PhD grant number SFRH/BD/141381/2018]. The authors also acknowledge the support provided by the project EcOffShorBe—Eco Offshore Built Environment, n. 37417, R&D cores in Copromoção, 14/SI/2017, NORTE-01-0247-FEDER-037417, supported by ANI (FEDER).

References

1. Castelle, B., Guillot, B., Marieu, V., Chaumillon, E., Hanquiez, V., Bujan, S., Poppeschi, C.: Spatial and temporal patterns of shoreline change of a 280-km high-energy disrupted sandy coast from 1950 to 2014: SW France. Estuar. Coast. Shelf Sci. **200**, 212–223 (2018). https://doi.org/10.1016/j.ecss.2017.11.005
2. Narra, P., Coelho, C., Sancho, F., Palalane, J.: CERA: An open-source tool for coastal erosion risk assessment. Ocean Coast. Manag. **142**, 1–14 (2017). https://doi.org/10.1016/j.ocecoaman.2017.03.013
3. United Nations, "Transforming Our World: The 2030 Agenda for Sustainable Development A/RES/70/1," UN General Assembly (2015)
4. Weinberg, J.: "The Big Squeeze: Coastal megacities face growing pressure from sea and land," *Stockholm Waterfront*, no. 1, Stockholm, Sweden, pp. 5–7 (2015)
5. Gilbert, J., Vellinga, P.: Climate change: the IPCC response strategies—Chapter 5. In: Report prepared for Intergovernmental Panel on Climate Change by Working Groups III, Digitization and Microform Unit (2010), p. 330, UNOG Library (1990)
6. Rangel-Buitrago, N., Williams, A., Anfuso, G.: Hard protection structures as a principal coastal erosion management strategy along the Caribbean coast of Colombia. A chronicle of pitfalls. Ocean Coast. Manag. **156**, 58–75 (2018). https://doi.org/10.1016/j.ocecoaman.2017.04.006
7. Williams, A., Rangel-Buitrago, N., Pranzini, E., Anfuso, G.: The management of coastal erosion. Ocean Coast. Manag. **156**, 4–20 (2018). https://doi.org/10.1016/j.ocecoaman.2017.03.022
8. SWAN: SWAN manual (2018). http://swanmodel.sourceforge.net/online_doc/swanuse/node3.html (Accessed 08 Feb 2019)
9. Deltares: XBeach (2019) https://www.deltares.nl/en/software/xbeach (Accessed 12 Feb 2019)

10. Deltares: Delft3D 4 Suite (structured) (2019) https://www.deltares.nl/en/software/delft3d-4-suite (Accessed 07 Feb 2019)
11. Capitão, R., Fortes, C.: Análise comparativa entre estimativas do modelo SWAN e medições de agitação marítima efectuadas na Praia da Amoreira, Portugal. Rev. da Gestão Costeira Integr. **11**(3), 283–296 (2011). https://doi.org/10.5894/rgci269
12. Bolle, A., Mercelis, P., Roelvink, D., Haerens, P., Trouw, K.: Application and validation of Xbeach for three different field sites. In: Proceedings of 32nd Conference on Coastal Engineering, 30th to 5th July 2010, pp. 1–14 (2010). https://doi.org/10.9753/icce.v32.sediment.40
13. Roelvink, D., McCall, R., Mehvar, S., Nederhoff, K., Dastgheib, A.: Improving predictions of swash dynamics in XBeach: The role of groupiness and incident-band runup. Coast. Eng. **134**, 103–123 (2018). https://doi.org/10.1016/j.coastaleng.2017.07.004
14. Roelvink, D.J.A., van Dongeren, A., McCall, R., Hoonhout, B., van Rooijen, A., van Geer, P., de Vet, L., Nederhoff, K., Quataert, E.: XBeach Technical Reference: Kingsday Release (2015)
15. Mendonça, A., Fortes, C.J., Capitão, R., Neves, M.G., Moura, T., Antunes do Carmo, J.S.: Wave hydrodynamics around a multi-functional artificial reef at Leirosa. J. Coast. Conserv. **16**(4), 543–553 (2012). https://doi.org/10.1007/s11852-012-0196-1
16. Vieira, B.F.V.: Wave hydrodynamics in coastal stretches influenced by detached breakwaters. MSc Thesis in Civil Engineering, University of Minho (2014)
17. Razak, M.S.A., Nor, N.A.Z.M.: XBeach process-based modelling of coastal morphological features near breakwater. MATEC Web Conf. **203**(1007) (2018). https://doi.org/10.1051/matecconf/201820301007
18. Hanson, H.: Wave transformation (2019). http://www.tvrl.lth.se/fileadmin/tvrl/files/vvr040/3_Wave_transformation_3pp.pdf (Accessed 28 Oct 2019)

Concrete Drying Modelling in a Variable Temperature Environment

Jean-Luc D. Adia, Herman Koala, Justin Kinda, Julien Sanahuja, and Laurent Charpin

Abstract EDF studies the behavior of double walls Concrete Containment Buildings (CCB) in nuclear power plant, for which the inner wall is prestressed and has no metallic liner. The inner wall of CCB is subjected to delayed strains of concrete (which are highly related to concrete drying) causing prestress loss and increased leakage with time. Moreover, the permeability of concrete itself is vastly dependent on its water saturation ratio. For these reasons, a good prediction of strains and leaktightness of CCB must start with a good estimate of the moisture profile on concrete. On the EDF Lab Les Renardières site in France, the VERCORS mock-up, a 1/3 scale CCB without a liner, has been built to improve the understanding and the modelling capabilities of ageing and leakage of double walls CCB. The inner wall of the VERCORS mock-up is subjected to a variable temperature drying environment, both in space and in time. In this work, a simplified drying model which take into account effects of temperature on concrete desorption isotherm and drying kinetics is proposed. In order to define a strategy for fitting model parameters, a sensitivity analysis based on Sobol' indices is performed. Next, the model is calibrated against independent sets of experimental data available on VERCORS concrete specimens up to temperatures of $70\,°C$. The calibrated model is validated by predicting the mass loss kinetics of several specimens which experience variable temperature drying conditions. Then, the fitted model is compared to direct RH profiles measurements of $40\,cm$ thick blocks installed at different locations in the VERCORS mock-up. Finally, the prediction of the calibrated model is compared to the saturation profile of VERCORS mock-up measured by Time Domain Reflectometry (TDR) method.

Keywords Drying · Desorption isotherm · Temperature effects · Saturation profile · Concrete · VERCORS mock-up

J.-L. D. Adia (✉) · H. Koala · J. Kinda · J. Sanahuja · L. Charpin
EDF Lab Les Renardières, Avenue des Renardières, 77250 écuelles, France
e-mail: jean-luc.adia@edf.fr

H. Koala
École CentraleSupelec, 3 Rue Joliot Curie, 91190 Gif-sur-Yvette, France

J. Kinda
École normale supérieure Paris-Saclay, 61 Avenue du Président Wilson, 94230 Cachan, France

© The Author(s), under exclusive license to Springer Nature Switzerland AG 2022
J. Sena-Cruz et al. (eds.), *Proceedings of the 3rd RILEM Spring Convention and Conference (RSCC 2020)*, RILEM Bookseries 34,
https://doi.org/10.1007/978-3-030-76465-4_5

47

1 Introduction

Electricité de France (EDF) operates 58 nuclear power plants (NPP) in France. On each plants (Pressurized Water Reactors technology, PWR), the final barrier against accidental release of radioactive fission products to the environment consist in a prestressed concrete containment building (CCB). This structure is not replaceable, and the issue of ageing management has to be addressed with specific care for Long Term Operation (LTO). However, as for any other civil prestressed structure, these tendons forces decrease slowly with time, due to the effects of shrinkage and creep in concrete and to the steel tendons relaxation. In addition, for the CCB without steel liner, containment leak tightness seems to be closely linked to the available prestressing forces if the internal pressure rises. It is well known that shrinkage and creep of concrete are depending on concrete moisture or water content [5, 15]. In France, CCB are about 1 m thick, or more, and then the drying is slow. Therefore effects of shrinkage and creep are still sensitive and can entail significant prestressing losses even near the end of operation time. A correct assessment of water content throughout the life of the CCB is then a critical issue for LTO.

At EDF R&D, the VERCORS mock-up, a 1/3 scale CCB, is built to improve the understanding and the modelling capabilities of ageing and leakage of double walls CCB. The monitoring system of the VERCORS mock-up includes two innovative technologies which are tested to measure water content: pressure pulse decay technique "pulse" [2] and Time Domain Reflectometry (TDR) method [30]. A calibration on concrete tests specimen is necessary for both methods to establish the relation between the water content in the material and the measured value (apparent gas permeability or apparent relative permittivity). The characterization of the drying behavior of VERCORS concrete has been performed with two main different kinds of tests. First, the desorption isotherms have been measured at different temperatures (20, 25 and 70 °C) using a testing protocol described in [4, 22, 23, 29]. Second, weighing of full samples at different humidity (80, 50, 30%) and temperature (20, 25, 40 and 70 °C) conditions with various sample dimensions (around 3, 10 and 11 cm) test campaigns is also performed [8–10].

The aim of this paper is to propose a simplified drying model which take into account effects of temperature on concrete desorption isotherm and drying kinetics. In order to analyse this experimental database on VERCORS concrete drying at various temperatures, a modelling approach based on a single diffusion-type equation by introducing a "water diffusivity" which combines the transport of liquid water by capillarity and of water vapor by diffusion with temperature effects is used. For define model parameters fitting strategy, a sensitivity analysis based on Sobol' indices [27] is carried out. In addition, the validation process of the proposed drying model consists of calibrating the material properties on the mass loss kinetics tests and comparing drying model predictions, with others mass loss kinetics which are not used for model calibration for various hygrothermal conditions in VERCORS mock-up. Finally, model applications are proposed and discussed. These are, prediction of direct RH profile measurements by capacity sensors on 40 cm thick blocks installed in the VERCORS mock-up and of the on-site measurements by TDR technique of saturation profile on the VERCORS mock-up.

2 Moisture Transport Modelling: A Richards Fick with Temperature Effects (RFT) Approach

2.1 Gouverning Equations

Under non-steady-state conditions, in concrete assumed as a porous medium partially saturated by the liquid water phase, the moisture transport is usually described at the macroscopic scale by using single diffusion-type equation (also known in this case as Richard's equation [25]). The various formulas proposed in the literature [3, 6, 18] yield writing a single non-linear diffusion-type Eq. (1), which governs the evolution of the liquid water saturation ratio S_r [-]:

$$\phi \frac{\partial S_r}{\partial t} - \text{div} \left[D_{lv} \mathbf{grad}\,(S_r) \right] = 0 \tag{1}$$

with $D_{lv} = D_l + D_v$. This drying model is based on the main assumptions [18], the gas pressure remains close to atmospheric pressure and Darcean transport of the vapor phase is negligible compared to diffusive transport. In this paper, these assumptions are assumed to be valid for moderate temperatures up to 70 °C. The apparent coefficients, D_l and D_v are respectively, liquid water and vapor water diffusivities. The coefficient, D_l, which governs the liquid water transport by capillarity is given by an apparent Darcy law:

$$D_l(S_r, T) = \frac{K_l(T) k_{rl}(S_r)}{\eta_l(T)} \frac{\partial P_c}{\partial S_r}(S_r, T) \tag{2}$$

while, the coefficient D_v which leads the water vapor diffusion is expressed by an apparent Fick law:

$$D_v(S_r, T) = \frac{D_{v0}(S_r, T)}{P_v(T)} \left(\frac{M}{\rho_l(T) R(T + 273.15)} \right)^2 \frac{\partial P_c}{\partial S_r}(S_r, T) \tag{3}$$

where T[°C] is the absolute temperature, R the universal gas constant ($R = 8.3144$ [J·mol^{-1}K^{-1}]), M the molar mass of water ($M = 18 \cdot 10^{-3}$ [Kg·mol^{-1}]). In Eqs. (2) and (3), ϕ [-] is the porosity, $K_l(T)$ [m^2] is the liquid water intrinsic permeability, $D_{v0}(S_r, T)$ [m^2/s] is the effective vapor water diffusion coefficient in the porous medium, $P_c(S_r, T)$ [Pa] is the water desorption isotherm, $\eta_l(T)$ [Pa·s] is the dynamic viscosity of liquid water, $\rho_l(T)$ [Kg/m^3] is the liquid water density, $P_v(T)$ [Pa] is the vapor pressure, and $\rho_v(T)$ [Kg/m^3] is the vapor density.

2.2 Constitutive Laws with Temperature Effects

The temperature effects (up to 70 °C) on concrete drying are mainly governed by two mechanisms. First, there is a dependence of concrete desorption isotherm on temperature [22]. When temperature increases, the general isotherm curve shape is modified (more pronounced non-linearity), the saturation at equilibrium with an arbitrary humidity h is reduced and the saturation decrease is observed over the whole h range and the higher the temperature, the reduction is greater. Second, increasing of kinetic mechanisms of concrete drying with temperature [11, 12]. In fact, all the water properties, concrete permeability, vapor diffusivity in air and local water capacity retention (desorption isotherm derivative relative to saturation ratio $\partial P_c / \partial S_r$) depends on temperature. Below, constitutive laws with respect to temperature effects are described. The liquid and vapor water properties, η_l, ρ_l, γ_{lv} [N/m](surface tension between liquid and vapor phases) and P_v, depend on the absolute temperature T. The empirical expressions given in Eqs. (4)–(9) are used.

$$\eta_l(T) = 0.6612\,(T+44.15)^{-1.562} \tag{4}$$

$$\rho_l(T) = 314.4 + 685.6\left(1 - (T/374.14)^{1/0.55}\right)^{0.55} \tag{5}$$

$$\gamma_{lv}(T) = 0.1558\,(1 - ((T+273.15)/647.1))^{1.26} \tag{6}$$

$$P_v(T) = h(S_r, T) \cdot P_{vs}(T) \tag{7}$$

where h is the relative humidity obtained by desorption isotherm Eqs. (11)–(13) and the Kelvin law:

$$h(S_r, T) = \exp\left(Pc(S_r, T)\frac{M}{\rho_l(T)RT}\right) \tag{8}$$

while, P_{vs} is vapor saturated pressure given by the Rankine equation:

$$Pvs(T) = 101325\exp\left(13.7 - (5120/(T+273.15))\right) \tag{9}$$

The porosity is assumed to be constant for temperature up to 70 °C. In the same condition, the increasing of temperature may have a coarsening effect on the pore structure. This effect is negligible on porosity but can modify the intrinsic permeability. To take into account this effect, an Arhenius law [21] is used:

$$K_l(T) = K_0 \left(\frac{T+273.15}{T_0+273.15}\right)\exp\left(-\left(\frac{E_a}{R}+273.15\right)\left(\frac{1}{T+273.15} - \frac{1}{T_0+273.15}\right)\right) \tag{10}$$

where, K_0 [m^2], T_0 [°C] and E_a/R [°C] are, the intrinsic permeability relative to liquid water at the reference temperature, the reference temperature and an energy activation. The desorption isotherm in temperature [11, 17] is expressed as follows:

$$Pc(S_r, T) = -\frac{\rho_l(T_0)RT_0}{\alpha M}\left(S_r^{-1/\beta} - 1\right)^{1-\beta}\frac{\gamma_{lv}(T_0)}{\gamma_{lv}(T)}\sqrt{\frac{K(T_0)}{K(T)}} \tag{11}$$

or,

$$S_r(h, T) = \left(1 + \left(-\alpha\frac{\rho_l(T)}{\rho_l(T_0)}\frac{\gamma_{lv}(T_0)}{\gamma_{lv}(T)}\frac{T}{T_0}\sqrt{\frac{K(T)}{K(T_0)}}\ln(h)\right)^{1/(1-\beta)}\right)^{-\beta} \tag{12}$$

with,

$$\frac{K(T)}{K(T_0)} = 10^{A_d\left(2\cdot10^{-3}(T-T_0)-10^{-6}(T-T_0)^2\right)} \tag{13}$$

In these equations, α [Pa], β [-] are the van Genuchten material parameters at the reference temperature T_0 and A_d [-] is the material parameter which leads the temperature effects [13]. The water retention capacity which controls the ability of the material to locally retain or release water is given by,

$$\frac{\partial P_c}{\partial S_r}(S_r, T) = \frac{-\beta+1}{\beta}\frac{\rho_l(T_0)RT_0}{\alpha M}\frac{\gamma(T)}{\gamma(T_0)}\sqrt{\frac{K(T)}{K(T_0)}}S_r^{-\frac{1+\beta}{\beta}}\left(S^{-1/\beta} - 1\right)^{-\beta} \tag{14}$$

Then, the relative permeability is assumed to be temperature independent and its expression is obtained following the Mualem-Van Guenucthen method [14, 20]:

$$k_{rl}(S_r) = S_r^p\left(1 - \left(1 - S_r^{1/\beta}\right)^\beta\right)^2 \tag{15}$$

here, the pore interaction factor [24] is introduced to take into account the much more pronounced tortuosity effects on permeability in concrete compared to sand. The Fickian (diffusive) transport of water vapor governed by the effective water vapor diffusion coefficient in the porous medium is,

$$D_{v0}(S_r, T) = D_0(T)R(\phi, S_r) \tag{16}$$

where D_0 [m^2/s] is the free (out of the porous medium) water vapor diffusion coefficient in the air,

$$D_0(T) = 0.217\cdot10^{-4}\left(\frac{T+273.15}{273.15}\right)^{1.88} \tag{17}$$

and $R(S_r, \phi)$ is the so-called resistance factor which accounts for both, the tortuosity effects and the reduction of space offered to gas diffusion in a partially saturated porous medium, compared to free diffusion in the air [19],

$$R(S_r, \phi) = \phi^a D_{rv}(S_r) \tag{18}$$

$$D_{rv}(S_r) = (1 - S_r)^b \tag{19}$$

with D_{rv}, a, b, the relative vapor diffusivity and the Millington parameters [19].

2.3   Initial State and Boundary Conditions

Like any transient phenomena, for solving concrete drying problem, it is mandatory to know the initial condition and the boundaries conditions. For concrete in endogenous or saturated state, the homogeneous initial state can be prescribe. Depending of the environment conditions, direct equilibrium saturation or flux conditions can be imposed.

3   Sobol' Sensitivity Analysis of the RFT Model

A sensitivity analysis is carried-out to define a strategy for fitting the parameters of the model. Depending on the model and the experiment data used, sometimes fitting problem of the drying models parameters on the mass loss measurements can be ill-posed [7]. So, sensitivity analysis can give us some indications about influence of each model parameters and potential interactions between them. This information is very important in order to proposed an adapted calibration methodology. The Sobol' indices based sensitivity analysis in the following is carried-out according to the methodology proposed by [28]. In this methodology, the Sobol' indices are calculated by the following expressions:

$$\mathbf{S}_i = \frac{\mathrm{Var}\left(E(\mathbf{Y}/X_i)\right)}{\mathrm{Var}\left(\mathbf{Y}\right)} \tag{20}$$

$$\mathbf{S}_{ij} = \frac{\mathrm{Var}\left(E(\mathbf{Y}/X_i X_j)\right) - \mathrm{Var}\left(E(\mathbf{Y}/X_i)\right) - \mathrm{Var}\left(E(\mathbf{Y}/X_j)\right)}{\mathrm{Var}\left(\mathbf{Y}\right)} \tag{21}$$

$$\mathbf{S}_{T_i} = \mathbf{S}_i + \mathbf{S}_{ij} + \mathbf{S}_{ijk} + \dots \tag{22}$$

where, X_i, $\mathbf{Y}$, Var and E, are the random input parameters, the model outputs, the statistic variance and mean respectively.

Table 1 Probability distribution functions (pdf) for RFT drying model parameters

Parameters	Pdf	Values	Units
ϕ	Constant	0.16	–
T_0	Constant	25	°C
S_{r0}	Constant	0.98	–
α	Constant	9.33427661	Pa
β	Constant	0.38923339	–
A_d	Constant	10.160105601	–
$\log(K_0)$	Normal	$\mathcal{N}\,(\mu = -20, \sigma = 1)$	$\log(\text{m}^2)$
p	Normal	$\mathcal{N}\,(\mu = 5.5, \sigma = 2)$	–
a	LogNormal	$\mathcal{LN}\,(\mu = 1.39, \sigma = 0.26)$	–
b	Uniform	$\mathcal{U}\,(inf = 1, sup = 4)$	–
E_a/R	Uniform	$\mathcal{U}\,(inf = 2226.85, sup = 5226.85)$	°C

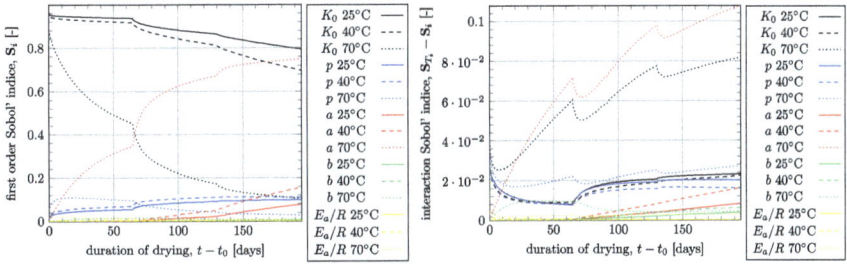

Fig. 1 First order and interactions Sobol' indices, of RFT model kinetics parameters on mass loss for cyclic relative humidity environment (80% → 50% → 30%) at several temperatures (25, 40 and 70 °C)

In the following, first order Sobol' indices, $\mathbf{S}_i$, and interactions Sobol indices, $\mathbf{S}_{T_i} - \mathbf{S}_i$, are analysed. The first order Sobol' indices measures the influence of each model parameter on the outputs and the interactions one, gives all the influence of the potential correlations between them. The Table 1 summed-up the probability distribution functions chosen for the analysis. The studied model output concerns a mass loss of concrete cylinder of 11.8 cm of diameter drying in 1D (along its diameter) in cyclic humidity environment, 85% between 0 and 65 days, 50% between 65 and 130 days and 30% between 130 and 195 days, at three temperatures 25, 40 and 70 °C.

The Fig. 1 shows the first order (left) and interactions (right) Sobol' indices as a function of the drying time. One can deduce a parameters or mechanisms hierarchy. For drying at 25 °C with RFT model, liquid water transport phenomena is the predominant mechanism regardless to the drying regime (K_0 and p are the most important parameters). At 40 °C around 170 days of drying (in drying regime between 50 and

30% of RH), the diffusive vapor becomes the mechanisms which controls drying. Whereas, at 70 °C, diffusive vapor becomes leading phenomena from 65 days of drying only (between 85 to 50% of RH). On the Fig. 1, one can also see a very weak influence of the interactions between the parameters on the loss of mass.

4 RFT Model Calibration on VERCORS Experiment Database

A very large experimental program has been designed and performed at EDF, in partner labs, or in the project MACENA.[1] Also, the VERCORS project interacts with the COST action TU1404[2] [26] where the VERCORS concrete has been used. The Fig. 2 shows that, the proposed desorption isotherm and RFT models fit very well the experimental data. The models fitted parameters are given in Table 2.

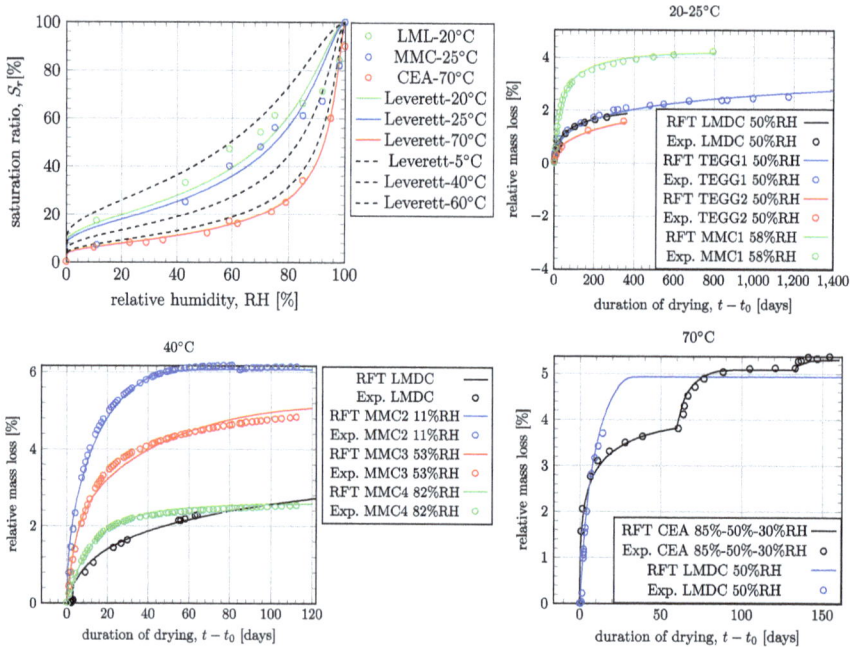

Fig. 2 VERCORS desorption isotherm and mass loss fitting at T = 20–25, 40 and 70 °C

[1] https://www.agence-nationale-recherche.fr/ProjetIA-11-RSNR-0012.
[2] https://www.tu1404.eu/.

Table 2 Fitted parameters RFT model

RFT model		
Parameters	Values	Units
K_0	$1.175 \cdot 10^{-20}$	m^2
p	2.91	–
a	2.607	–
b	7.0	–
Ea/R	-273.15	°C
T_0	25	°C
$h_{conv_{MMC1}}$	$2.73 \cdot 10^{-9}$	m/s
$h_{conv_{MMC2}}$	$4.297 \cdot 10^{-8}$	m/s
$h_{conv_{MMC2}}$	$1.133 \cdot 10^{-8}$	m/s
$h_{conv_{MMC2}}$	$8.548 \cdot 10^{-9}$	m/s

5   Validation and Application of RFT Model

In order to validated the fitted RFT model, predictions of the mass loss kinetics of several specimens, EEE, EIbas, EImi-fut and EIdome, which experience variable temperature drying conditions is curried out. These concerns cylindrical specimens, 16×32 cm, placed at 90 days of age at different locations in the VERCORS mock-up so as to measure the spatio-temporal variation effects of the ambient conditions in the VERCORS mock-up on the concrete drying. The Fig. 3 shows the predictions of these mass loss kinetics. One can see the ability of the RFT model to predict drying in variable conditions.

After the models validations on mass loss data, the aim in Fig. 4 is to compared models saturation profile predictions to direct RH measurements by capacity sensor and saturation measurements by TDR technique on 40 cm thick blocks [1] and VERCORS wall [16]. For 40 cm thick block, the RFT model is not able to predict the correct amplitude of drying profile, however those predicted without too much difficulty in the loss of mass of 16×32 cm specimens which are in the same drying conditions. An explanation could be that the humidity profile measurement by capacitive probe is not perfectly waterproof or that the endogenous initial saturation state measured on the $30 \times 30 \times 15$ cm samples at EDF DI TEGG is not the same as that of the 40 cm blocks simulated here. This second hypothesis is likely to the extent that the slopes of the simulated profiles and those of the measurements seem to be comparable.

For the TDR measurements on PACRA zone in VERCORS mock-up, the RFT model is not able to reproduce either the slope or the amplitude of the drying profile given by the TDR sensor. As explained in the introduction to this article, the TDR technique is based on the measurement of the permittivity which depends on the water saturation in concrete and this is converted into saturation by means of a calibration curve connecting the two physical quantities. This same calibration curve

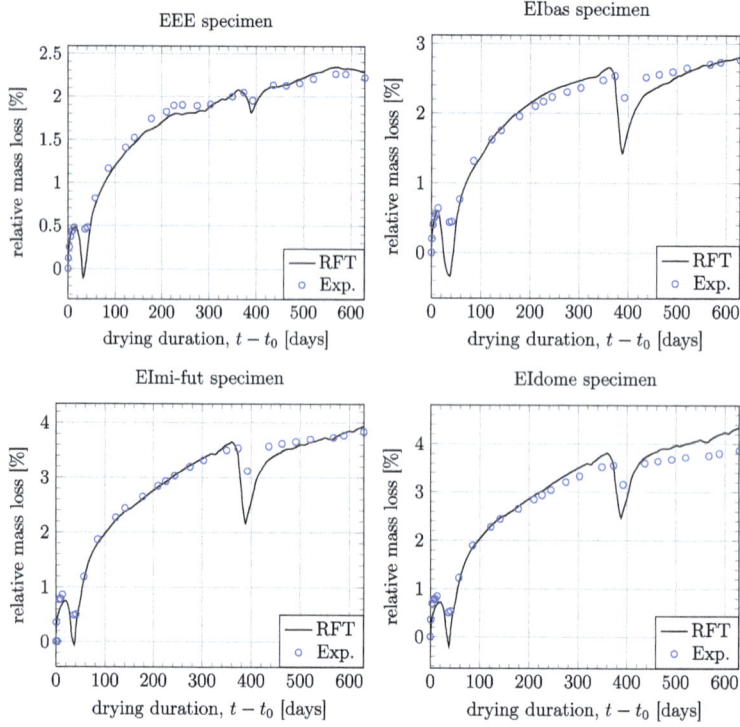

Fig. 3 EEE, EIbas, EImi-fut and EIdome relative mass loss predictions

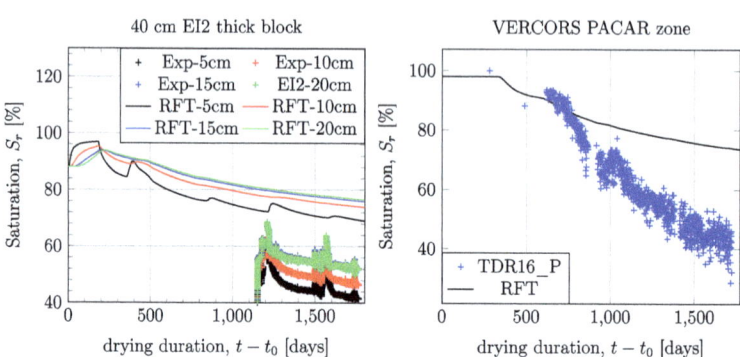

Fig. 4 VERCORS 40 cm EI2 thick block and PACAR zone drying profile in variable temperature environment

probably depends on the temperature, but in our case the calibration curve used is that measured in the works of [16]. The latter was measured at 20 °C. So, part of the difference between the models and the measurements could perhaps be explained by not taking into account the effects of temperature on this calibration curve.

6 Conclusion

In this contribution, a simplified drying model RFT was proposed and studied by comparison to experimental data. A sensitivity analysis, based on Sobol' indices give some informations about drying mechanism hierachy in different regime in temperature up to 70 °C. The calibrated RFT model was validated for mass loss predictions in variable temperature under 40 °C and in variable humidity conditions. The thick wall drying profile predictions shows that there is still progress to be made for improvement on drying modelling and on techniques for measuring saturation profiles.

References

1. Adia, J.L., Charpin, L., Martin, B., Leroy, D., Masson, B., Courtois, A.: In-situ RH measurements in concrete in a variable temperature environnement. In: FraMCoS-X, Bayonne, France (2019)
2. Agostini, F., Clauzon, T., Courtois, A., Skoczylas, F.: Monitoring of gas permeability and water content in large concrete structures: a new method based on pressure pulse testing. In: TINCE 2016, Paris (2016)
3. Baroghel-Bouny, V.: Water vapour sorption experiments on hardened cementitious materials. Part II: Essential tool for assessment of transport properties and for durability prediction. Cement Concr. Res. **37**, 438–454 (2007)
4. Baroghel-Bouny, V., Mainguy, M., Lassabatère, T., Coussy, O.: Characterization and identification of equilibrium and transfer moisture properties for ordinary and high-performance cementitious materials. Cement Concr. Res. **29**, 1225–1238 (1999)
5. Bazant, Z., Jirasek, M.: Creep and hygrothermal effects in concrete structures (chapter 10). Solid Mech. Appl. **225**, (2018)
6. Carette, J., Soleilhet, F., Benboudjema, F., Xiaoyan, M., Nahas, G., Abahri, K., Darquennes, A., Bennacer, R.: Identifying the mechanisms of concrete drying: an experimental numerical approach. Constr. Build. Mater. **230**, 117001 (2020)
7. Charpin, L., Courtois, A., Taillade, F., Martin, B., Masson, B., Haelewyn, J.: Calibration of Mensi/granger constitutive law: evidences of ill-posedness and practical application to VeRCoRs concrete. In: TINCE 2018 proceedings, Paris-Saclay, France (2018)
8. Charpin, L., Le Pape, Y., Coustabeau, E., Masson, B., Montalvo, J.: EDF study of 10-years concrete creep under unidirectional and biaxial loading: evolution of Poisson coefficient under sealed and unsealed conditions. In: CONCREEP 10, Vienna, Austria (2015)
9. Charpin, L., Le Pape, Y., Coustabeau, E., Toppani, E., Heinfling, G., Le Bellego, G., Masson, B., Montalvo, J., Courtois, A., Sanahuja, J., Reviron, N.: A 12 year EDF study of concrete creep under uniaxial and biaxial loading. Cement Concr. Res. 140–159 (2015)
10. Chhun, P.: Modélisation du comportement thermo-hydro-chemo-mécanique des enceintes de confinement nucléaire en béton armé-précontraint. Ph.D. thesis, Université de Toulouse (2017)
11. Davie, C., Pearce, C., Kukla, K., Bićanić, N.: Modelling of transport processes in concrete exposed to elevated temperatures—an alternative formulation for sorption isotherms. Cement Concr. Res. **106**, 144–154 (2018)
12. Drouet, E., Poyet, S., Torrenti, J.M.: Temperature influence on water transport in hardened cement pastes. Cement Concr. Res. **76**, 37–54 (2015)
13. Gawin, D., Pesavento, F., Schrefler, B.: Simulation of damage-permeability coupling in hygrothermo-mechanical analysis of concrete at high temperature. Numer. Methods Eng. Commun. **18**, 113–119 (2002)

14. van Genuchten, M.T.: A closed-form equation for predicting the hydraulic conductivity of unsaturated soils. Soil Sci. Soc. Am. J. **44**, 892–898 (1980)
15. Granger, L.: Comportement différé du béton dans les enceintes de centrales nucléaires: analyse et modélisation. Ph.D. thesis, Ecole Nationale des ponts et Chaussées, France (1996)
16. Guihard, V.: Homogénéisation de grandeurs électromagnétiques dans les milieux cimentaires pour le calcul de teneur en eau. Ph.D. thesis, Université Toulouse 3, France (2018)
17. Leverett, M.: Capillary behaviour in porous solids. Trans. Am. Inst. Mining Metall. Eng. **142**, 152–169 (1941)
18. Mainguy, M., Coussy, O., Baroghel-Bouny, V.: Role of air pressure in drying of weakly permeable materials. Eng. Mech. **127**(6), 582–592 (2001)
19. Millington, R.: Gas diffusion in porous media. Science **130**, 100–102 (1959)
20. Mualem, Y.: A new model for predicting the hydraulic conductivity of unsaturated porous media. Water Resour. Res. **12**, 513–522 (1976)
21. Powers, T.: The specific surface area of hydrated cement obtained from permeability data. Mater. Struct. **12**, 159–168 (1979)
22. Poyet, S.: Experimental investigation of the effect of temperature on the first desorption isotherm of concrete. Cement Concr. Res. **39**, 1052–1059 (2009)
23. Poyet, S., Charles, S.: Temperature dependence of the sorption isotherms of cement-based materials: heat of sorption and Clausius-Clapeyron formula. Cement Concr. Res. **39**, 1060–1067 (2009)
24. Poyet, S., Charles, S., Honoré, N., L'hostis, V.: Assessment of the unsaturated water transport properties of an old concrete: determination of the pore-interaction factor . Cement Concr. Res. **41**, 1015–1023 (2011)
25. Richards, L.A.: Capillary conduction of liquid through porous mediums. J. Appl. Phys. **1**, 318 (1931)
26. Serdar, M., Staquet, S., Gabrijel, I., Cizer, O., Nanukuttan, S., Bokan-Bosiljkov, V., Rozière, E., Šajna, A., Schlicke, D., Azenha, M.: COST Action TU1404: recent advances of the extended round robin test RRT+. In: Early Age Cracking and Serviceability in Cement- Based Materials and Structures, Brussels, Belgium (2017)
27. Sobol', I.: Sensitivity estimates for nonlinear mathematical models. Math. Model. Comp. Exp. **1**, 407–414 (1983)
28. Sudret, B.: Global sensitivity analysis using polynomial chaos expansions. Reliab. Eng. Syst. Safety **93**(7), 964–979 (2008)
29. Sémété, P., Février, B., Le Pape, Y., Delorme, J., Sanahuja, J., Legrix, A.: Concrete desorption isotherms and permeability determination: effects of the sample geometry. Eur. J. Environ. Civil Eng. (2015)
30. Vautrin, D., Taillade, F., Courtois, A., Clauzon, T., Bore, T., Daout, F., Placko, D., Lesoille, S., Sagnard, F.S.G.: Adaptation of a TDR probe design for the estimation of water content in concrete. In: TINCE 2016, Paris (2016)

Hydric Characterisation at Different Temperature of Nuclear Waste Package's Concrete

François Soleilhet, Patrick Sémété, Laurent Charpin, and Ginger El Tabbal

Abstract In the frame of its electro-nuclear activity, EDF has to deal with nuclear wastes. Containment of waste packages must be demonstrated. The latter is strongly influenced by the delayed strains of cement materials that can lead to cracking. Moreover, the permeability of concrete itself is vastly dependent on its saturation degree. For these reasons, a good prediction of strains and leak-tightness of concrete containment must start with a good estimate of the moisture profile on concrete. In this framework a large panel of experiments and studies are performed in order to demonstrate the package containment. The temperature experienced by waste packages is closed to 50 °C in the concrete shell and 60 °C in the filling slurry. Moreover, traditionally, the identification of the parameters of concrete drying constitutive laws is based on mass measurement of samples exposed to a constant humidity and temperature. However, it has been observed for different drying laws that this methods leads to a multiplicity of solutions, potentially corresponding to different moisture profiles and also to different extrapolations of the mass-loss beyond the end-time of the experiment used for the identification [1]. As an attempt to calibrate drying model, this study is interested in both obtaining drying experimental data in temperature and also original experimental data set that can discriminate between numerical parameter sets. Thus, data regarding relative mass variation, desorption isotherm at various temperature and relative humidity are carried out. Finally an easy way of achieving cup test is presented in order to assess water vapor permeability for different RH levels.

Keywords Drying · Drying shrinkage · Cup test

F. Soleilhet (✉) · P. Sémété · L. Charpin
EDF Lab Les Renardières Ecuelles, Moret-Loing-et-Orvanne, France
e-mail: francois.soleilhet@edf.fr

L. Charpin
e-mail: laurent.charpin@edf.fr

G. El Tabbal
EDF Lab Saclay, Palaiseau, France
e-mail: ginger.el-tabbal@edf.fr

© The Author(s), under exclusive license to Springer Nature Switzerland AG 2022
J. Sena-Cruz et al. (eds.), *Proceedings of the 3rd RILEM Spring Convention
and Conference (RSCC 2020)*, RILEM Bookseries 34,
https://doi.org/10.1007/978-3-030-76465-4_6

1 Introduction

Nowadays, EDF is involved on multiple projects concerning the durability of various concrete structures. This research focuses on the long term behaviour of a high performance concrete used in nuclear waste packages. These packages will be conserved in a surface storage facility and then disposed at CIGEO[1][2].

The objective of this research is to characterize in detail the hydric properties of the concrete mix used for the shell's package. This concrete is part of a structure which is subjected to restrained drying shrinkage at high temperature and low humidity (the most severe condition being 50 °C and 30 % RH). The study aims to collect experimental results in order to efficiently calibrate drying, shrinkage and creep models developed at EDF R&D, to be able to predict the conditions of cracking of concrete due to restrained shrinkage.

The concrete used is a high performance concrete with a porosity around 9 % and a compressive strength around 100 MPa. The maximum aggregate size is 16 mm. A larger experimental campaign is undergoing.

The present paper will focus and highlight some experimental tests related to drying and drying shrinkage. To characterise drying properties, EDF R&D has undertaken different tests.

1. Drying tests at different RH and two temperatures (20 °C and 50 °C) for the determination of the kinetics of drying and the desorption isotherms (static methods and dynamic methods);
2. Drying and shrinkage tests at 20 °C/50 % RH and 50 °C/30 % RH;
3. Cup tests at 20 °C and 50 °C for different RH steps.

Some insight regarding test methods will be developed and then results will be presented and discussed in the following section.

2 Methods

2.1 Drying Test

The main goal of the drying test is to assess, for the considered material, the evolution of relative mass variation over the time. To carry out these tests, a gravimetric method was used. The $7 \times 7 \times 28$ cm^3 specimens stored under endogenous conditions for 28 days, were weighed at the beginning of the test (time t_0) and then placed under the two different controlled conditions, respectively in a climate room (20 °C and 50 % RH) and in a regulated climate chamber (50 °C and 30 % RH).

---

[1]Forcasted underground facility disposal for radioactive waste in France.

Moreover, in order to obtain drying kinetics and saturation degree within a short time, slices of material 1 and 3 cm thick were machined in the prisms with a circular saw and placed in the same two conditions. In the drying environments, the sample are mass measured daily at the beginning and then over longer periods of time until the relative mass variation reaches the equilibrium. All samples are weighed using a precision balance (Kern type) with a resolution of 0.01 g. Finally, the relative mass variation ($\frac{\Delta m(t)}{m}$) is computed according to the expression (1).

$$\frac{\Delta m(t)}{m} = \frac{m(t) - m(t_0)}{m(t_0)} \tag{1}$$

with $m(t)$ the mass measured at time t in (g), $m(t_0)$ the initial mass in (g) and $\frac{\Delta m(t)}{m}$ in (%). To assess samples equilibrium, a criterion (given by the Eq. 2) is constructed.

$$\varepsilon_\delta = \frac{dm_i}{dt_i} \frac{1}{m(t_0)} \tag{2}$$

with dm_i in (g) the mass variation over two weighings spaced by a time dt_i and $m(t_0)$ the initial mass in (g). The equilibrium is assumed to be reached when the criterion is less than 10^{-3} % d^{-1}.

2.2 Drying Shrinkage Test

In addition to the drying test, the shrinkage of the six prisms were monitored. Before the prisms were stored in controlled climatic condition (c.f Sect. 2.1), an initial comparator reading was performed. During each measurement, the same protocol is performed (Fig. 1). First, a reference bar, made with Invar, is placed in the comparator and the comparator dial is set to zero. Then the specimen is measured. The shrinkage strain is then determined as :

$$\varepsilon_{sh}(t) = \frac{l(t) - l(t_0)}{l(t_0)} \tag{3}$$

where $\varepsilon_{sh}(t)$ is the drying shrinkage strain at the time t, $l(t)$ and $l(t_0)$ are the values given by the comparator respectively at time t and t_0 (initial gage length) in (m).

For each measurement, the readings are taken in the same order, one sample at the time. First the comparator measurement and then the mass measurement. These precautions are taken in order not to be disturbed by thermal strains.

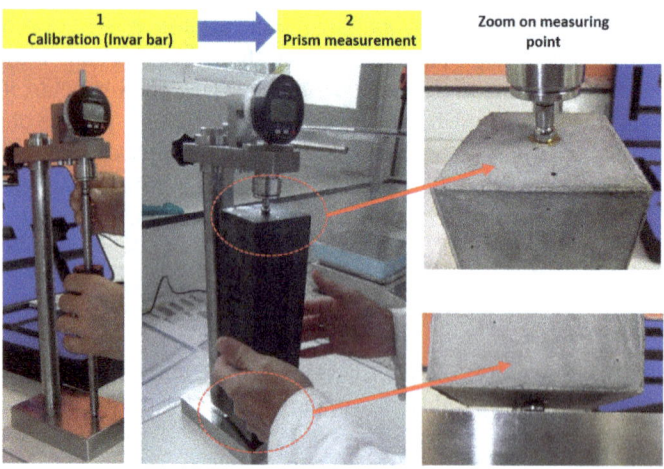

Fig. 1 Drying shrinkage setup

2.3 Cup Test

In order to characterize the water vapor permeability δ, a steady state vapor diffusion test is performed. This test is called the cup test. The cup test method has been used in different studies such as [4]. The main goal of this test is to determine the vapor moisture flux a sample in an unidirectional and steady state conditions which is easily computed under these assumptions as :

$$q_v = -\delta \frac{dP_v}{dx} \tag{4}$$

with q_v the vapor moisture flux (kg m^{-2} s^{-1}), δ the water vapor permeability (kg m^{-1} Pa^{-1} s^{-1} or in S.I units) and dP_v/dx the water vapor gradient in 1D (Pa m^{-1}). The principle of the test is to place a concrete sample between two different drying environments. This sample, a concrete prism of dimensions $7 \times 7 \times 3$ cm^3 (sawed in a $7 \times 7 \times 28$ cm^3 prism) is first equilibrated at the highest RH value of the step (RH2) in a desiccator placed in a climatic chamber (for temperature control). Then the concrete is assembled on a glass and sealed with aluminium tape on its sides to prevent multi-directional drying. The glass contains a satured salt solution which gives a RH1 (Fig. 2a). Finally, this setup is placed in a desiccator containing the saturated salt solution RH2 used for equilibrium phase (Fig. 2b).

During the first equilibration step and the second steady state step, the mass of the sample is monitored. The same weighting device used in the drying test is used. When the steady state is reached, the water vapor moisture flux through the sample is equal to the moisture transport.

Finally, the water vapor permeability depends on the saturation state of the material [5, 6]. In order to assess the evolution of the parameter for different relative humidity

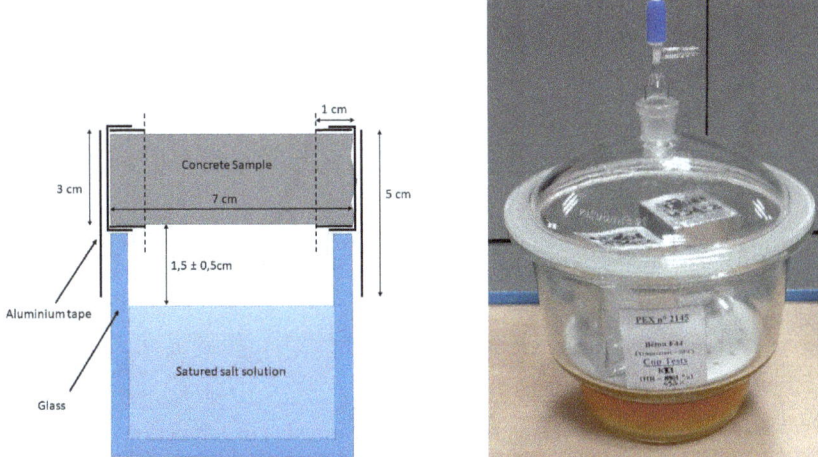

(a) Schematic illustration of the glass cup (b) Setup with the desiccator setup

Fig. 2 Cup test setup for assessing the moisture transport coefficient

Table 1 Theoretical values of the RH steps and corresponding salt solutions

Temperature	20 °C		50 °C	
RH steps	RH1 ⟵ RH2		RH1 ⟵ RH2	
High	81 %	96 %	85 %	98 %
	(KCl)	(K$_2$SO$_4$)	(KCl)	(K$_2$SO$_4$)
Intermediate	50 %	64 %	58 %	69 %
	(NaBr)	(KI)	(NaBr)	(KI)
Lower	11 %	31 %	11 %	33 %
	(LiCl)	(MgCl$_2$)	(LiCl)	(MgCl$_2$)

steps, three relative humidity levels are investigated. The values are summarized in the Table 1. Two samples for each RH steps are used.

3   Results and Discussions

3.1   Relative Mass Variation

The measurement of relative mass variation was carried out over a period of 254 days for the drying samples and 526 days for the sample regarding drying shrinkage. The variation in relative mass loss shows little scattering for the three samples investigated

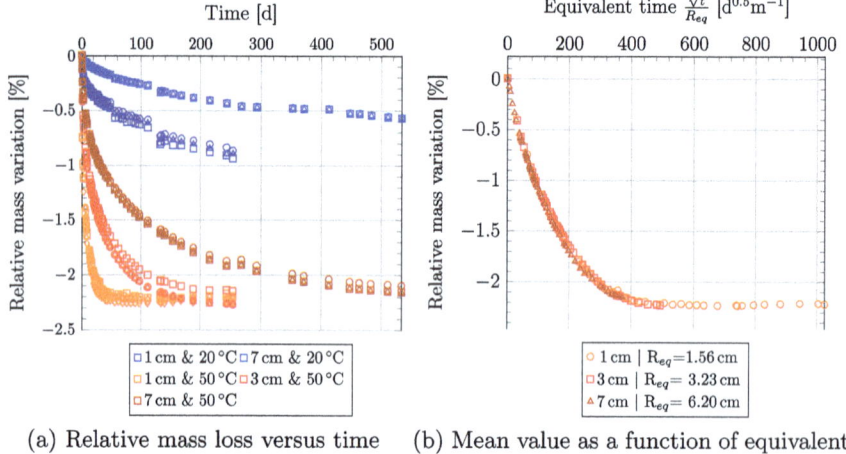

(a) Relative mass loss versus time   (b) Mean value as a function of equivalent time

Fig. 3 Evolution of drying for different sample geometries. Samples at 20 °C are conserved at 50 % RH and those at 50 °C at 30 % RH

at both temperatures (Fig. 3a). For the batch of specimens kept under the conditions at 20 °C and 50 % RH, the value reached for the 1 cm specimens is 0.9 % and for the 7 cm specimens 0.55 %. For both types of sample, the water balance is not reached at the end of the measurements. For specimens stored at 50 °C and 30 % RH, water equilibrium is reached for 1 cm and 3 cm samples after respectively 70 and 254 days. It is reached at a relative mass loss value closed to 2.22 %. For the largest specimens (7 cm), equilibrium is almost reached after 524 days. On these three samples, the influence of the sample size on the drying kinetic is highlighted.

To compare the different sample sizes at 50 °C, the equivalent drying radius is introduced (c.f Eq. 5).

$$R_{eq} = \frac{4V}{S} \qquad (5)$$

As a result, the evolution of relative mass variation for the three sample size are overlapping (Fig. 3b). They are following the same drying kinetic. These evolutions ensure that the smaller samples are representative of the behaviour of the larger ones. It also speeds up the drying process by reducing the sample size. Finally, it is noticeable that the 7 cm samples are close to equilibrium.

3.2 Drying Shrinkage

The measurement on prisms were carried out over a period of 526 days. To monitor the drying shrinkage, it is plotted as a function of the mass loss over a relevant period

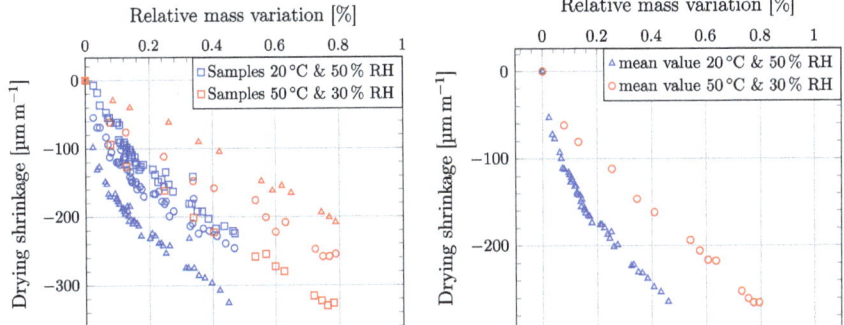

(a) Drying shrinkage versus relative mass variation at two drying ambiances

(b) Mean value of drying shrinkage

Fig. 4 Evolution of drying shrinkage for $7 \times 7 \times 28$ cm^3 samples. Drying environments respectively equal to 20 °C & 50 % RH for blue color and 50 °C & 30 % RH for red color

(Fig. 4a). For samples at 20 °C, the mass loss after 526 days is close to 0.55 % and the drying shrinkage is between 210 and 320 μm m^{-1}. The specimens show some scattering but much less than the samples at 50 °C. The latter reach 0.75 % loss in mass for a drying shrinkage between 208 and -326 μm m^{-1}.

The averages of these curves are then plotted (Fig. 4b). As it was intuitively predictable regarding the different evolutions, the two curves do not overlap. The curves are linear (almost bilinear) as a function of the relative mass variation but do not have the same slopes. There are different modelling approaches in the literature to account for drying shrinkage. Among these are the phenomenological methods proportional to water content [3]. In an isothermal framework this approach seems to be suitable. According to this result, in a non isothermal test, it no longer seems to be valid.

3.3 Cup Test

For the cup test results, not all drying curves are shown. The behaviour obtained for the cup tests is illustrated (Fig. 5a). A first phase of sample equilibration which lasts for these samples 400 days. Then between 400 and 550 days linear phase appears. In this linear part (Fig. 5b), the derivative value over the time of the mass is equal to the water vapor moisture flux.

For the specimens at 50 °C, the evolution of relative mass variation over the time is plotted (Fig. 5a). After 400 days they were almost all at equilibrium. The value obtained for the lower range was -2.3 %, for the intermediate -1.48 % and for the highest range $+0.14$ %. The value at 30 % RH is closed to the one observed for samples of similar size and conditions (c.f Fig. 3a). In the linear part, the reference value for the evolution of mass loss is the mass of the glass setup.

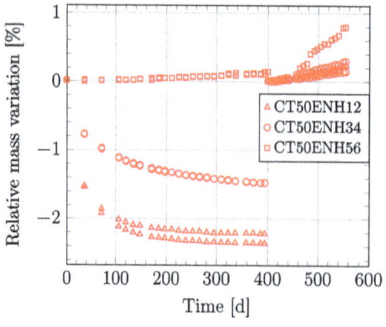 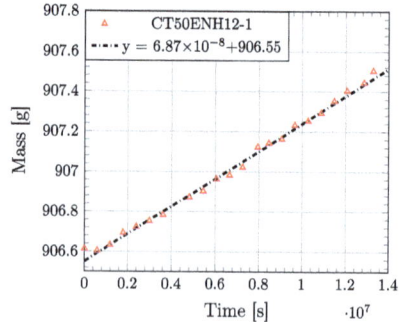

(a) Evolution of relative mass variaton: samples at 50 °C (two samples for each curves)

(b) Evolution of mass in the linear phase for "CT50EN12-1" sample

Fig. 5 Evolution of relative mass variation and mass at 50 °C

The moisture flux is thus post-processed for all the samples which exhibit the increasing linear part and compiled (Fig. 6a). Over the target humidity ranges, it spreads over a decade for the highest temperature. It varies from 1.19×10^{-8} to 7.54×10^{-8} kg m^{-2} s^{-1} at 50 °C and is equal to 2.96×10^{-9} kg m^{-2} s^{-1} for the one at 20 °C. Even it is not possible to conclude at this stage of the experiment, it seems that there is a clear separation between the two temperatures (c.f samples at higher RH step). Since the moisture flux is dependent on both the saturation state and the temperature. This difference seems to be justified. With the raise of the temperature, the water content (and the saturation degree) at equilibrium for a given RH is reduced [7]. Thus, the porous network at 50 °C is less saturated than the one at 20 °C, which favors transport by diffusion. Likewise, the raise of the intrinsic permeability with temperature [8] driven by the expansion of the porous network also benefits to raise transport by permeation. Both processes increase the amount of moisture flux.

Finally, the evolution δ the water vapor permeability coefficient over the different RH ranges is plotted (Fig. 6b). Computed values is equal to 2.62×10^{-12} s at 20 °C and are between 2.12×10^{-12} to 1.42×10^{-11} s at 50 °C. Remarks are likely the same as the one for moisture flux. The data at the highest temperature show a higher coefficient of water vapor permeability for all moisture levels (at 50 °C).

To characterize the obtained values, mass conservation can be applied to the porous sample in a steady state (Eq. 6). For this purpose a drying model[2] such as the one proposed by [9] is used.

$$\text{div}\left(\rho_l \frac{K_{int} \times k_{rl}}{\mu_l} \nabla P_l + \frac{D_{int} \times d_{rl} M_v}{RT} \nabla P_v \right) = 0 \tag{6}$$

---

[2]The model assumptions are not adressed in this paper.

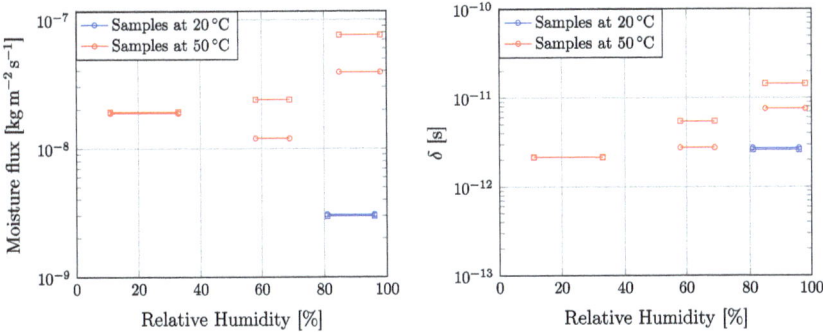

(a) Moisture flux: two samples for each steps

(b) Water vapor permeability coefficient as a function of relative humidity

Fig. 6 Evolution of moisture flux and water vapor permeability over the RH range. Color blue corresponding to sample at 20 °C and red at 50 °C

That equation is rewritten with a water vapor pressure gradient as a driving potential, spatially integrated and projected in an unidirectional case. It gives the same equation as the Eq. 4 with the δ parameter. This allows to identify the δ parameter (Eq. 7).

$$q_v = \underbrace{\left[\rho_l^2 \frac{RT}{M_v P_v} \frac{K_{int}(T) \times k_{rl}(S_w)}{\mu_l} + \frac{D_{int}(T) \times d_{rl}(S_w)M_v}{RT} \right]}_{\delta} \frac{dP_v}{dx} \qquad (7)$$

In ranges of relative humidity allow to assume that the transport by permeation (first term) is limited or even null either for values of RH lower than 40 % according to [11]. The δ parameter is identified using the relative diffusivity $d_{rl}(S_w)$ carried by Millington and Quirk [10].

$$\delta = D_{int}(T) \times \phi^a (1 - S_w)^b \times \frac{M_v}{RT} \qquad (8a)$$

$$D_{int}(T) = 2.17 \times 10^{-5} \left(\frac{T}{273.15} \right)^{1.88} \qquad (8b)$$

The a and b parameters are then fitted in order to obtain the mean value of δ (2.13×10^{-12} s) at 50°C for the lowest humidity (Eq. 8a). The parameters used for the identification are a temperature of (T= 323.15 K), porosity of ($\phi = 0.09$) and a saturation of ($S_w = 0.21$). The fitted values are a = 1.55 and b = 3.43. These two values are within the order of magnitude of the parameters proposed by [10] a = 1.33 and b = 3.33 for a more permeable granular medium and the one proposed by [9] for mortars a = 2.74 and b = 4.2. One would expect to obtain values closer to those of mortars, even lower for a HPC. One reason could explain this result. It is possible that the assumption regarding permeation transport at this humidity level is not totally valid. As a result, a part of the identified diffusion could be permeation.

4 Conclusion

In the context of radioactive waste management, it is necessary to model and therefore characterise the hydric behaviour of waste packages. Drying modelling is complex and involves a large number of parameters. It results sometimes in a multiplicity of solutions for a given problem. It becomes necessary to have various tests to discriminate some sets of parameters.

The purpose of the present works was to characterize the hydric properties of a HP concrete. Drying, drying shrinkage and cup method tests were undertaken under conditions of 20 °C and 50 % and 50 °C and 30 %.

Drying tests for different sample sizes highlight differences in kinetics, providing various data for the characterization of properties governing drying rates. Tests of drying shrinkage at two different temperatures allowed to find a linearity between relative mass loss and shrinkage. However this linearity is not similar for the two temperatures. This result prevents the use of phenomenological approaches for non isothermal modelling. Finally, the cup test allowed to define the moisture flow and to determine for humidity levels the permeability coefficient of the studied concrete. The value evolves as a function of the humidity (and therefore of the state of saturation). Using a drying model, the parameters of the Millington and Quirk model [10] are identified. For the considered concrete they were equal to $a = 1.55$ and $b = 3.43$.

References

1. Charpin, L., Courtois, A., Taillade, F., Martin, B., Masson, B., Haelewyn, J.: Calibration of Mensi/Granger constitutive law: evidences of ill-posedness and practical application to VeRCoRs concrete, TINCE 2018 proceedings. Paris-Saclay, France (2018)
2. CIGEO Homepage. https://www.andra.fr/cigeo. Accessed 6 Jan 2020
3. Benboudjema, F., Meftah, F., Torrenti, M.: Interaction between drying, shrinkage, creep and cracking phenomena in concrete. Eng. Struct. **27**(2), 239–250 (2005)
4. Jooss, M., Reinhardt, H.W.: Interaction between drying, shrinkage, creep and cracking phenomena in concrete. Cement Concr. Res. **32**(9), 1497–1504 (2002)
5. Villain, G., Baroghel-Bouny, V., Kounkou, C., Hua, C.: Measuring the gas permeability as a function of saturation rate of concretes. French J. Civ. Eng. **5**(2–3), 251–268 (2001). [Transfer in concrete and durability, In French]
6. Sercombe, J., Vidal, R., Adenot, F.: Gas diffusivity in cement, Transfert, Lille, Ecole Centrale (2006). [Transfer, In French]
7. Poyet, S.: Experimental investigation of the effect of temperature on the first desorption isotherm of concrete. Cement Concr. Res. **39**(11), 1052–1059 (2009)
8. Drouet, E., Poyet, S., Torrenti, J.M.: Temperature influence on water transport in hardened cement pastes. Cement Concr. Res. **76**, 37–50 (2015)

9. Thiery, M., Baroghel-Bouny, V., Bourneton, N., Villain, G., Stéfani, C.: Concrete drying modelling: analysis of the different modes of water transfer. French J. Civ. Eng. **11**(5), 541–577 (2007)
10. Millington, R.J., Quirk, J.P.: Permeability of porous solids. Trans. Faraday Soc. **57**, 1200–1207 (1959)
11. Kameche, Z.A., Ghomari, F., Choinska, M., Khelidj, A.: Assessment of liquid water and gas permeabilities of partially saturated ordinary concrete. Constr. Build. Mater. **65**, 551–565 (2014)

Modelling the Three-Stage of Creep

Vitor Dacol⬥ and Elsa Caetano⬥

Abstract Over the last years, several studies have addressed the time-dependent mechanical behaviour of polymeric composites. When subjected to a constant stress, viscoelastic materials experience a time-dependent increase in strain. This phenomenon is known as viscoelastic creep and manifests as a tendency of a solid material to deform permanently under the influence of constant stress: tensile, compressive, shear or flexural. When applied to polymers, creep is the result of the inherent viscoelastic nature that causes time dependency of behavior. As well known, the initial strain in a material is roughly predicted by its stress-strain modulus. Then, the material will continue to deform slowly with time indefinitely or until rupture or yielding causes failure. A typical creep curve reveals a three-stage behavior, (1) the transient stage, where the deformation rate decreases with time, (2) the steady-state, characterized by a "relatively" uniform rate gradient and (3) the accelerated phase, where the strain rate increases until rupture. In polymers at low strains (nearly to 1%), creep is essentially recoverable after unloading. However, in certain cases, creep failure is the most important degradation mode of a structure (turbine blades, aircraft parts). Furthermore, in civil engineering works, this kind of deformations may be substantial throughout the required service life. The investigation of the creep response of selected engineering materials should integrate the design of structures subjected to mechanical loads over a long time of operation (self-weight, static loads). The aim of "creep modeling for structural analysis" is the development of methods to simulate and analyze the time-dependent changes of stress and strain states in engineering structures up to the critical stage of creep rupture, passing through service state. In particular, the key in identifying these three stages above described lies in the location of the transition points between stages. This work presents a study conducted to estimate the 4 instants of time of the creep curve: (1) the first transition point, that is the transition point between transient and steady creep; (2) the secondary point, that is the inflexion point of the creep curve where the strain rate reaches its minimum; (3) the second transition point, that is the transition point between steady and accelerated creep; and (4) the instability point. This research follows the work

V. Dacol (✉) · E. Caetano
Faculty of Engineering (FEUP), CONSTRUCT (ViBEST), Porto, Portugal
e-mail: vitor.dacol@fe.up.pt

© The Author(s), under exclusive license to Springer Nature Switzerland AG 2022   71
J. Sena-Cruz et al. (eds.), *Proceedings of the 3rd RILEM Spring Convention
and Conference (RSCC 2020)*, RILEM Bookseries 34,
https://doi.org/10.1007/978-3-030-76465-4_7

publish by Crevecoeur [3] and is based on a combination of an exponential and a power law approach to the creep test data of HDPE pipe sample.

Keywords Viscoelasticity · Creep · Creep behaviour · Stages of creep · Polymer Ageing

1 Introduction

Creep is a slow, continuous deformation of a material under a constant stress and temperature. As a result of its inherent viscoelastic nature, polymers present a time-dependent deformation that occurs under any temperature. Creep of polymeric solids can be subdivided into several stages, where instantaneous creep takes place upon loading, followed by transient creep where the strain-rate reaches its minimum. It then increases and approaches an asymptotic limit and rupture occurs.

Therefore, the study on the process of creep is essential and significant for engineering applications concerning temperature and stress and has been a challenge in the design of structures subjected to loads for long periods of time. According to Monfared [14], predicting creep behaviour is of utmost importance on advanced reinforced or non-reinforced materials design, such as polymeric composite.

In fact, to define parameters to prevent unsafe and undesired events such as progressive deformations and local reduction of material strength, may be an important tool in development of more slender and durable structures. In this context, estimating of transition points between stages of creep may help in linear and non-linear analysis of the structure behaviour.

On the other hand, Crevecoeur [6] shows a non-linear kinetics in the ageing of biological systems, which is characterized by three successive stages such as observed in a creep curve of polymeric composite. Crevecoeur shows that a model useful for the follow-up of operating inert systems allows to find back typical curves and laws related to the ageing of biological systems.

Based on the work of Crevecoeur and through a combination of an exponential and a power law approach to the creep test data of FRP sample, this paper proposes a simple approach to modelling the three-stages of polymer creep.

2 Theoretical Background

The combination of mechanical behavior of elastic and viscous materials characterizes the response of a viscoelastic material.

According to Barra [1], due its molecular nature, the polymeric mass may be seen formed from a twisted or tangled chains in its undeformed state. Subjected to constant tension, molecules exert a retractive response to this tension in order to maintain their stable (folded) conformation. Under continuous tension, the chains

Fig. 1 Stress-strain relationship of viscoelastic materials

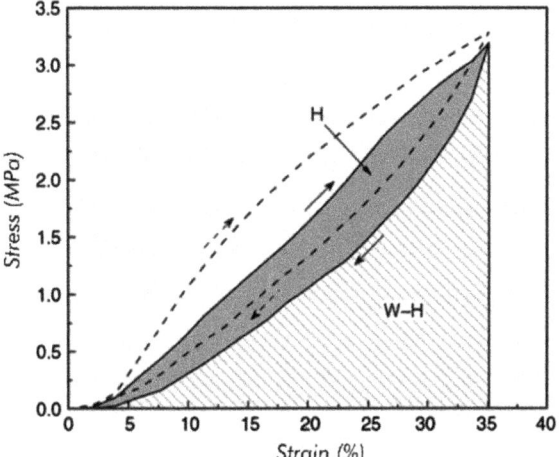

slide relatively to each other and the tangled structure is deformed into an extended conformation. When the applied load is released before the creep rupture occurs, an immediate elastic recovery equal to the elastic deformation, followed by a period of slow recovery is observed.

Within certain limits, the graphic tension versus deformation will take the shape of an ellipse. This format tends to be maintained to the extent that the amplitudes of tension and deformation grow (Fig. 1).

The viscoelastic behaviour manifests itself in several ways, which comprise: (a) creep under constant load; (b) relaxation of tension under constant deformation; (c) recovery of time dependent deformation after complete removal of load; (d) rupture due to time-dependent creep; (e) fatigue resistance dependent on the excitation frequency.

Basically, there are three principal ways to describe the viscoelastic behavior [19] in time domain. The first one is the Boltzmann superposition principle so-called the integral representation expressed by the convolution integral (1).

$$\varepsilon(t) = \int_{-\infty}^{t} J(t - \tau_n) d\sigma(\tau_n) \tag{1}$$

The second way, represented by a differential equation (2), uses assemblages of Hookean springs and Newtonian viscous elements (dashpots) as models through several forms of combinations. For a Maxwell rheological model, total strain yields:

$$\frac{\partial \varepsilon}{\partial t} = \eta \dot{\varepsilon} \tag{2}$$

The third method is based on assumptions about the molecules themselves which dynamic tests are more convenient to be used for. This last method will not be described in this paper.

2.1 Creep

Depending on the level of excitation applied, the material may not recover to its original shape, and then a permanent deformation becomes installed. The magnitude of this permanent deformation will depend on the duration and the amount of tension applied, as well as the temperature [8].

For a constant temperature, this creep description is illustrated by Fig. 2 and translated by expression (3).

$$J(t) = \epsilon\ (t)/\sigma_0 \qquad\qquad (3)$$

For most applications, composite structures may be designed using linear elastic analysis. However, due to the effects of environment, there are instances where time dependent phenomena such as creep may be important long-term design considerations.

In these cases, creep failure is the most important degradation mode of a structure (turbine blades, aircraft parts) and its behavior must be known until its failure. An example showing importance of the creep phenomenon is the creep behavior of epoxy-based structural adhesives. Compared to other joining methods, welding or bolting for example, epoxy-based structural adhesives provide advantages, such as its high strength-to-weight ratio [10].

Fig. 2 Creep and recovery description

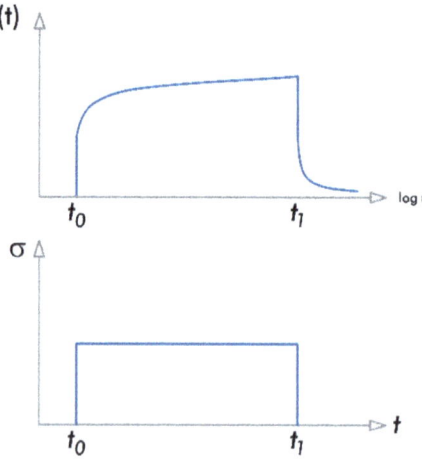

2.2 The Three-Stages of Creep

As known, viscoelastic material is a rheological material that exhibits time-temperature rate of loading dependence, with the response being a function of the current input and also of the current and past input history.

Under a sustained load over a long period of time, viscoelastic materials undergo three stages of strain evolution (Fig. 3).

The typical creep curve showed in Fig. 3 may be described as follow:

(a) First stage: so-called Primary or Transient creep, in which creep occurs at a decreasing rate;
(b) Second stage: so-called Second or Steady-state creep, in which strain rate reaches its minimum and shows a nearly constant value;
(c) Third stage: so-called Tertiary or Accelerated creep, where the strain rate increases quickly and its end the rupture occurs.

Figure 4 shows the strain rate of creep behavior and the total relationship of an ideal creep curve, the so-called creep bathtub. The bathtub curve usually gives the failure rate for an operating system.

According to Naumenko and Altenbach [15], the secondary creep is for many applications the most important creep model. The relative equilibrium, i.e., a uniform rate of strain gradient, enables the long-term behavior of the structure to be analyzed

Fig. 3 Three-stages of creep

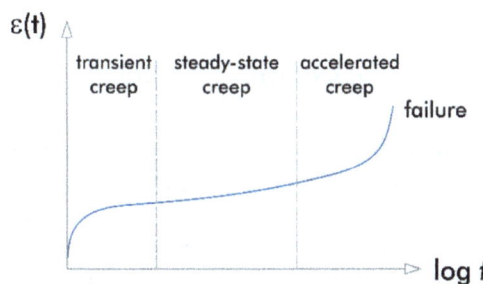

Fig. 4 Strain rate bathtub

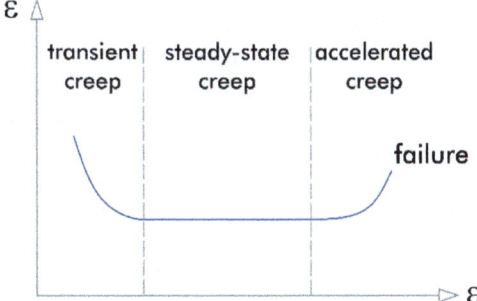

assuming stationary creep processes. The extent of this stage depends on the combination of applied stress and system temperature. For relatively low stresses, the secondary stage may have a considerable extent before the onset of tertiary creep [2].

However, for components with a close fit, the primary creep needs to be taken into account. Bolts and copper canisters for the storage of spent nuclear fuel are an example of this [16].

On its turn, the tertiary creep leads to an unstable crack propagation and materials failure at end. The point where the accelerated creep starts is known as flow time and is defined as the time at which the shear deformation under constant volume starts, i.e., wherein the volume does not change. In an idealized curve of creep, tertiary stage only occurs at high stress or for ductile materials.

Furthermore, stress impacts significantly the strain rate where the increase of the stress by one order causes an increase of strain rate about four orders approximately [11].

Thus, to provide a better understanding of the structure behavior over time, these three-stages of creep should be known in order to avoid undesired behavior and even irreversible failure.

In particular, the key to determining these three-stages is in the fact that how to locate the transition points between stages along the creep curve. At last instance, it translates in the determination of secondary stage boundaries and in the predicting a creep safe envelope.

2.3 Biological Ageing Analogy

As defined in Luder [13], biological ageing is usually considered the continuous accumulation of damage or deterioration, at the level of the cells, tissues, organs or organism, which reduces vitality and ultimately leads to death.

Crevecoeur [6] shows that ageing could be seen as a multifactorial process. According to that, there is a widely observed non-linear kinetics in the ageing of biological systems, which is characterised by three successive stages, (1) firstly the ageing rate is high, but it decreases quickly to a minimum, from which (2) it remains nearly constant during the major part of the process until (3) it starts increasing again up to the final collapse of the system.

Concerning to this, taking ageing in the broad sense, as the whole process of going from birth to death, non-linearities appear. Those can be found in nature as reflecting successive periods in the life of biological systems, such as in the "survival, mortality or growth curves" (see Van Voorhies [20], Luder [13] and Fotsis et al. [9], respectively).

Selye [17] described the three stages of a system reaction, termed General Adaptative Syndrome (GAS) where "there is an integrated syndrome of closely interrelated adaptive reactions to non-specific stress itself". The first stage of GAS is the "Alarm Reaction", which can be corresponding to an elastic response of polymer under static load; the second stage is called "Stage of Resistance", in which "a certain resistance is built up against the damaging stimulus" that could be translated in terms of steady creep rate; at last, the tertiary stage, called "Stage of Exhaustion", where the system loses its resistance and fails with several internal changes.

Thus, the parallel between biological ageing and polymeric creep behavior can be explained by assuming that a polymeric structure is composed of interconnected microstructures operating together under an input stress.

In this way, Crevecoeur [4] shows the similarity between creep behavior and mechanical system ageing, where the failure rate of mechanical repairable systems that deteriorate with time due to ageing can usually be visualized by a bathtub curve, as shown in the creep rate curve.

3 Modeling Approach

Crevecoeur [5] suggests a differential equation to synthesize the strain rate behavior of creep expressed by the bathtub curve:

$$\frac{\dot{\varepsilon}(t)}{\varepsilon_e(t)} = \alpha + \frac{\beta}{t_i} \tag{4}$$

where:

$\dot{\varepsilon}(t)$ *is the strain rate*
$\varepsilon_e(t)$ *is the measured strain from creep test*
α, β *are constants to be fitted, where* $\alpha > 0$ *and* $\beta < 1$ *and* $\alpha \ll \beta$.

Once (4) is integrated, it gives a combination of an exponential and a power law function (5).

$$\varepsilon_e(t) = k.e^{\alpha.t}.t^{\beta} \tag{5}$$

where:

k *is a constant for a given temperature and load.*

The three parameters α, β and k are constant as long as the physical constraints on the system remain constant, i.e., while the stress and temperature remain constant, these three parameters stay constant.

Crevecoeur [6] introduces a third system constraint which is called the internal "*organisation*" of the system. In the case of a polymer, this can be understood as the molecular arrange of the polymer and the micromechanical assemblage of the composite material.

According to Smith [18], the values of the parameter β will typically range between 0.15 and 0.65. However, this quantity can have values between 0 and 1 [3].

In (5) the quantity $(1/\alpha)$ gives the order of magnitude of the time scale for the creep process. Physically, this quantity corresponds to a time called *"instability time"*, t_i, [6], i.e.:

$$\frac{1}{\alpha} = t_i \tag{6}$$

The *instability time* is one that from it there is a risk that the system starts to fail. At t_i, if the load is not removed or operating conditions are not relaxed, the system will become impaired. It means that the system will usually continue after t_i but the process leading to the collapse is irreversibly started.

According to Crevecoeur [6], it is noticeable that the *instability time* occurs in tertiary creep and it is possible to divide the third stage into two parts, (1) before t_i (the predictable ageing) and (2) after t_i (the statistically predictable ageing).

The system failure takes place about 20–40% of life after t_i.

Applying (6) into (5), yields (7).

$$\varepsilon_e(t) = k.e^{\left(\frac{1}{t_i}\right)}.t^\beta \tag{7}$$

where:

$\varepsilon_e(t)$ *is the creep strain measured at* t.

From Eq. (7), at some given time t_n, it is expectable that:

$$t = t_i \ldots \text{that means} \ldots \varepsilon_e(t) = \varepsilon(t_i) \tag{8}$$

From the equality (8), Eq. (7) yields (8).

$$\varepsilon_i(t_i) = k.e^1.t_i^\beta \tag{9}$$

where:

$\varepsilon_i(t_i)$ *is the strain at* t_i.

The quotient between (7) and (9) yields (10):

$$\frac{\varepsilon_e(t)}{\varepsilon_i(t_i)} = \exp\left(\frac{t}{t_i} - 1\right).\left(\frac{t}{t_i}\right)^\beta \tag{10}$$

From data pairs (ε_e, t), it is expectable that $(\varepsilon_i, t_i, \beta)$ converge from a successive iteration of (10).

Moreover, according to Crevecoeur [7] the use of $\varepsilon_e(t)/\varepsilon_i(t_i)$ and t/t_i allows to normalize the shape of the creep curve, independently of how long the process lasts. In fact, for a given β and making $\varepsilon_e(t) = \varepsilon_i(t_i)$, the creep curve will look similar for a small t_i (tests at higher stress or temperature) as well as for a larger t_i (tests at lower stress or temperature).

To find the time of steady creep, Crevecoeur [6] suggests the following relation:

$$t_{2ary} = t_i \cdot \left(\sqrt{\beta} - \beta \right) \tag{11}$$

Expression (11) gives the point where the strain rate reaches its minimum and corresponds to the point of inflexion on the creep curve.

From (11), the secondary strain point, ε_{2ary}, is estimated using (7).

It must to be warned that the 2ary stage (the steady-state stage) has visually the aspect of a straight line but in fact it is an approximate linearization for practical applications and shows a quasi-straight linear zone.

The point $(\varepsilon_{2ary}, t_{2ary})$ is somewhere in the 2ary stage where the creep curve turns from concave (from below) to convex (from below) form.

Furthermore, it is possible to mark several periods of time on the creep curve. Comparing biological and inert systems, Crevecoeur [6] distinguishes some periods of time:

(1)   the instability time—t_i;
(2)   the actual time of collapse—t_r (the 'lifespan');
(3)   the maximum possible lifetime—t_m (the 'longevity'); and
(4)   the statistical average of lifetimes—t_μ, (the 'life expectancy').

These periods of time are shown in Fig. 5

where:

t_{1ary} is the transition point between transient and steady creep
t_{2ary} is the inflexion point of the creep curve
t_{3ary} is the second transition point (flow time)

Fig. 5 Periods of time on a hypothetical creep curve

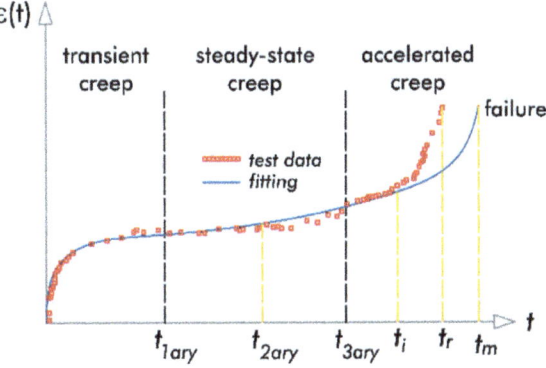

Fig. 6 Creep envelope for a
given temperature

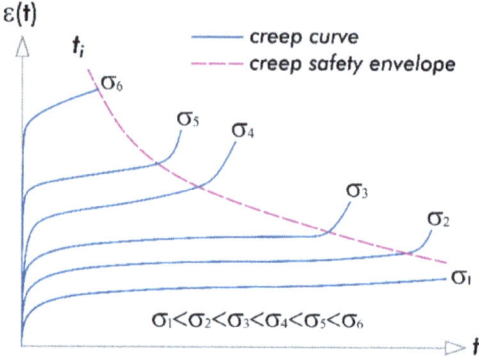

t_i is the instability time
t_r is the actual time of failure
t_m is the maximum possible lifetime.

Applying t_{2ary} to the derivative of (5) and comparing its straight line to creep data curve, it is expectable to determine t_{1ary} and t_{3ary} taking the difference between the straight line and creep data less than to **1%**.

3.1   Creep Safety Envelope

Once defined the critical points on the creep curve (for a given temperature and load), it is possible to determine different critical points for a spectral of different values of stress for a specific temperature or different levels of temperature for a range of stress values. Fig. 6 shows a creep safety envelope for an aleatory polymeric sample at a given temperature.

4   Material Testing

In order to simulate the above described fitting, sampled points from the creep curves identified by Liu et al. [12] are used as the actual experimental values to be matched.

4.1   Liu's Creep Data

The creep experimental data are obtained for creep testing of a sample extracted from HDPE (high-density polyethylene) pipes at 5 stress levels. From a Generalized

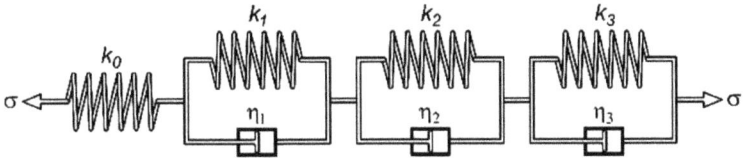

Fig. 7 Generalized kelvin model

Table 1 Fitted values from Eq. (12)

Stress (MPa)	k_0 (MPa)	k_1 (MPa)	k_2 (MPa)	k_3 (MPa)
2.97	650	797.3889	2320.3566	925.0882
5.97	580	913.5936	1212.2605	695.0461
7.71	520	1224.7911	1104.9922	385.8572
10.31	500	1034.2045	694.1084	226.4555
12.19	470	1128.4448	806.0972	140.6875

Adapted from Liu et al. [12]

Kelvin Model (Fig. 7), a Dirichlet Series with three terms were used to represent the resulting strain (12). The tests were conducted at room temperature of about 22 °C. Table 1 shows the fitted values of experimental data from Eq. (12).

$$\varepsilon_D(t) = \frac{\sigma(t)}{k_0(\sigma)} + \sum_{i=1}^{3} \frac{\sigma(t)}{k_i(\sigma)}\left[1 - \exp\left(-\frac{t}{\tau_i}\right)\right] \tag{12}$$

where the respective relaxion times are:

$\tau_1 = 500\,s$
$\tau_2 = 10000\,s$
$\tau_3 = 200000\,s$

The sampled points from the creep curves identified in Table 1 are used to simulate the actual experimental values to be matched and the resulted strain-time curves are depicted in Fig. 8.

5 Application of Crevecoeur Fitting Methodology

From the data spectra generated by Eq. (12), Crevecoeur equation coefficients are fitted for each stress level and presented in Table 2.

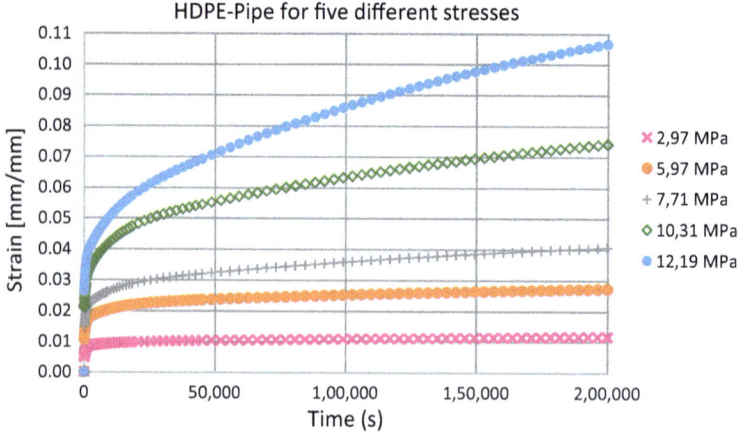

Fig. 8 Creep curves until end of measurements

Table 2 Crevecoeur equation coefficients

Stress (MPa)	k	β	t_i (s)	error (%)	r-Person
2.97	0.00558	0.055	3.403.676	0.60	0.9763
5.97	0.00880	0.085	1.709.402	0.80	0.9901
7.71	0.00900	0.113	1.459.854	0.90	0.9990
10.31	0.01340	0.124	924.214	1.10	0.9982
12.19	0.01320	0.150	740.741	1.90	0.9975

5.1 Fitting Quality

In order evaluate the quality of the fitted values, the relationship diagram between $log\ \sigma$ versus $log\ t_i$ and $\varepsilon(t_i)$ versus $log\ t_i$ is shown in Fig. 9.

Fig. 9 Evaluation diagram

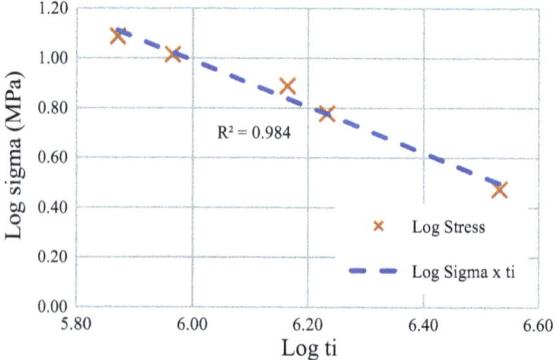

The linear relation plotted is similar to a typical failure mode of HDPE Pipe, the so-called Mode–1, where the pipe fails mainly in a ductile manner under mechanical overload at higher stresses (see Xu et al. [21]), that indicates an acceptable correlation between data test and the approach.

6 Results and Comments

Figure 10 and Table 3. Strain Correlation show the relationship between the strain at t_i and t_{2ary}, $\varepsilon_i(t_i)$ and $\varepsilon_{2ary}(t_{2ary})$. This one reveals a constant ratio between the instability time and the time of steady creep, over the stress increment.

6.1 Transition Points

In order to exemplify the strain rate bathtub the strain rate for 12,19 MPa is depict in Fig. 11.

From derivative of (5) and applying the t_{2ary}, it is expectable to find the values of t_{1ary} and t_{3ary} where the difference between the strain rate line from (4) and the creep data curve is less than or equal to 1%. Tables 4 and 5. Pairs of Transition Points (2/2) shows the values of each pairs of fitted points.

Fig. 10 Strain correlation

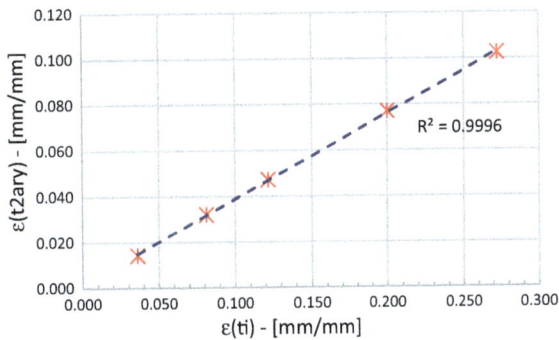

Table 3 Strain correlation

Stress (MPa)	$\varepsilon(t_i)$ (mm/mm)	$\varepsilon(t_{2ary})$ (mm/mm)	Strain ratio
2.97	0,036	0,014	0,398
5.97	0,081	0,032	0,396
7.71	0,122	0,047	0,388
10.31	0,200	0,077	0,385
12.19	0,272	0,102	0,376

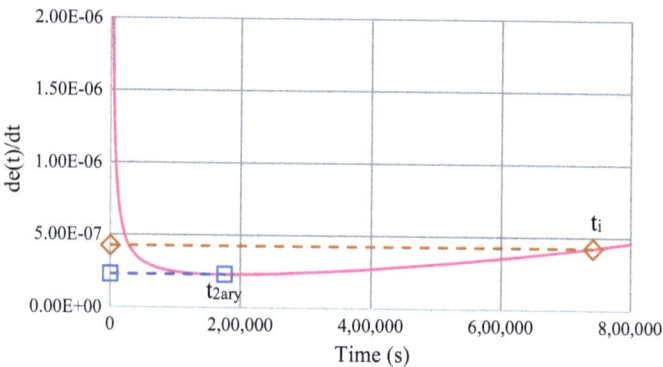

Fig. 11 Strain bathtub for 12,19 MPa.

Table 4 Pairs of transition points (1/2)

Stress (MPa)	$\varepsilon(t_{1ary})$ (mm/mm)	t_{1ary} (s)	$\varepsilon(t_{2ary})$ (mm/mm)	t_{2ary} (s)
2.97	0,011	77.400	0,014	611.031
5.97	0,026	142.200	0,032	353.073
7.71	0,044	260.000	0,047	325.773
10.31	0,059	77.400	0,077	210.847
12.19	0,087	109.800	0,102	175.777

Table 5 Pairs of transition points (2/2)

Stress (MPa)	$\varepsilon(t_{3ary})$ (mm/mm)	t_{3ary} (s)	$\varepsilon(t_i)$ (mm/mm)	t_i (s)	$\varepsilon(t_m)$ (mm/mm)	t_m (1s)
2.97	0,023	2.150.000	0,036	3.403.676	0,055	4.424.779
5.97	0,042	710.000	0,081	1.709.402	0,124	2.393.162
7.71	0,051	404.000	0,122	1.459.854	0,189	2.043.796
10.31	0,120	518.000	0,200	924.214	0,311	1.293.900
12.19	0,139	330.000	0,272	740.741	0,428	1.037.037

From expression (5) and the values from Tables 4 and 5. Pairs of Transition Points (2/2), the theoretical creep curves for each step of stress are depicted in Fig. 12

Once defined the critical points on the creep curve shown in Table 4, creep safety envelope is created and the shape of the curve is given in Fig. 13.

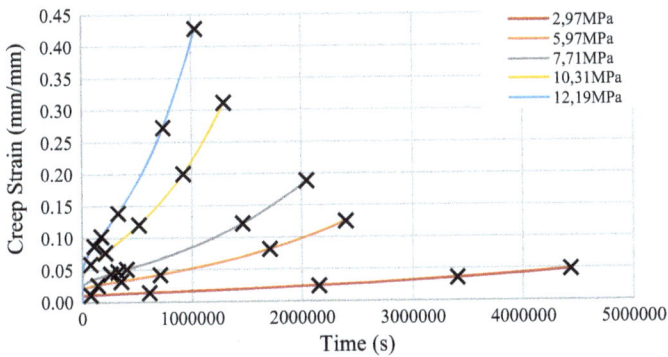

Fig. 12 Transition points

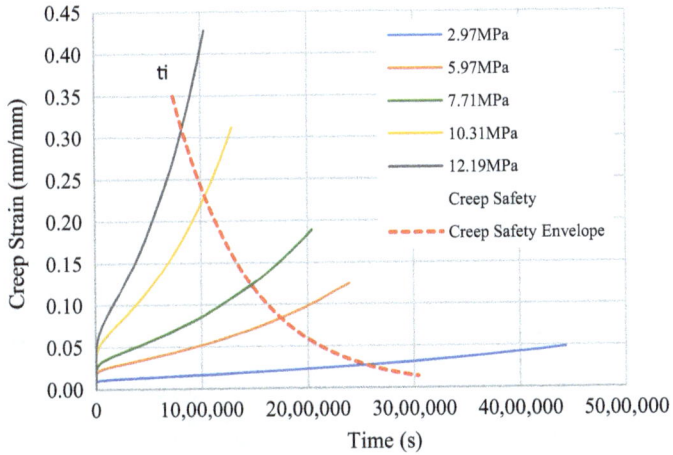

Fig. 13 Creep safety envelope

7 Conclusions

This paper presented a study conducted to estimate the 4 instants of time character-izing the creep curve: (1) the first transition point, that is the transition point between transient and steady creep; (2) the secondary point, that is the inflexion point of the creep curve where the strain rate reaches its minimum; (3) the second transition point, that is the transition point between steady and accelerated creep; and (4) the instability point.

The creep curve used in this study was constructed using data resulting from a numerical simulation of an experiment of Liu et al. [12], where a sample cut from a 24" HDPE-Pipe was tested for five different stress levels.

A brief review of the fitting methodology was first presented concerning the creep behavior and its three-stages. This methodology was applied then to the numerical description of the creep behavior of a PE. The time-dependent and stress-dependent approach was easily applied to a viscoelastic material. The transition points were calculated and a creep safety envelope was defined and a lifetime prediction could be made.

Although there was no comparative analysis of the useful life of the material based on standards, the creep curves obtained by the Crevecoeur adjustment methodology evidenced a good fitting.

Acknowledgements The authors wish to express their utmost thanks to Professor Guibert Ulric Crevecoeur (Federal Public Service Economy—Department of Energy—Brussels, Belgium) for his useful remarks having led to a significant improvement of the present contribution.

This work was financially supported by: Base Funding—UIDB/04708/2020 of the CONSTRUCT—Instituto de I&D em Estruturas e Construções—funded by national funds through the FCT/MCTES (PIDDAC).

References

1. Barra, G.: Fundamentos de reologia de materiais poliméricos. Florianópolis–SC (2010). http://emc5744.barra.prof.ufsc.br/Reologiaparte1.pdf
2. Cao, P., Short, M.P., Yip, S.: Understanding the mechanisms of amorphous creep through molecular simulation. Proc. Nat. Acad. Sci. U. S. A. **114**(52), 13631–13636 (2017). https://doi.org/10.1073/pnas.1708618114
3. Crevecoeur, G.U.: Quelques réflexions autour de la courbe de fluage—Une nouvelle perspective. Journal des Ingénieurs, bimestral (1992)
4. Crevecoeur, G.U.: A Model for the integrity assessment of ageing repairable systems. IEEE Trans. Reliab. **42**(1), 148–155 (1993). https://doi.org/10.1109/24.210287
5. Crevecoeur, G.U.: Reliability assessment of ageing operating systems. Eur. J. Mech. Environ. Eng. **39**(4), 219–228 (1994)
6. Crevecoeur, G.U.: A system approach modelling of the three-stage non-linear kinetics in biological ageing. Mech. Ageing Dev. **122**(3), 271–290 (2001). https://doi.org/10.1016/s0047-6374(00)00233-5
7. Dacol, V.: Prof. Guibert Ulric Crevecoeur interview, ResearchGate, 19 Oct 2019, Porto–Portugal (2019)
8. Findley, W.N., Lai, J.S., Onaram, K.: Creep and Relaxation of Nonlinear Viscoelastic Materials—With an Introduction to Linear Viscoelasticity. Dover Publications Inc., New York (1976)
9. Fotsis, T., et al.: The endogenous oestrogen metabolite 2-methoxyoestradiol inhibits angiogenesis and suppresses tumour growth. Nature **368**(6468), 237–239 (1994). https://doi.org/10.1038/368237a0
10. Guedes, R.M.: *Creep and Fatigue in Polymer Matrix Composites, Woodhead Publishing.* Woodhead Publishing Limited, Cornwall, UK (2011). https://doi.org/10.1007/s13398-014-0173-7.2
11. Iskakbayev, A., Teltayev, B., Rossi, C.O.: Steady-state creep of asphalt concrete. Appl. Sci. **7**(2) (2017). https://doi.org/10.3390/app7020142
12. Liu, H., Polak, M.A., Penlidis, A.: A practical approach to modeling time-dependent nonlinear creep behavior of polyethylene for structural applications. Polym. Eng. Sci. **48**, 9 (2008). https://doi.org/10.1002/pen

13. Luder, H.U.: Onset of human aging estimated from hazard functions associated with various causes of death. Mech. Ageing Dev. **67**(3), 247–259 (1993). https://doi.org/10.1016/0047-6374(93)90003-a
14. Monfared, V.: Review on creep analysis and solved problems. Intech **i**(10), 31 (2016). http://dx.doi.org/10.5772/57353
15. Naumenko, K., Altenbach, H.: Modeling of creep for structural analysis (2007). https://doi.org/10.1007/978-3-540-70839-1
16. Sandström, R.: Basic model for primary and secondary creep in copper. Acta Mater. **60**(1), 314–322 (2012). https://doi.org/10.1016/j.actamat.2011.09.052
17. Selye, H.: Stress and the general adaptation syndrome. Br. Med. J., 1383–1392 (1950). https://doi.org/10.1159/000227975
18. Smith, D.J.: Reliability and maintainability in perspective—practical, contractual, commercial and software aspects. J. Chem. Inform. Model. (1988). Third Edit. Edited by Macmillan. https://doi.org/10.1017/cbo9781107415324.004
19. Vincent, J.: Structural Biomaterials, Structural Biomaterials. Springer International Publishing Switzerland, London (1982)
20. Van Voorhies, W.A.: Production of sperm reduces nematode lifespan. Nature **359**, 710–713 (1992)
21. Xu, C., Xu, P., Shi, J.: Investigation on creep-rupture failure time of HDPE pipe under hydrostatic pressure. American Society of Mechanical Engineers, Pressure Vessels and Piping Division (Publication) PVP, 6(PARTS A AND B), pp. 913–918 (2011). https://doi.org/10.1115/pvp2011-57840

Effects of the Thermal Conditioning on the Mechanical Properties of an FRCM (Fiber Reinforced Cementitious Matrix) Strengthening System

L. Ombres, P. Mazzuca, and S. Verre

Abstract Fiber reinforced composite materials are starting to have very widespread use for rehabilitation and strengthening of existing concrete structures. As demonstrated by several experimental results, the use of composites made with fibers and inorganic matrix (such as cement based or lime based matrix) is very useful from a mechanical point of view. The performances of existing reinforced concrete structures strengthened in bending and shear with FRCM systems evidenced a significant improvement both in strength and ductility. The durability of FRCM strengthening systems are not adequately analysed; due to a limited number of experimental studies and research carried out. Moreover, the effects on the environmental conditions on the mechanical properties of strengthening systems are not well investigated. In this paper, the effects of the thermal conditioning on the mechanical properties of the FRCM system was experimentally investigated. The examined strengthening system consists of PBO (short of Polyparaphenylenebenzobisthiazole) fabric meshes embedded into a cement-based matrix. The system was exposed to thermal cycles at different temperature values (100 and 200 °C) and, then, tested to evaluate its mechanical properties (ultimate strength, ultimate strain and elastic modulus). Five thermal cycles were developed by daily exposure of PBO fabric mesh, matrix and PBO-FRCM specimens at constant temperature value, over six hours and subsequently cooling down freely to ambient temperature (20 °C). After the thermal treatment, specimens were tested until failure, at room temperature. Results of tests analysed in terms of stress–strain relationships allowed influence evaluation of the thermal treatment on the structural response of PBO fibers, matrix and PBO-FRCM specimens.

L. Ombres (✉) · P. Mazzuca · S. Verre
Department of Civil Engineering, University of Calabria, Calabria, Italy
e-mail: luciano.ombres@unical.it

P. Mazzuca
e-mail: pietro.mazzuca@unical.it

S. Verre
e-mail: salvatore.verre@unical.it

© The Author(s), under exclusive license to Springer Nature Switzerland AG 2022
J. Sena-Cruz et al. (eds.), *Proceedings of the 3rd RILEM Spring Convention and Conference (RSCC 2020)*, RILEM Bookseries 34,
https://doi.org/10.1007/978-3-030-76465-4_8

Keywords FRCM · Strengthening · Thermal conditioning · Mechanical properties

1 Introduction

In the last decades, Fiber Reinforced Polymer (FRP) and Fiber Reinforced Cementitious Matrix (FRCM) started to be widely used in the repair and strengthening intervention of degraded structures. The first use of FRP material was observed in 1980; at that time this new system represented a development of a strengthening system called "Beton plaque" [1]. At the beginning of the XXI century, an engineering company called Ruregold developed the FRCM system, proposing this system as an alternative of the FRP system. The main difference between the two systems consists of the matrix. FRCM are characterized by an inorganic matrix (i.e. cement-based) while the FRP are based on organic matrix (i.e. polymeric and epoxy resins).

Moreover, the fibers sheets used in the FRP system are replaced with open fabric meshes in the FRCM system. This choice allows the cementitious matrix to penetrate and impregnate the fibres allowing to obtain a higher area at the interface matrix-reinforcement [2]. The advantages of using the FRCM system instead of the FRP system are the higher durability and deformability observed in the former mentioned. Moreover, the use of the inorganic matrix allows overcoming some drawbacks related to the use of the FRPs: (i) the mechanical behaviour when subjected to high temperature and (ii) inapplicability in high humidity environment [3]. With this in mind, the FRCM system represents an excellent alternative to the traditional strengthening system made of either hybrid steel–concrete or timber concrete or reinforced concrete (RC) structures. In fact, in repair and rehabilitation interventions, this solution allows increase of the ductility of the existing structure without adding substantial weight on it. Several studies are available regarding the tensile response of FRCM systems [4–7]. In [4] three FRCM systems, comprising carbon, glass, and steel fabrics embedded in cement, lime, and geopolymer mortar matrices respectively, were studied. The results obtained show in terms of tensile stress-tensile strain a tri-linear behaviour of the strengthening system. The first linear stage is characterized by a phase in which the specimen is un-cracked, then with the increase of the load, perpendicular cracks start to appear on the specimens, which become more significant in the last linear phase as the fabric mesh deforms. The observed failure mode is due to tensile failure of the fabrics. The FRCM system comprised of the steel fabric mesh embedded in a geopolymer matrix shows the highest value in terms of tensile strength. Similar results were observed in [5–7]. Moreover, [5] evaluated the influence of the textile coating on the stress–strain response showing an improvement of the mechanical properties of the FRCM system due to coating treatment. In [7], authors also studied the influence of the loading rate; tests were conducted at displacement control using a rate of 0.1 and 0.6 mm/min. This parameter seems not to have effects on the tensile response of the system. Despite the good results obtained with this strengthening system, the use of it is still hindered due to a lack of information related to its behaviour when subjected to elevated temperatures. For this reason, this study aims to characterize the tensile response of FRCM system that has been thermally conditioned up to 200 °C.

2 Experimental Investigation

The experimental campaign was focused on the evaluation of the mechanical properties at different temperature of the single part (fiber and matrix) and on the strengthening system. The specimen preparation and the experimental procedure are also discussed in this section.

2.1 Fiber and Cement Based Mortar

The PBO fabric mesh has the following dimensions 500 mm in length and 50 mm in width. In particular, the strips of the PBO fabric mesh are characterized by five yarns. The cement based mortar specimens are characterized by a cross section of 40×40 mm, with a length equal to 160 mm. The dimensions used for the cement based mortar are in accordance to [8]. Four specimens were tested for both cement based mortar and fiber.

2.2 Strengthening System

Concerning the strengthening system, three steps were carried out to develop the FRCM specimens. The first step consisted of placing the cement-based mortar in the mould (see Fig. 1a) in order to develop the internal matrix layer, then the PBO fibers were applied and pressed into the matrix (see Fig. 1b). The last step was covering the PBO fiber with the external matrix layer of a cement-based matrix (see Fig. 1c). The dimensions used for the FRCM system are similar to the dry PBO fiber as reported in Sect. 2.1. As showed in Fig. 1a cardboard strip was used to ensure a constant thickness and a flat surface of the matrix. The thickness adopted is equal to 4 mm for each mortar layer.

2.3 Experimental Procedure

All samples (dry fibers, cement based mortar and FRCM system) were thermally conditioned at 100 and 200 °C in a thermal chamber with a heating rate of 2 °C/min as shown in Fig. 2.

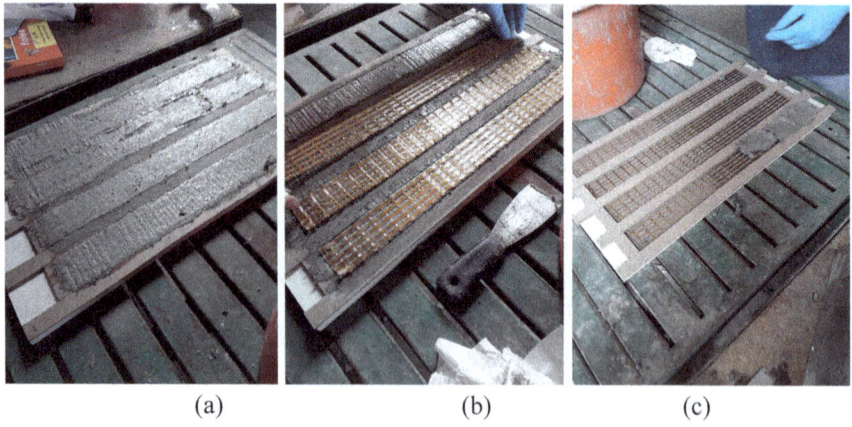

(a)   (b)   (c)

Fig. 1 FRCM system preparation: **a** internal matrix layer, **b** PBO fiber mesh and **c** external matrix layer

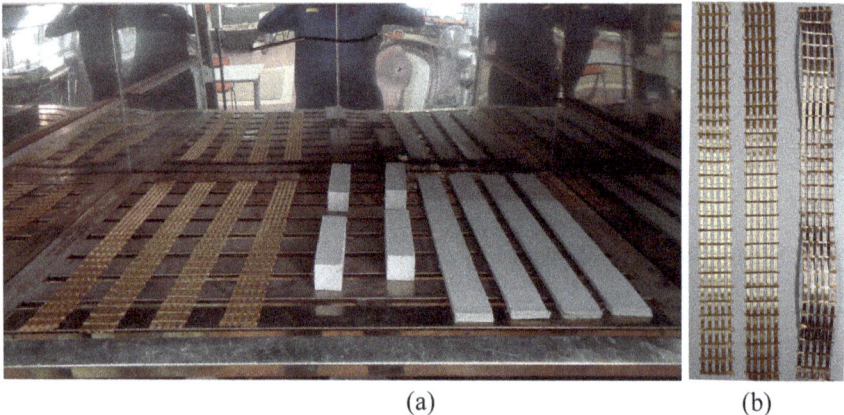

(a)   (b)

Fig. 2 Specimen involved: **a** Thermal treatment on the specimens and **b** Dry PBO fiber after the thermal treatment

After the set point of the temperature was reached, it was kept constant for six hours. The specimens were left cooling down in the thermal chamber, and then after 24 h, another cycle started again. The heating process took into account five cycles. After the thermal treatment was completed, the specimens were taken out from the oven and left to cool at room temperature. With this procedure it is possible to check if the specimens report some degradation due to the heat treatment. The PBO fibers reported a change in the surface colour (see Fig.2b) and became stiffer after being thermally conditioned. No signs of degradation were found either on the cement-based mortar and FRCM system. The PBO fabric mesh and the FRCM system were tested in direct tension up to failure according to [9] while the cement based mortar

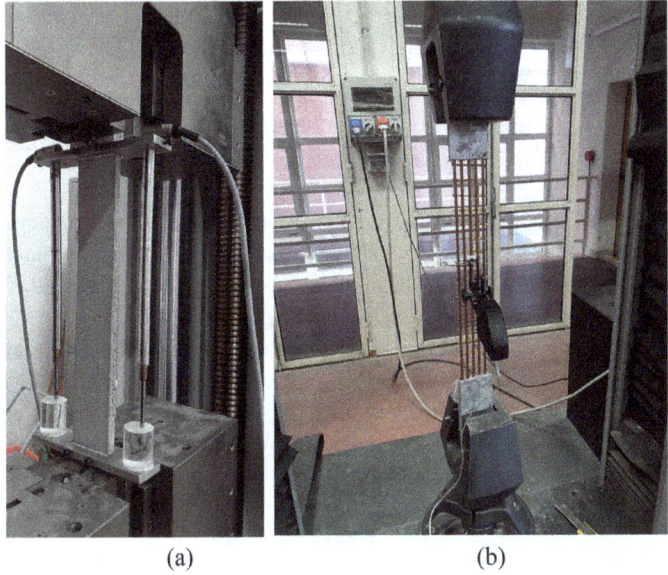

(a)   (b)

Fig. 3 Test setup adopted for: **a** FRCM specimen (DT-20-3) and **b** dry fiber (DF-20-1)

was subjected to three points bending test in order to evaluate the tensile strength (indirectly) and compressive test to obtain the compressive strength [8, 10]. On the PBO fibers and FRCM system, tests were carried out using a standard tensile test configuration, in which the ends of the specimens were clamped. The tests were conducted at displacement control with a rate of 0.5 and 0.2 mm/min, respectively. To guarantee a uniform load distribution, 3 mm thick aluminium tabs and 2 mm thick PVC tabs were bonded on the PBO fiber mesh and FRCM specimen, respectively (using an epoxy resin). In order to evaluate the strain, an extensometer with a gauge length of 50 mm placed in the midsection of the fabric mesh was adopted, while for FRCM specimen two linear variable differential transducers (LVDT) with a gauge length of 300 mm were placed in the vertical direction on the two opposite sides of the FRCM system. The adopted test set up are shown in Fig. 3.

FRCM and dry fibers specimens were named according to the following designation: DT (or DF)-T-Z, where DT (or DF) indicates the direct tensile test on FRCM specimens (or dry fibers), T indicates the temperature investigated, and Z indicates the specimen number.

3   Results and Discussion

In Table 1 the results obtained trough tensile and compressive tests on the cement based mortar are summarized.

Table 1 Results of cement based mortar

	Temp. 20 °C	Temp. 100 °C	Temp. 200 °C
Compressive strength (MPa) (C.o.V.)	43.11 (0.032)	37.80 (0.044)	39.28 (0.027)
Flexural tensile strength (MPa) (C.o.V.)	6.73 (0.053)	5.73 (0.110)	5.23 (0.150)

As reported in Table 1, a minimal reduction in terms of tensile and compressive strength is observed with the increase of the temperature. After the thermal treatment, a single yarn of dry PBO fiber with a length of 300 mm was weighed, showing a reduction of the weight with the increase in temperature.

The weight *(p)* at room temperature is equal to 0.271 g., while at 100° and 200 °C the values observed are 0.222 and 0.153 g. showing a reduction of 22 and 43%, respectively. In Fig. 4a–c the results in terms of tensile strength versus axial strain for the dry fibers are reported. For the evaluation of the stress, the area of a single yarn was evaluated according to Eq. 1.

$$A_f = 1000 \cdot p/\rho_f \cdot l \qquad (1)$$

where ρ_f is the density of fibers, l is the length of the yarn and p the weight. In Table 2 the average values of elastic modulus and tensile strength with the corresponding coefficient of variation, are reported.

As shown in Table 2, significant variations are observed in terms of elastic modulus, in addition, with regard to the tensile strength. At 200 °C the specimens exhibited (lower) value than the specimens tested at 20 and 100 °C.

In Fig. 5a, b and c the results in terms of tensile strength versus axial strain for the FRCM specimen, of which four specimens are tested for each temperature are reported. From each figure it is possible to see an idealised behaviour made of three main phases: (i) Stage I-un-cracked zone, (ii) Stage II-crack development zone and (iii) Stage III-cracked stage up to the failure zone. Moreover, it is possible to observe a drastic reduction in terms of tensile strength and axial strain with the increase in temperature. In particular, compared to the values observed at room temperature, a loss of axial strength equal to 38.1 and 32.5% at 100 and at 200 °C, respectively, can be observed.

Table 2 Results of dry fiber

	Temp. 20 °C	Temp. 100 °C	Temp. 200 °C
Elastic modulus (GPa) (C.o.v.)	218.56 (0.13)	225.92 (0.06)	172.85 (0.06)
Peak tensile strength (MPa) (C.o.v.)	1734.60 (0.05)	1725.23 (0.10)	1673.38 (0.04)

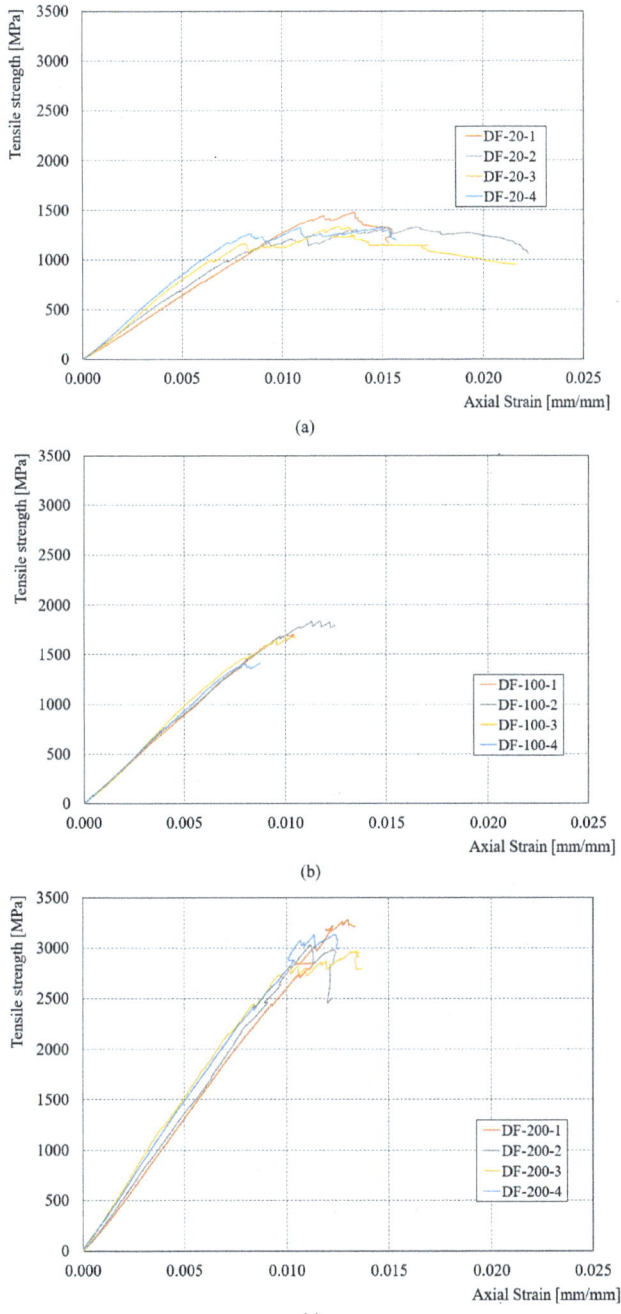

Fig. 4 Result of dry fiber at: **a** 20 °C, **b** 100 °C and **c** 200 °C

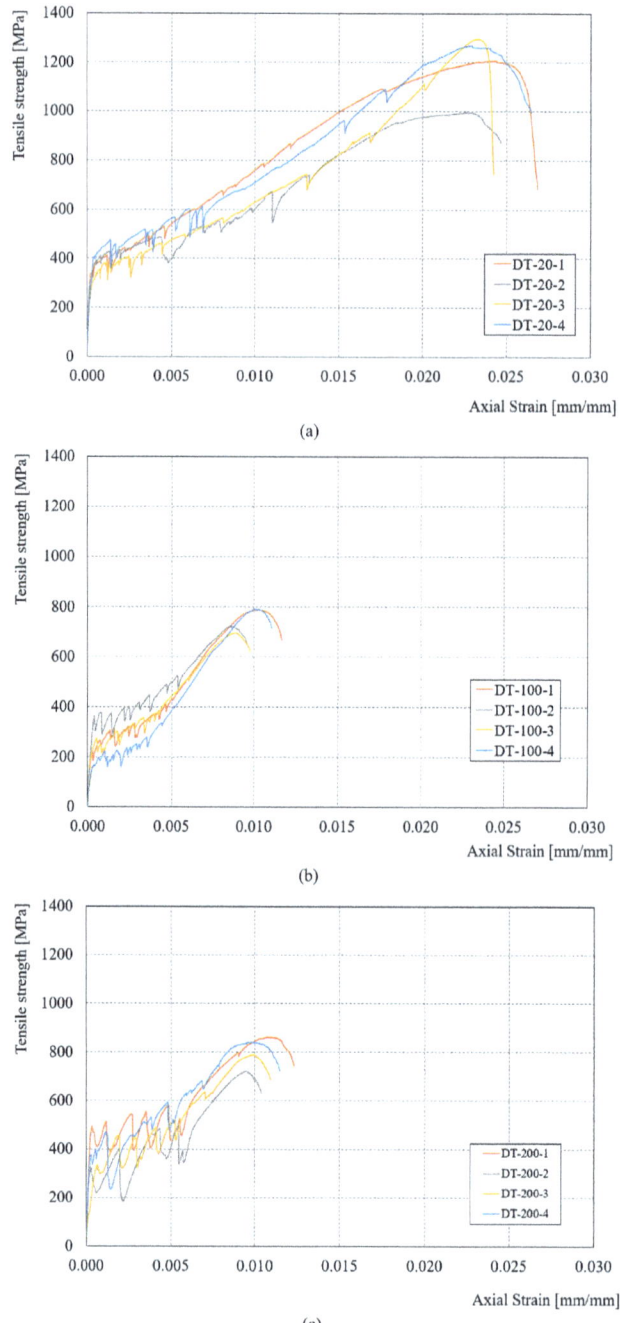

Fig. 5 Result of FRCM specimens at: **a** 20 °C, **b** 100 °C and **c** 200 °C

4 Conclusions

This paper presented the results of an experimental investigation on the effects of the temperature on cement based mortar, dry fiber and FRCM specimen. Based on the obtained results, the following conclusions can be drawn:

- The temperature has minimal influence on the cement based mortar in terms of flexural tensile strength and compressive strength. The specimen reports a compression strength reduction of 12% its ambient temperature at 100 °C and a tensile strength reduction of 22% when temperature was increased from 20 to 200 °C.
- There are significant variations on the elastic modulus of the PBO fibers with the increase of the temperature. The effect of the heat treatment can be observed on the tensile strength with the highest values observed at 200 °C.
- In the FRCM specimens, it is possible to observe severe reductions of the mechanical properties when the temperature increases. In particular, in terms of axial strength compared to the values observed at room temperature, there is a reduction equal to 38.1 and 32.5% at 100 and at 200 °C, respectively.

This preliminary work needs further studies with different temperature and number of layers.

Acknowledgements The authors would like to express their appreciation to Ruregold s.r.l. Italy, that provided the composite materials in this study. The second author would like to acknowledge POR CALABRIA 2014–2020 for funding his research.

References

1. Trimboli, A.: Guida ai materiali FRCM per il rinforzo delle strutture in c.a. e in muratura, 1st edn., Ruregold (2019)
2. https://www.structuremag.org/?p=5446. Accessed 09 Jan 2020
3. https://www.ruregold.it/rinforzi-strutturali/. Accessed 04 Jan 2020
4. De Santis, S., Hadad, H.A., De Caso y Basalo, F., De Felice, G., Nanni, A.: Acceptance criteria for tensile characterization of fabric-reinforced cementitious matrix systems for concrete and masonry repair. J. Comp. Constr. **22** (2018)
5. D'Antino, T., Papanicolaou, C.C.: Comparison between different tensile test set-ups for the mechanical characterization of inorganic-matrix composites. Constr. Build. Mater. **171**, 140–151 (2018)
6. Li, B., Xiong, H., Jiang, J., Dou, X.: Tensile behavior of basalt textile grid reinforced engineering cementitious composite. Compos Part B: Eng. **156**, 185–200 (2019)
7. Iorfida, A., Verre, S., Candamano, S., Ombres, L.: Tensile and direct shear responses of basalt-fibre reinforced mortar based materials. RILEM Bookseries **15**, 544–552 (2018)
8. UNI-EN, 1015-11.: Methods of Test for Mortar for Masonry Part 11: Determination of Flexural and Compressive Strength of Hardened Mortar. UNI, Rome (2016)

9. UNI EN 2561.: Materie plastiche rinforzate con fibre di carbonio. Laminati unidirezionali. Prova di trazione parallelamente alla direzione delle fibre. UNI, Rome (1995)
10. UNI EN 12190.: Product and System for the Protection and Repair Concrete Structures—Test Methods—Determination of Compressive Strength of Repair Mortar. UNI, Rome (2000)

Durability Under Thermal Actions of Concrete Elements Confined with an Inorganic Matrix Fiber-Reinforced Composites

L. Ombres, P. Mazzuca, and S. Verre

Abstract Fiber reinforced composite materials are starting to have very widespread use for rehabilitation and strengthening of existing concrete structures. As demonstrated by several experimental results, the use of composites made with fibers and inorganic matrix (such as cement based or lime based matrix) is very effective from a mechanical point of view. Nevertheless, there are legitimate concerns with the durability aspects of strengthened elements, which have hindered the widespread use of these composite materials in structural applications. The exposure of confined concrete elements to environmental actions could reduce the beneficial effects of the strengthening. For a better understanding of this aspect, through an experimental campaign, the paper aims to investigate the effects of thermal actions on the structural response of concrete elements confined with an inorganic matrix fiber-reinforced composite system, consisting of high strength fibers in the form of fabric embedded into a cement-based matrix. Cylindrical concrete specimens confined with one-layer of PBO (Polyparaphenylenebenzobisthiazole)-FRCM (Fiber Reinforced Cementitious Matrix) system were exposed to several (five) thermal cycles at different temperatures (100 and 200 °C) and, then tested under compression loads at ambient temperature (20 °C). When the desired temperature was reached, it was kept constant for 6 h and subsequently, the concrete specimens were cooled down freely to ambient temperature. The obtained results in terms of failure modes, peak strength, axial and radial strains were analysed, in order to evidence the effects of the thermal actions on the mechanical properties of confined concrete elements.

Keywords Durability · Composites · Strengthening · Concrete · Thermal actions

L. Ombres (✉) · P. Mazzuca · S. Verre
Department of Civil Engineering, University of Calabria, Calabria, Italy
e-mail: luciano.ombres@unical.it

P. Mazzuca
e-mail: pietro.mazzuca@unical.it

S. Verre
e-mail: salvatore.verre@unical.it

© The Author(s), under exclusive license to Springer Nature Switzerland AG 2022
J. Sena-Cruz et al. (eds.), *Proceedings of the 3rd RILEM Spring Convention and Conference (RSCC 2020)*, RILEM Bookseries 34,
https://doi.org/10.1007/978-3-030-76465-4_9

99

1 Introduction

During its nominal life, the load-bearing capacity of structures can be affected by several factors, namely degradation of constituent materials, change of use or catastrophic events such as earthquake or fire. Thus, the structures can be deteriorated or reach a substandard grade requiring urgent intervention. To reconstitute the original section, traditional interventions are usually carried out by replacing or reinforcing the degraded system using substantial solutions such as reinforced concrete or steel-concrete systems. These types of intervention, challenging in the application phase and highly invasive to the static and aesthetics of the buildings, also present low durability in keeping the efficacy of the strengthened system during the years. The durability problems aforementioned coupled with the needs of higher speeds of construction lead to the development of innovative structural solutions using composite materials such as fiber reinforcement polymers (FRP) and fiber fabric reinforced cementitious mortar (FRCM). FRCM composite materials comprised of fibers coupled with an inorganic matrix, unlike the organic matrix usually used in the more common FRP. Moreover, the FRCM allows overcoming some limits showed by the FRP such as the inapplicability in high humidity environment and the low fire resistance [1]. In this context, FRCM system represents a good alternative providing an increase of the seismic capacity by increasing the structural ductility of the existing structure without adding substantial weight on it.

Nevertheless, legitimate concerns are related to the behaviour of FRCM systems when subjected to thermal actions, hindering their development in structural applications. Several researchers have investigated the compressive behaviour of PBO-FRCM confined concrete cylinder [2–4], reporting a good efficacy of this strengthening system on the mechanical performances of the reinforced element. Although the potential of this solution is relevant, a lack of knowledge regarding the behaviour of these elements, when subjected to high temperatures, can be observed in literature. In literature, several authors studied the behaviour at elevated temperatures of a concrete element reinforced with FRP [5–8], but only two studies are available regarding the effect of the temperature on FRCM confined concrete specimens [9, 10]. In [9] concrete cylinder wrapped with CFRP sheets and FRCM mesh were placed in a thermal chamber and exposed to elevated temperatures: 40, 60, 80 °C for 24 h. After the heat treatment finished, the specimens were immediately placed in the testing machine and tested. From the test results, it can be observed that the CFRP confined specimens are more sensitive to the temperature than the FRCM confined specimens. For the FRCM strengthening system, upon temperature increase from 40 °C to 80 °C, a reduction of 5–11% was observed in load-bearing capacity and compressive strength, respectively. Instead, the CFRP strengthening system, in the same range of temperatures, showed a higher percentage of loss in terms of load-bearing capacity and compressive strength equal to 20% and 50%, respectively.

In [10], FRCM confined concrete cylinders with different fiber reinforcement ratio have been subjected to thermal cycles from 20 °C to 250 °C, and then tested up to failure at room temperature. With the increase of the temperature, the compression strength of the confined specimens decreases and the axial strain increased. Moreover, the specimens characterised by a higher percentage of fiber reinforcement are more sensitive to the temperature in terms of ductility loss compared to the specimens with a lower fiber reinforcement ratio. Concerning the failure mode, no significant changes were observed in the specimens tested. This paper aims to study the effects of the temperature on compressed PBO confined and unconfined concrete cylinders. The tests carried out on the unwrapped specimens are useful for a more accurate analysis of the confinement effect at elevated temperature. Test results are analysed in terms of compressive strength, failure mode, axial and lateral strains.

2   Experimental Investigation

The experimental campaign comprises in 18 concrete cylinders of 150 mm diameter and 300 mm in height. The specimens were characterised by nine unwrapped cylinders and nine cylinders wrapped with one layer of PBO fabric mesh. Before testing, all samples were thermally conditioned (up to 200 °C) to simulate the behaviour of an industrial building when subjected to high service temperature and of a strengthened concrete column under fire actions. In Table 1 the tested specimen in terms of temperature adopted and the reinforcement ratio (ρ_f) are reported. Confined cylinders were named according to the following designation: C (or U)-T-Z, where C (or U) indicates the confined specimen (or un-confined), T indicates the temperature investigated, and Z indicates the specimen number. For cylindrical specimens, the reinforcement ratio was calculated with the following Eq. (1):

$$\rho_f = 4\,t_f/D \qquad (1)$$

where D is the specimen diameter and t_f is the thickness of the PBO-FRCM reinforcing system; and the ρ_f is equal to 0.121%.

Table 1 Specimens involved

Room temperature		Temperature 100 °C		Temperature 200 °C	
Unconfined	Confined	Unconfined	Confined	Unconfined	Confined
U-20-1	C-20-1	U-100-1	C-100-1	U-200-1	C-200-1
U-20-2	C-20-2	U-100-2	C-100-2	U-200-2	C-200-2
U-20-3	C-20-3	U-100-3	C-100-3	U-200-3	C-200-3

Fig. 1 PBO fabric mesh

2.1 Specimen Preparation and Test Setup

The FRCM system adopted in this experimental campaign consists of unidirectional PBO fabric mesh embedded in an inorganic matrix designed to be chemically bonded together. The fibers are characterised by an equivalent thickness of 0.0455 mm and 0.0115 mm in the principal and transversal direction, respectively (see Fig. 1).

The manufacturer provides the mechanical properties of both fibers and matrix. According to the material's datasheet, the PBO mesh presents a density of 1.56 g/cm^3. Tests were carried out, according to [11] by the authors to define the tensile strength and elastic modulus, obtaining values of 1083 MPa (C.o.V. = 0.05) and 122 GPa, (C.o.V. = 0.13), respectively. With regards to the cementitious matrix, specimens exhibited tensile strength and compressive strength equal to 6.73 MPa (C.o.V. = 0.053) and 43.11 MPa (C.o.V. = 0.032) (values at 28 days) according to [12]. Before confining the concrete cylinder with PBO fabric mesh, the surface of the concrete specimens was sandblasted in order to remove any possible defects on it and to create a roughness profile in order to assure a perfect adhesion between the concrete substrate and inorganic matrix (see Fig. 2a). The strengthened system was applied at the cylinders by performing three steps. The first step was applied to the internal matrix layer, with a thickness equal to 3 mm on the concrete surface (see Fig. 2a). The second step, the PBO fabric mesh was slightly pushed inside the internal matrix layer (see Fig. 2b) and then, as the third step, the external matrix layer with a thickness of 3 mm (see Fig. 2c) was applied. Two circular plexiglass plates placed at the top and bottom of the cylinder with a diameter of 15.6 and 16.2 mm, were used to control the thickness of the internal and external matrix layers. Moreover, the principal direction of the PBO fabric mesh was applied with the principal direction parallel at the transversal cross section. Finally, to avoid a premature rupture in the fibers, an overlap equal to 100 mm was used.

Before the thermal treatment, the specimens were in ambient condition-after the preparation they were kept in the laboratory enviroment. Before testing, all specimens were exposed to thermal cycles at 100 and 200 °C at a heating rate of 2 °C/min. Once the desired temperature was reached, isothermal conditions were maintained for 360 min at the end of each cycle, the speciemens were left cooling in the oven. The heat treatment took five days and took in account five cycles (one cycle per

(a)   (b)   (c)

Fig. 2 Confined concrete cylinder preparation: **a** internal matrix layer and concrete surface, **b** PBO fiber mesh and **c** external matrix layer

day). After heating, the specimens were taken out from the oven and left to cool in the lab temperature before testing. This procedure was chosen to investigate any permanent deterioration caused by elevated temperatures in the strengthened system. As shown in Fig. 3, a superficial crack can be observed on the external matrix due to the heat treatment. The concrete cylinders were tested in uniaxial compression load up to failure. The compressive loading, measured by a load cell, was applied at a displacement controlled rate of 0.0005 mm/s. The vertical displacement by means of two linear variable differential transducers (LVDTs) were placed at the two opposite sides of the sample (see Fig. 3) in order to evaluate the axial strain. Moreover, in each cylinder, a lateral displacement was evaluated through nine LVDTs placed in the horizontal direction, in particular, three at the top part, three at mid-height and three at the bottom of the wrapped specimen (see Fig. 3).

Fig. 3 Test setup

3 Results and Discussion

The obtained results are summarized in Tables 2 and 3. In particular, in Table 2 the results for the un-confined cylinder and the average value in terms of strength, f_{c0}, and corresponding axial strain, ε_{c0} are reported.

In Table 3, in addition to the peak stress f_{cc} and the peak strain ε_{cc} the ultimate strain ε_{ccu} (evaluated as the strain recorded at 80% of the peak strength in the descending branch), the confinement ratio in terms both of strength, f_{cc}/f_{c0} and axial strain, $\varepsilon_{cc}/\varepsilon_{c0}$ and the strain efficiency factor k defined as the ratio between the hoop strain at failure and the ultimate strain of the FRCM system, are, also, reported.

The axial strength recorded during the tests is reported in Tables 2 and 3. In both unconfined and confined cylinders, it should be noted that a slight decrease in terms of axial strength occurred between the specimens tested at 20 and 100 °C.

With the increase in temperature, the strength's reduction becomes more relevant. As shown in Fig. 4, concerning the ratio f_{cc}/f_{c0} and $\varepsilon_{cc}/\varepsilon_{c0}$, it is possible to observe that in terms of axial strength, the effect of the confiment are not so much affected

Table 2 Unconfined cylinders: test results

Specimens	Temperature (°C)	f_{c0} (MPa)	Avg f_{c0} (MPa)	ε_{c0} ($\times 10^{-3}$)	Avg ε_{c0} ($\times 10^{-3}$)
U-20-1	20	28.98		3.86	
U-20-2	20	26.54	28.73	3.66	3.67
U-20-3	20	30.69		3.49	
U-100-1	100	27.38		3.83	
U-100-2	100	29.56	28.1	3.45	3.65
U-100-3	100	27.35		3.66	
U-200-2	200	26.58	24.65	5.52	5.21
U-200-3	200	21.3		5.02	

Table 3 Confined cylinders: test results

Speci	Temp (°C)	f_{cc} (MPa)	ε_{cc} ($- \times 10^{-3}$)	ε_{ccu} ε_{cc} ($- \times 10^{3}$)	f_{cc}/f_{c0} (–)	$\varepsilon_{cc}/\varepsilon_{c0}$ (–)	k (–)
C-20-1	20	33.83	8.47	10	1.18	2.3	0.896
C-20-2	20	38.13	8.33	10.59	1.33	2.27	0.732
C-20-3	20	42.84	7.35	9.56	1.49	2	0.451
C-100-1	100	38.12	7.15	9.84	1.36	1.96	0.544
C-100-2	100	33.75	6.76	9.11	1.2	1.85	0.484
C-100-3	100	33.78	6.72	11.02	1.2	1.84	0.754
C-200-1	200	26.35	6.98	9.05	1.07	1.33	0.553
C-200-2	200	28.04	6.99	8.73	1.14	1.34	0.464
C-200-3	200	22.98	9.77	12.52	0.93	1.87	0.725

Fig. 4 **a** f_{cc}/f_{c0} vs temperature and **b** $\varepsilon_{cc}/\varepsilon_{c0}$ vs temperature

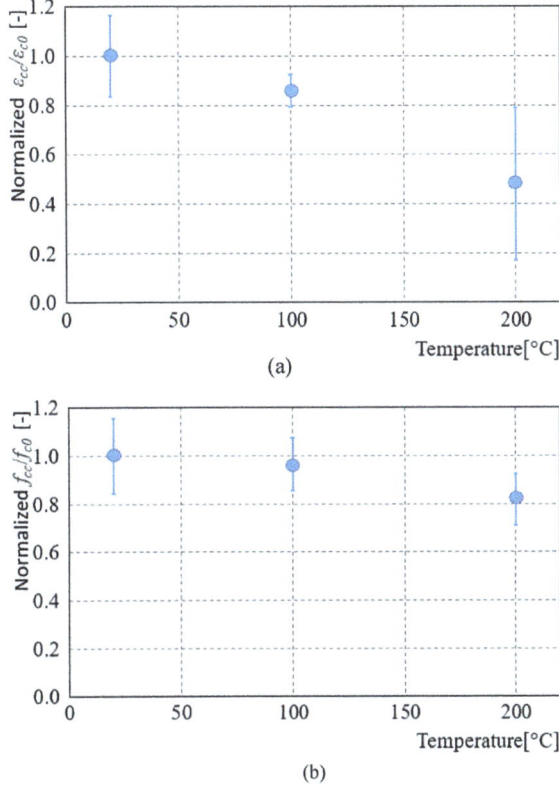

(a)

(b)

by the temperature, showing a loss lower than 18% of the ambient temperature value at 200°C. Nevertheless, in the ratio $\varepsilon_{cc}/\varepsilon_{c0}$, temperature plays an important role. In fact, at 200°C the effect of the confinement is reduced of almost 50% of the room temperature value.

In Fig. 5 the results in terms of axial stress versus axial and lateral strain for unconfined (see Fig. 5a) and confined (see Fig. 5b) cylinders, are reported. All curves showed a loss in stiffness with the increase in temperature. The target temperature and the exposition time are factors that affect this loss in strength. The curves also exhibited an increase of both axial and lateral strain with the increase of the temperature, this may be due to the damage reported by the mortar at high temperature. With regards to the efficiency factor, all specimens showed values of k lower than 1, this means that a localized brittle failure causes a premature rupture of the reinforcement.

In Fig. 6 the typical failure mode observed for both unconfined and confined cylinder, is reported. In the unconfined cylinders, vertical cracks occurred with the increase of the load (see Fig. 6a). Tests were carried out up to the complete crushing of the concrete. Concerning the wrapped specimens, several vertical cracks were observed on the external layer (see Fig. 6b).

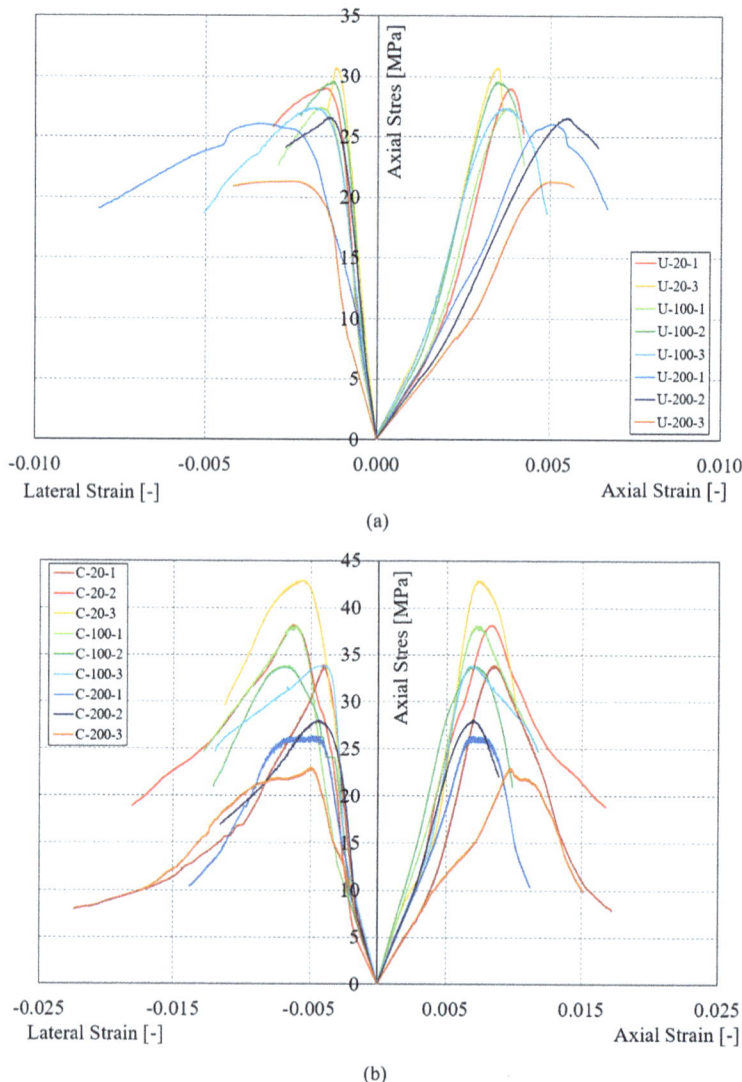

Fig. 5 Stress–strain curves: **a** unconfined and **b** confined cylinders

Before the peak load was reached, a wide vertical crack appeared, which became more substantial with the increase of the load (see Fig. 6c). This failure mode may be related to the slippage of the PBO sheets from the matrix. As reported in [10], this type of behaviour is typical for concrete elements reinforced with PBO fibres.

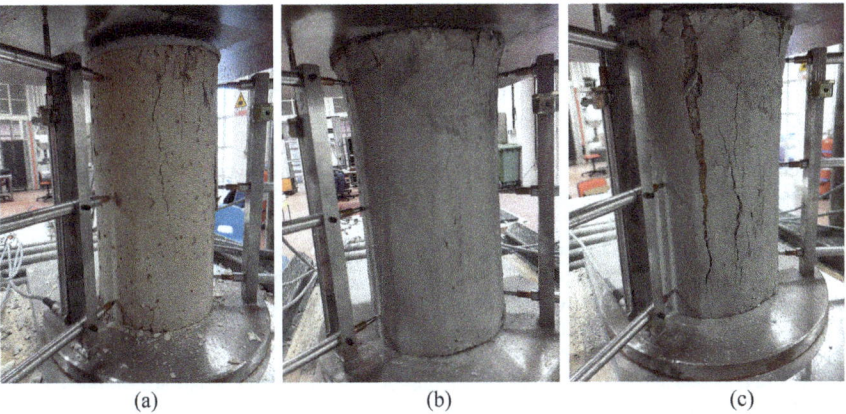

| (a) | (b) | (c) |

Fig. 6 Failure mode: **a** U-200-1, **b** and **c** C100-3

4 Conclusion

This paper presented the results of an experimental investigation on the effects of the temperature on unconfined and confined concrete cylinders with FRCM system. Based on the results, the following can be concluded:

- Reinforcing the cylinders with PBO fabric meshes has a positive effect in terms of mechanical properties. An increase in axial strength equal to 22% (33%) and 20% (25%) for the specimens tested at 20°C and 100°C, respectively, is observed. At 200°C no relevant improvements are noted.
- The effects of the temperature is more relevant for temperatures above 100°C. At 200°C, the unconfined and confined cylinders show a percentage of axial strength loss equal to 17% and 35%, respectively. Concerning the tests carried out at 100°C, both unconfined and confined cylinders, suffer little loss in axial strength, with percentages of loss equal to 4% and 8%, respectively.
- Both unconfined and confined cylinders show a slight decrease in terms of initial stiffness with the increase of the temperature.
- The heat treatment seems not to have effects on the failure mode of both unconfined and confined cylinders; in all tests, failure occurs due to the slippage of the fibres from the cementitious matrix.

This preliminary work needs further studies with different temperature and number of layers.

Acknowledgements The authors would like to express their appreciation to Ruregold s.r.l, Italy, that provided the composite materials in this study. The second author would like to acknowledge POR CALABRIA 2014–2020 for funding his research.

References

1. https://www.ruregold.it/rinforzi-strutturali/. Accessed 04 Jan 2020
2. Trapko, T.: Confined concrete elements with PBO-FRCM composites. Constr. Build. Mater. **73**, 332–338 (2014)
3. Colajanni, P., De Domenico, F., Recupero, A., Spinella, N.: Concrete columns confined with fibre reinforced cementitious mortars: experimentation and modeling. Constr. Build. Mater. **52**, 375–84 (2014)
4. Carloni, C., Bournas, D., Carrozzi, G.C., D'Antino, T., Fava, G., Focacci, F., Giacomin, G., Mantegazza, G., Pellegrino, C., Perinelli, C., Poggi, C.: Fiber Reinforced Composites with Cementitious (Inorganic) Matrix, p. 349–91. Springer, Berlin (2015). ISBN 978-94-017-7335-5. Chapter 9, ID 336067_1_En
5. Rosa, I.C., Firmo, J.P., Correia, J.R., Barros, J.A.O.: Bond behaviour of sand coated GFRP bars to concrete at elevated temperatureDefinition of a bond vs slip relations. Compos. B **160**, 329–340 (2019)
6. Rosa, I.C., Firmo, J.P., Correia, J.R., Mazzuca, P.: Influence of elevated temperatures on the bond behaviour of GFRP bars to concrete—pull-out tests, IABSE Symposium Guimarães, Portugal (2019)
7. El-Gamal, S., Al-Jabri, K., Al-Mahri, A., Al-Mahrouqi, S.: Effects of elevated temperatures on the compressive strength capacity of concrete cylinders confined with FRP sheets: an experimental investigation. Int. J. Poly. Sci. (2015)
8. Al-Salloum, Y.A., Elsanadedy, H.M., Abadel, A.A.: Behavior of FRP-confined concrete after high temperature exposure. Constr. Build. Mater. **25**(2), 838–50 (2011)
9. Trapko, T.: The effect of high temperature on the performance of CFRP and FRCM confined concrete elements. Compos. B **54**, 138–45 (2013)
10. Ombres, L.: Structural performances of thermally conditioned PBO FRCM confined concrete cylinders. Comp. Struct. **176**, 1096–1106 (2017)
11. UNI EN 2561.: Materie plastiche rinforzate con fibre di carbonio. Laminati unidirezionali. Prova di trazione parallelamente alla direzione delle fibre. UNI, Rome (1995)
12. UNI-EN, 1015-11.: Methods of Test for Mortar for Masonry Part 11: Determination of Flexural and Compressive Strength of Hardened Mortar. UNI, Rome (2016)

Accelerating the Hardening of Lime Mortar on Addition of Organics for Repair in Heritage Structures

Celesta Issac and Simon Jayasingh

Abstract Repair and restoration of heritage structures have become important as they are the evidence of cultural and historical significance. Strength and durability of lime mortar makes it a good repair material for achieving movement with the old mortar and therefore compatible. Hardening of lime mortar may take centuries to complete and therefore accelerating the hardening process helps in achieving the desired strength earlier. This work aims at studying the impact of organic addition on the physical, mechanical, chemical and durability properties of hydraulic lime mortar to accelerate the hardening. A comparative study of lime mortars with addition of three different organics namely Vitis vinifera (Raisins), Vita vinifera (Zante currant) and Phoenix dactylifera (Dates) were performed. The lime mortar was prepared with binder to aggregate ratio of 1:3, water to binder ratio of 0.75 and with addition of organics in various concentrations (0, 2, 4 and 6%). Bulk density, porosity and water absorption were the physical properties analyzed. The compressive strength was determined for the specimens. The rate of carbonation was determined through the phenolphthalein test. The mineralogical composition of the mortar was identified by using analytical techniques like XRD and FTIR. Capillary rise test was conducted to study the durability of the lime mortar. It was observed that organic addition to lime mortar has enhanced its properties along with an increase in the rate of carbonation and thereby accelerating the hardening. The carbohydrate rich organic additives increased the CO_2 content throughout the mortar. Hence increased the strength and durability of lime mortars used in the repair of heritage structures.

Keywords Heritage · Lime · Mortar · Organic additive · Carbonation

C. Issac · S. Jayasingh (✉)
Vellore Institute of Technology, Vellore, Tamil Nadu, India
e-mail: simon.jayasingh@gmail.com

© The Author(s), under exclusive license to Springer Nature Switzerland AG 2022
J. Sena-Cruz et al. (eds.), *Proceedings of the 3rd RILEM Spring Convention and Conference (RSCC 2020)*, RILEM Bookseries 34,
https://doi.org/10.1007/978-3-030-76465-4_10

1 Introduction

Lime mortar was prominently used as a construction material in historic times [1]. Lime was used as a binding material along with aggregates since the fourth century which replaced the ancient construction materials clay and gypsum [1]. Historic buildings serve to provide aesthetical and cultural significance in a society. Conservation of historic buildings has gained importance in recent times due to the damages and deterioration of heritage structures by physical and chemical factors [3]. The restoration works on these heritage structures are carried out considering the compatibility of ancient and modern construction materials [2]. Hence the demand for lime based mortars have increased as cement based mortars prove to be incompatible with the old mortars. The higher strength of cement based materials makes it inappropriate for repair as it limits the ability to allow movement with the old mortar [2].

Lime mortar exhibits properties like fresh state properties and hardened state properties. The fresh state properties include workability and better water retention. The improved workability depends on the plasticity of the mortar and makes the construction easier. Good water retention property of lime enhances its ability to resist water penetration while being the more breathable and environment friendly material [17]. The hardened state properties of lime mortar like strength and durability makes it a better repair material. Hardening of lime mortar is a gradual process after the setting of mortar, where the mortar attains its strength progressively [1]. Hardening of lime mortar occurs on drying by crystallization and also through carbonation process that precipitates calcium carbonate crystals [18]. This process helps in the subsequent strength gain and resistance to stresses. However this is a slow process and the desired outcome is obtained after long time. The increased durability of lime mortar was evident from the ancient structures all over the world. The durability was enhanced by using additives which act as water repellent agents that protects the structures from harmful weathering agents [9]. Calcite, vaterite and aragonite are the commonly known polymorphs of calcium carbonate of which calcite being the most stable and vaterite the least [5]. The delay in the transformation of vaterite to calcite prolongs the drying process and also increases the durability [14]. The porosity of lime mortar also plays a major role in protection from harmful agents and also moisture control [17]. It controls the water absorption and gas permeability characteristics which aids in the carbonation process and hence in the strength and durability of lime mortar [19]. The performance and the properties of lime mortar depend on the morphology and microstructure of calcium carbonate crystals [4]. Calcium carbonate crystals are formed as a by product of carbonation process.

Carbonation of lime mortar is a chemical reaction which involves the diffusion of atmospheric carbon dioxide and reaction with calcium hydroxide resulting in the formation of calcium carbonate crystals [6]. The diffusion of CO_2 gas occurs through the pores of the lime mortar and adequate water is required for the dissolution of CO_2 [19]. The chemical reaction is represented in "Eq. (1)" as follows:

$$Ca(OH)_2 + CO_2 + nH_2O = CaCO_3(n+1)H_2O \qquad (1)$$

The precipitated insoluble calcium carbonate crystals occupy the pore spaces present in the hardened mortar which makes it more compact and increases the strength [18]. An increase in volume is also observed during this process. Carbonation affects the microstructure properties of lime mortar and its porosity [19]. It plays a major role in the hardening of lime mortar and hence its durability. This process primarily depends on the carbon dioxide content in the air and occurs on the surfaces in contact with atmosphere. Natural carbonation is a slow process due to reduced CO_2 concentration in the air. Thus the complete carbonation of lime mortar may take centuries to occur [7]. This explains the reason for reduction in use of lime based materials nowadays.

Many researchers have done study on the increase in strength of lime mortar using various organic additives. Organic additives are used as source of carbohydrates, proteins, polysaccharides and fats since ancient times as it helps in enhancing the binding capability and also prevents cracking in lime mortar [8]. Characterisation of ancient mortars by chemical and mineralogical analysis shows that organic additives were used to enhance the properties like mechanical strength and durability. Various organic admixtures like cactus plant extract, latex from rubber, resins, banana plant leaves, oils, egg whites, jute, rice husk, sugarcane bagasse, areca nut, blood lime and many other fruits and flowers were used for studying the influence on physical, mechanical and durability properties [1, 9]. The availability of raw materials and traditional production technology differs across the world. The water proofing property of Euphoria lacter was utilized by the African and south American region [10]. The juice extracts of fruits, keratin, egg whites and blood of bullocks were used as polymers in Egypt during ancient times [11]. Tung oil was used in early wooden ships and special structures in China as it acts as a sealant and provides better moisture resistance. Shiqiang Fang et al. (2014) highlights the performance of tung oil lime mortar and shows increased mechanical properties and resistance to moisture and weather compared to the reference lime mortar [12]. Increase in mechanical strength was observed with the use of sticky rice as an admixture in lime mortar [4]. Cissus Glauca Roxb extract greatly influenced the compressive and flexural strength of the hydraulic lime mortar. The hydrophobic nature of the polymer led to the blockage of capillary pores which protects the structure from moisture penetration and weathering agents [16]. The various organic additives used in India include pulp of fruits, flowers, beans and leaves soaked in oils [13]. The studies conducted by Thirumalini et al. provide information on the influence of organic additives on the mechanical strength and durability of the structure. The presence of carbohydrates and proteins greatly influenced the strength and longevity of the structure [14, 15].

Studies have revealed that addition of organics to lime mortar improves its properties whereas detailed information regarding the influence of organics and its ability to accelerate the carbonation process and enhance hardening is not available. The present study aims in accelerating the hardening process with the help of organic additives. The influence of organics on the physical, mechanical, chemical and durability properties of lime mortar was studied. The organic additives used in this study include Vitis vinifera (Raisins), Vita vinifera (Zante currant) and Phoenix dactylifera (Dates). They commonly belong to the group of dry fruits and are rich in nutrients like carbohydrates, proteins, calories and other minerals. Since they are rich in carbohydrate content which helps in increasing the rate of carbonation, it accelerates the hardening process and improves its strength. Comparative study on the effect of each organic type is conducted. The lime mortar was prepared with lime to aggregate ratio of 1:3, 0.75 water-binder ratio and with the addition of organic admixtures in various concentrations (0%, 2%, 4% and 6%). The physical properties like porosity, water absorption and bulk density were studied. Compressive strength was the mechanical property analyzed. The durability and resistance of lime mortar specimens were determined through the capillary rise test. The carbonation rate was identified through the phenolphthalein test. Analytical study on the mineralogical composition of the modified lime mortar was conducted through techniques such as XRD and FTIR. This validation provides detailed information on the chemical reactions and compositions in organic modified lime mortar which contributes to the improved properties.

2 Materials and Experiments

2.1 Raw Materials

The procurement of hydrated air lime powder was from Valasaravakkam, Chennai. Lime powder sieved under 90 μ sieve was used for the preparation of mortar specimens. River sand of specific gravity 2.67 was used as fine aggregates for lime mortar. The sieve analysis was performed as per IS: 2386 (Part 1) 1963 [20]. The fine aggregates sieved under the sieve size 4.75 mm was used for the study. The mineralogical composition of lime powder and sand was determined using XRD analysis. The workability test of lime was performed according to IS: 6932 (Part 8) 1973 [21] and the water to binder ratio of 0.75 was adopted from the test. The initial setting time and final setting time of lime was determined according to the code provisions IS 6932 (Part 11) 1983 [22].

The utilization of organic additives in ancient construction was determined through characterisation of historic mortars and materials. This study aims in using carbohydrate rich organic additives like Vitis vinifera (Raisins), Vita vinifera (Zante currant) and Phoenix dactylifera (Dates). These organics are categorised as dry fruits, which contains high amount of carbohydrates and proteins. The presence of polysaccharides, proteins and amides and the chemical bonding was determined through FTIR analysis of these organic additives. All the three organics were soaked in water and then grinded to paste form. These were added to the lime mortar mix in various concentrations (0, 2, 4 and 6%) of water during the casting process.

2.2 Mortar Specimen Preparation

The workability of lime was determined through the flow table test and the confirmation was done through initial flow test. As per the results the optimum water/lime ratio of 0.75 was adopted for the preparation of lime mortar. The lime to aggregate ratio of 1:3 has been proven to be most appropriate for conservation of heritage structures, hence this ratio was adopted. The organic addition was carried out as a percentage of the water required for the mortar.

The required amount of binder, fine aggregate and water was mixed in a basin and filled into cube shaped moulds of size $50 \times 50 \times 50$ mm. Light compaction was provided by pressing with thumb and tamping. Three set of lime mortar cubes with different organics (in various percentages) and a reference mortar was prepared. The samples were given codes such as Reference (R), Zante Currant (Z- 2%, 4%, 6%), Dates (D- 2%,4%,6%), Raisins (R- 2%, 4%, 6%). The mortar cubes were casted and demoulded on the 5th day and kept for curing under room temperature for 28 days and 90 days.

2.3 Experimental Investigation

The compressive strength of the mortar cubes were determined through the compressive strength test according to IS: 6932 (Part 7) 1973 [23]. The physical properties like bulk density, water absorption and porosity was identified according to RILEM (1980). The core sample was taken after the compression test and powdered finely for the secondary tests. The mineralogical and chemical composition of the hardened mortar was determined through analytical techniques like XRD and FTIR. X-Ray Diffraction spectroscopy helps in identifying the mineral composition of lime mortar using Cu Ka radiations. The presence of calcite, vaterite, quartz and other minerals were obtained through the analysis which was interpreted using X-Pert High score software. The graphical representation of the results was plotted using Origin software. Fourier Transform Infrared spectroscopy was used as a supplementary to XRD and to identify the organic materials and chemical bonding of the sample. The carbonation test using phenolphthalein was conducted on lime mortar samples to identify the depth of carbonation. The difference in colour during the phenolphthalein test indicates incomplete carbonation. Durability test such capillary rise test was conducted on the mortar specimens after the curing period.

3 Results and Discussion

3.1 Test on Raw Materials

The mineralogical composition of lime was obtained from XRD analysis. The presence of high calcium content was evident from the graph shown in "Fig. 1 XRD pattern for air lime". The predominant compound present in the lime sample was $Ca(OH)_2$ which confirmed it can be used as air lime (Table 1). The particle size distribution of fine aggregate (river sand) is shown in "Fig. 2". The obtained curve indicates that the aggregates are well graded. The XRD analysis of fine aggregates shows high silica content in the form of quartz. The initial setting time of lime was found to be 2 h and 45 min and the final setting time was found to be 13 h.

The results obtained from FTIR analysis of the organic additives are shown in "Fig. 3a Dates, 3b Zante Currant, 3c Raisins". The interpretation of results shows that the absorption bands 2933, 1026 and 2931 cm^{-1} indicates the presence of polysaccharides. These strong bands represent C–H group and C–O–H or C–O–R group which are alkanes and secondary cyclic alcohol. The peaks 1631 cm^{-1} and 1629 cm^{-1} shows $C = C$ bond which indicates the presence of proteins and amides. Peaks representing $C = O$ and O–H bands are also present.

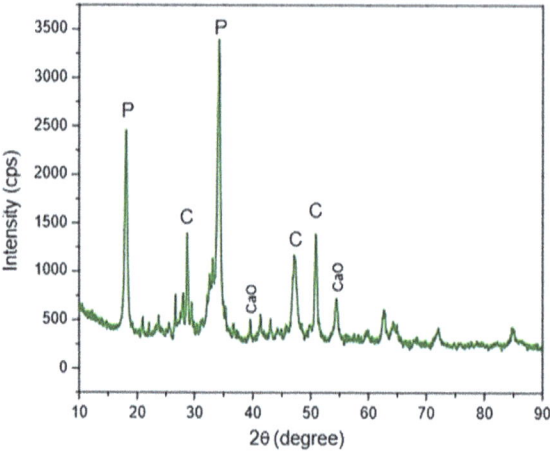

Fig. 1 XRD pattern of air lime

Table 1 XRD analysis of lime

Compounds	Proportion
Portlandite (**P**)—$Ca(OH)_2$	Predominant compound
Calcite (**C**)—$CaCO_3$	Low proportion
Calcium Oxide—CaO	Traces obtained

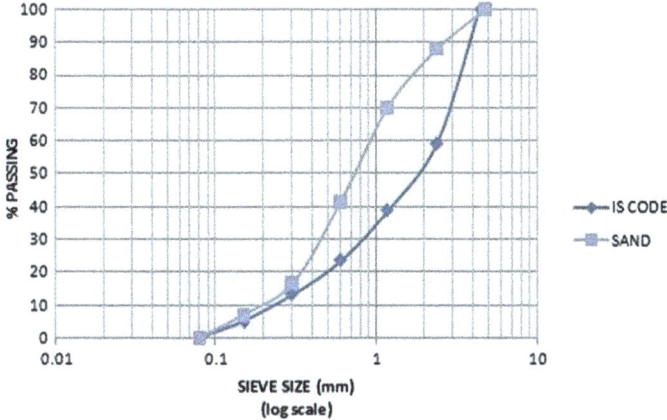

Fig. 2 Particle size distribution graph of river sand

3.2 Test on Hardened Mortar

Mechanical Properties. The mechanical property of the reference lime mortar and the organic modified lime mortar was determined through the compressive strength test conducted on the INSTRON testing machine. The lime mortar samples after 28 days of curing were subjected to compressive strength test. The results obtained have been shown below in "Fig. 4". The strength gain is not prominent in the lime mortar specimens with organics after 28 days whereas the strength has increased almost twice after 90 days of curing. A comparative study reveals that out of the three organics the addition of raisins in 4% concentration has shown the maximum strength gain. The compressive strength of lime mortar with raisins R-4% is 1.367 N/mm^2 at 28 days and 2.76 N/mm^2 at 90 days which was higher than the reference mortar. From the results it is evident that an increase in strength is attained due to the presence of polysaccharides and proteins in organics. The polysaccharides initially slow down the process of drying and helps in attaining strength in the later stages. The moisture content of the lime mortar is retained by the polysaccharides which facilitate the diffusion of CO_2 [4]. The reduction of polysaccharides to CO_2 due to fermentation helps in the carbonation process by providing more CO_2 content throughout the mortar and increase the calcite formation. Thus the organic addition enhances the properties of lime mortar.

Mineralogical Analysis. The mineralogical composition of the hardened mortar was identified through XRD analysis. The XRD results obtained for lime mortar with organic additives are shown in "Fig. 5". The high intensity peak of calcite was observed in both the reference mortar as well as the mortar with Raisins −4% addition. This shows the occurrence of carbonation process and conversion of portlandite to calcite. Portlandite peaks with medium intensity was also observed in organic modified samples which indicated the conversion process was to endure. These were only observed as traces after 90 days of curing. The presence of Quartz, Dioptase,

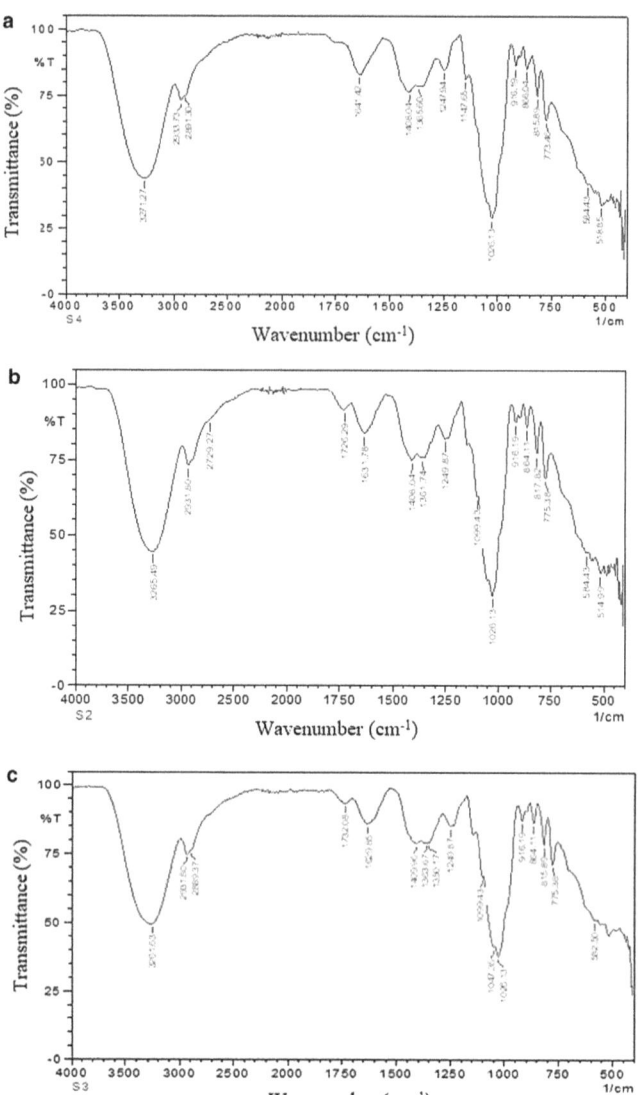

Fig. 3 a FTIR analysis of Dates (Phoenix dactylifera). **b** FTIR analysis of Zante Currant (Vita vinifera). **c** FTIR analysis of Raisin (Vitis vinifera)

Titanite, Genthelvite and Gismondine was observed in the organic added mortar. Gismondine (Calcium Aluminium Silicate Hydrate) was found in both the reference and organic modified lime mortar samples and Gryolite (Calcium Silicate Hydrate) was found in the reference sample. The presence of these compounds in lime mortar indicates the hydraulic reactions of lime with silicates (SiO_2) and aluminates (Al_2O_3) which helps in hardening of mortar and imparts strength to the mortar. The calcium

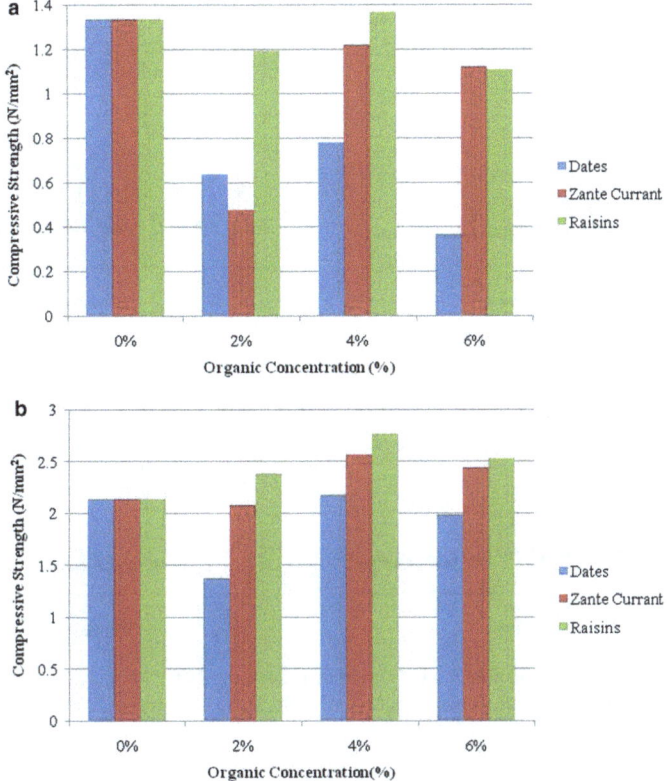

Fig. 4 **a** Compressive strength of lime mortar after 28 days. **b** Compressive strength of lime mortar after 90 days

hydroxide present in the hydraulic lime has been converted into calcium carbonate crystals in lime mortar confirming to the strength and hardening of the mortar.

The confirmation of XRD results was carried out using the FTIR spectroscopy and represented in "Fig. 6". The identification of organic was done through the FTIR analysis. The strong and medium bands at 1409, 873 and 711 cm^{-1} indicate the presence of calcite (calcium carbonate) showing CO_3 bending vibration. The intensities of peaks are higher for lime mortar with organic addition. The absorption band at 1033 cm^{-1} shows the presence of aluminium silicates and the sharp medium bands at 3641 cm^{-1} shows hydroxyl group and amides with O–H stretching. The broad peak at 1004 cm^{-1} indicates the $C = C$ bending.

Physical Properties. The bulk density, water absorption and porosity of the lime mortar samples were determined and tabulated below in "Table 2". The bulk density of the reference mortar was slightly less compared to the lime mortar with organics. The bulk density of lime mortar with raisins was found to be 2.31% higher than the reference mortar, lime mortar with zante currant has 2.03% and dates has 1.12% higher bulk density. The reduction in water absorption and porosity values for organic

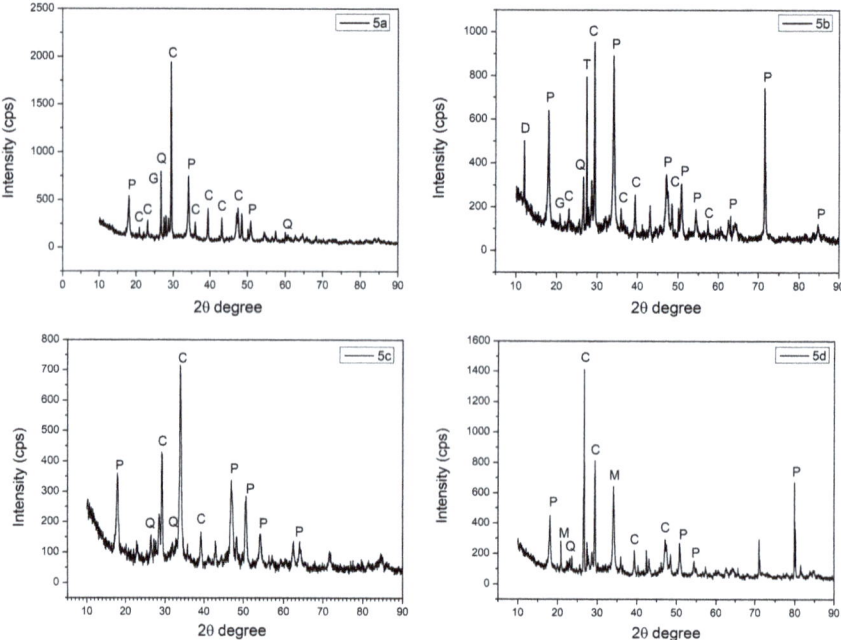

Fig. 5 XRD results of: **a** Reference lime mortar, **b** Lime mortar with Raisins, **c** Lime mortar with Dates, **d** Lime mortar with Zante Currant [C-Calcite, P-Portlandite, Q-Quartz, G-Gismondine, D-Dioptase, T-Titanite]

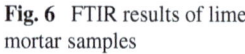
Fig. 6 FTIR results of lime mortar samples

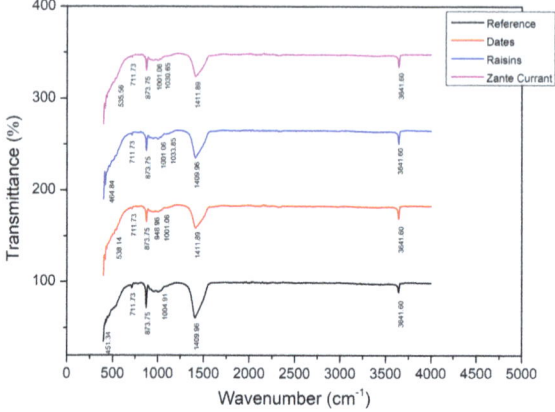

added lime mortar conveys the hydrophobic nature of the organics which consists of proteins and amides. The hydrophobic property of the organic prevents the absorption of water and other solvents and thus increases strength and durability. The water absorption was found to be the least for the lime mortar samples with raisins −4% addition.

Table 2 Physical properties of lime mortar

Sample	Water absorption(%)	Porosity (%)	Bulk density (g/cm^3)
Reference	14.4	25.8	1.77
Dates	14.3	25.7	1.79
Zante Currant	13.9	25.2	1.806
Raisins	13.4	24.4	1.811

Carbonation. The phenolphthalein test was performed to determine the depth of carbonation on the lime mortar samples. The carbonation test results show dark pink colour towards the inner core and light colourless portion on the exterior surface. This indicates that carbonation has occurred predominantly on the outer surface of the mortar which is in contact with atmospheric CO_2, while the interior portion shows less carbonation. The presence of carbohydrates in the organic aids in carbonation process which helps in conversion of $Ca(OH)_2$ to $CaCO_3$. The depth of carbonation was determined as 0.3, 0.15 and 0.2 cm for mortar with organics while 0.1 cm depth obtained for lime mortar without organic additives. The carbonation process helps in the hardening of lime mortar and consequently in the strength gain and durability (Fig. 7).

Fig. 7 Carbonation indication on lime mortar samples: **a** Reference, **b** with Raisins, **c** with Dates, **d** with Zante Currant

Durability Properties. The durability property was determined using the capillary rise test. The results obtained are listed in the "Table 3". The coefficient of capillary rise of water for the lime mortar samples with organics have decreased compared to the reference lime mortar sample. From the results the samples with organics Zante currant and Raisins have given maximum resistance to capillary rise.

Table 3 Capillary rise test results for lime mortar samples

Lime mortar sample	Capillary rise coefficient of water $(kgm^{-2}\ s^{-0.5})$
Reference	1.7
Zante Currant	0.8
Dates	1.5
Raisins	0.8

The presence of organics plays a vital role in resisting the capillary rise of water and other solutions. The hydrophobic property of the organic additives resitricts the capillary suction through the pores. This proves to increase the durabilty of the lime mortar and thereby prevention of structural damages from weathering agents and rain water.

4   Conclusion

The present study on the impact of organic addition on the hardening of lime mortar has enlightened the influence of organic in lime mortar and heritage structures. The physical, mechanical, mineralogical and durability properties of lime mortar with three different organic additives were analyzed and compared. A comparative study on the effect of the three organic additives on lime mortar has revealed that the addition of Raisins in 4% concentration have produced the maximum effect on the properties of lime mortar. Higher compressive strength was obtained at an earlier stage after 28 days of curing for the lime mortar samples with raisins. As these dry fruits are rich in carbohydrates, the polysaccharides act as moisturizing agents as well as reduce to form CO_2. Thus increasing the CO_2 concentration throughout the mortar and thereby increasing the rate of carbonation. The presence of calcite and CAH observed through XRD and FTIR analysis shows evidence for strength gain due to the hydraulic and carbonation reactions. The carbonation profile attained through the phenolphthalein test provided the depth of carbonation. The presence of proteins, amides and hydrophobic groups in the organics improved the physical and durability properties of the lime mortar. The water repellant property reduced the water absorption through the pores enhancing the strength and preventing damage. The capillary rise coefficient also decreased with organic addition indicating higher durability of the lime mortars. Hence it can be inferred that the addition of carbohydrate rich organics in lime mortar has improved its properties and aids in the hardening process. Improved durability and acceleration in mechanical strength proves to be functional for the repair works in heritage structures and its conservation.

References

1. Khalid, G.A., Ravi, R., Thirumalini, S.: Revamping the traditional air lime mortar using the natural polymer—areca nut for restoration application. Constr. Build. Mater. **164**, 255–264 (2018)
2. Scannell, S., Mike, L., Peter, W.: Impact of aggregate type on air lime mortar properties. Energy Proc. **62**, 81–90 (2014)
3. Fang, S., Zhang, K., Zhang, H., Zhang, B.: A study of traditional blood lime mortar for restoration of ancient buildings. Cem. Concr. Res. **76**, 232–241 (2015)
4. Yang, T., Ma, X., Zhang, B., Zhang, H.: Investigations into the function of sticky rice on the microstructures of hydrated lime putties. Constr. Build. Mater. **102**, 105–112 (2016)

5. Gopi, S., Subramanian, V.K.: Aragonite-calcite-vaterite: a temperature influenced sequential polymorphic transformation of $CaCO_3$ in the presence of DTPA. Mater. Res. Bull. **48**(5), 1906–1912 (2013)
6. Balen, K.V., Gemert, D.V.: Modelling lime mortar carbonation. Mater. Struct. **27**, 393–398 (1994)
7. Cultrone, G., Sebastia, E., Huertas, M.O.: Forced and natural carbonation of lime-based mortars with and without additives: Mineralogical and textural changes. Cem. Concr. Res. **35**, 2278–2289 (2005)
8. Thirumalini, S., Sekar, S., Bhuvaneshwari, B.: Bio-inorganic composites as repair mortar for heritage structures. J. Struct. Eng. **42**, 294–304 (2015)
9. Ramdoss, R., Thirumalini, P., Sekar, S.K.: Characterization of hydraulic lime mortar containing opuntia ficus indica as a bio admixture for restoration applications. Int. J. Architect. Herit. **10**(6), 714–725 (2015)
10. Clifton, J.: Preservation of historic adobe structures, US National Bureau Standards, Technical Note 934. US Department of Commerce, Washington DC (1997)
11. Moorehead, D.R.: Cementation by the carbonation of hydrated lime. Cem. Concr. Res. **16**(5), 700–708 (1986)
12. Fang, S., Zhang, H., Zhang, B., Li, G.: A study of Tung-oil–lime putty—a traditional lime based mortar. Int. J. Adhesion Adhesives **48**, 224–230 (2014)
13. Chandra, S., Aavik, J.: "Influence of proteins on some properties of portland cement mortar. Int. J. Cem. Compos. Lightweight Concr. **9**, 91–94 (1987)
14. Ravi, R., Thirumalini, S., Taher, N.: Analysis of ancient lime plasters—reason behind Longevity of the Monument Charminar, India a study. J. Build. Eng. **20**, 30–41 (2018)
15. Thirumalini, S., Ravi, R., Sekar, S.K., Nambirajan, M.: Knowing from the past – Ingredients and technology from ancient mortar used in Vadakumnathan temple, Trissur, Kerala, India. J. Build. Eng. **4**, 101–112 (2015)
16. Ravi, R., Thirumalini, S.: Effect of natural polymers from cissus glauca roxb on the mechanical and durability properties of hydraulic lime mortar. Int. J. Architect. Heritage **13**(2), 229–243 (2019)
17. Edwards, J.A.: Properties of hydraulic and non hydraulic limes for use in construction, A thesis submitted in partial fulfilment of the requirements of Napier University for the Degree of Doctor of Philosophy, Napier University School Of The Built Environment
18. Zhang, H.: Air hardening binding materials. In: Building Materials in Civil Engineering, vol. 423, Chap. 3, pp. 29–45 (2011)
19. Lawrence, R.M., Mays, T.J., Rigby, S.P., Walker, P., Dina, D.A.: Effects of carbonation on the pore structure of non-hydraulic lime mortars. Cem. Concr. Res. **37**, 1059–1069 (2007)
20. IS:2386 (Part 1) -1963—Methods of Tests for Aggregates for Concrete—Particle Size and Shape
21. IS:6932 (Part 8) -1973—Methods of Tests for Building Limes—Determination of Workability
22. IS:6932 (Part 11)-1983—Methods of Tests for Building Limes—Determination of Setting Time of Lime
23. IS:6932 (Part 7) -1973—Methods of Tests for Building Limes—Determination of Compressive and Transverse Strengths

Effect of Relative Humidity on Cement Paste: Experimental Assessment and Numerical Modelling

J. Kinda, L. Charpin, R. Thion, J.-L. Adia, A. Bourdot, S. Michel-Ponnelle, and F. Benboudjema

Abstract In this study, the objective was to verify whether or not the classical Richard-Fick model was able to take into account the size effect on drying. First, mass loss experiments are made on ordinary cement paste cylinders of 3.6×18 cm geometry that enable us to calibrate the model parameters. Second, the quality of identification and drying model was checked by predicting mass loss evolution of small prism of $1 \times 5 \times 10$ mm size dried at different steps of relative humidity. The results demonstrate that the present drying model is able to predict the drying of specimen for different sizes and levels of humidity at ambient temperature. And finally, drying shrinkage experiments for different rates of drying are performed on cement paste cylinders of 3.6×18 cm; then, the prediction of drying shrinkage evolution was made using a simple model where the drying shrinkage is supposed to be linear with respect to the relative humidity. The result appears to be satisfactory. Since the spatio-temporal evolution of water content is needed as input to shrinkage prediction, the latter result confirms that the identification of drying model parameters is trustworthy.

Keywords Drying · Drying rate · Mass loss · Drying shrinkage

J. Kinda (✉) · L. Charpin · R. Thion · J.-L. Adia
EDF R&D MMC, Paris, France
e-mail: justin-j.kinda@edf.fr

J. Kinda · A. Bourdot · F. Benboudjema
LMT—Laboratoire de Mécanique Et Technologie, Université Paris-Saclay, ENS Paris-Saclay, CNRS, 94235 Cachan, France

S. Michel-Ponnelle
EDF R&D ERMES, Grenoble, France

J. Sena-Cruz et al. (eds.), *Proceedings of the 3rd RILEM Spring Convention and Conference (RSCC 2020)*, RILEM Bookseries 34,
https://doi.org/10.1007/978-3-030-76465-4_11

1 Introduction

On the one hand, drying of cement based materials is a key factor of durability of concrete structures [1], therefore one needs to know its evolution on the long term for such structures [2]. On the other hand, drying is a very slow diffusion process, and may last for many decades for large structures such as large bridges or nuclear power plants [3]. Thus a modeling is needed to bridge that gap. Usually, models are calibrated on laboratory specimens [4] and then the parameters are used for prediction at structural level. Yet it is important to verify whether or not, this model accounts correctly for size effects. In this study, mass loss and sorption isotherm experiments have been conducted on ordinary cement paste specimens, for different sizes and at different rates of drying. Drying shrinkage measurement is also performed for some specimens. A drying model accounting for water permeation and vapor diffusion, [5] called Richards-Fick model in this paper is used to simulate the experiments. This model is calibrated by inverse analysis and then used for prediction. The prediction is shown to be very satisfactory. Therefore, we were confident to predict drying shrinkage. By using the well-known drying shrinkage $k\dot{h}$ model proposed by [6, 7], we were able to reproduce correctly the drying shrinkage at different drying rates with a single set of parameters. First the method of study in introduced. Second the experiment and simulation results are presented. Then the discussion on the results is undertaken on the following section. And last, we end this paper with conclusion on the whole study.

2 Method

2.1 Experiments

The material we studied is a cement paste, of 0.525 w/c ratio. A cement of type CEM I 52.5 N is used, corresponding to the VERCORS concrete, [8, 9]. All specimen are kept under endogenous conditions just after casting up to testing date. Below are listed the different experiments of the study:

- Test CP1: Sorption isotherm of the material is characterized first by means of dynamic vapor sorption (DVS).
- Test CP2: Mass loss measurement is made on specimen of $1 \times 5 \times 10$ mm geometry using the same DVS used to test CP1
- Test CP3: Mass loss and drying shrinkage tests are performed on cylinders of 3.6×18 cm size, for three humidity levels obtained with saline salt solutions, embedded in specific designed chambers.
- Test CP4: A final test is made in testing room on cylinder of 3.6×18 cm size, where humidity and temperature are kept constant (50% and 20 °C).

2.2 Modelling

Drying model: The drying model used herein was implemented in code aster (https://www.code-aster.org). A non-linear diffusion equation governs the evolution of liquid water saturation degree S, modeled by "Eq. (1)". The equivalent coefficient of diffusion D in "Eq. (2)" takes into account for permeation of liquid and diffusion of water vapor in the porous network of the material.

$$\frac{\partial S}{\partial t} = \nabla(D(S)\nabla S) \tag{1}$$

$$D = \frac{\partial P_c}{\partial S}\left[\frac{K_0 k_{rl}(S)}{\varnothing \mu_l\left(1 - \frac{\rho_v}{\rho_l}\right)} + D_{v0} d(S)\left(\frac{M_v}{\rho_l RT}\right)^2 P_{vsat} e^{\left(\frac{P_c M_v}{\rho_l RT}\right)}\right] \tag{2}$$

$$P_c = a\left(S^{-b} - 1\right)^{1-\frac{1}{b}} \tag{3}$$

$$k_{rl} = S^{nk}\left(1 - \left(1 - S^b\right)^{\frac{1}{b}}\right)^2 \tag{4}$$

$$D_v(s) = D_{v0}\varnothing^{a_{mq}}(1 - S)^{b_{mq}} \tag{5}$$

where:

- S is the liquid water saturation degree, chosen herein as the internal variable of the model
- K_0 is the intrinsic permeability of the material to be identified
- P_c is the capillary pressure which can be related to saturation degree using Van Genuchten [10] two parameters relation and reads "Eq. (3)"; where b is the same as in k_{rl} relation, and a, a parameter to be identified on desorption isotherm of the given material.
- k_{rl} is the relative permeability of the material, which depends on saturation degree. A choice is made here to use Mualem empirical relation "Eq. (4)" [10], where n_k is a material parameter to be identified.
- $D_v(s)$ is the vapor diffusion coefficient of the material, it decreases when the water content increases. It is chosen here to use Mualem [10] equation shown in "Eq. (5)"; where D_{v0} is the diffusion coefficient of the material at dry state; a_{mq}, b_{mq} are the parameters describing the tortuosity of the material and are to be identified by inverse analysis on mass loss experiment.
- $\varnothing$ is the porosity of the material measured experimentally,
- ρ_l, ρ_v, M_v are, respectively, the bulk density of liquid water, the bulk density of vapor, the molar volume of vapor.
- T is temperature, considered to be constant in this modelling and R is the universal gas constant.

- The boundary condition adopted here is of Robin type and reads "Eq. 6" where S_e is the equivalent saturation degree of water in air close to drying surface, S_i the saturation degree at drying surface level and C_s the surface *exchange factor to be identified*.

$$f_s = C_s(S_e - S_i) \tag{6}$$

Drying shrinkage:
The drying shrinkage is supposed to be proportional on the humidity change rate [7]:

$$\dot{\varepsilon} = k\dot{h} \tag{7}$$

2.3   Identification Method

Drying model First parameters a and b of capillary pressure in "Eq. (3)" are identified using sorption isotherm measurement (Test CP1, see "Fig. 1"). The calibration of the isotherm is very satisfactory.

Second parameters K_0 and n_k describing liquid water permeation in porous media are determined using Test CP4, and Dirichlet boundary condition, see "Fig. 2".

And last, the parameters a_{mq} accounting for tortuosity of the material and the surface factor C_s accounting for real boundary condition are fixed by performing optimization on mass loss measurement of Test CP3-1. Experimental results of CP3-2, and CP3-3 are only used in aim of prediction, see "Fig. 3".

Fig. 1 Fitting experimental isotherm of Test CP1 of cement paste w/c = 0.525 by Van Genuchten model

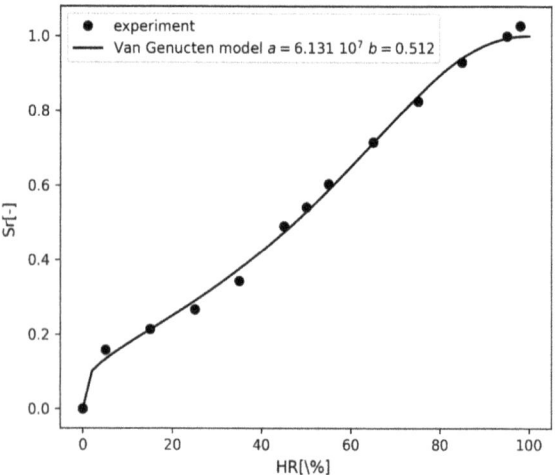

Fig. 2 Identification of intrinsic water permeability $K_0 = 1.02\ 10^{-21}$ and the relative permeability parameter $n_k = 2.52$ on experiment Test CP4

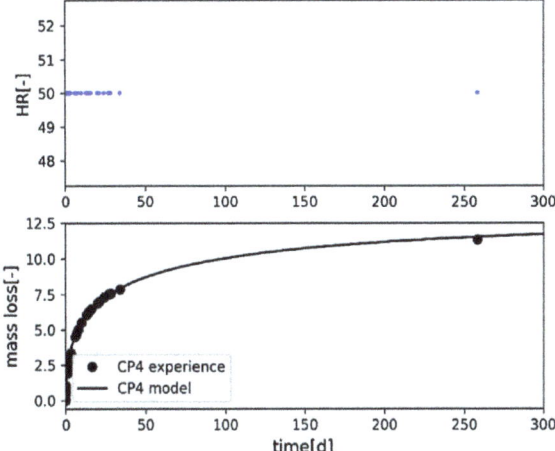

Fig. 3 Identification of amq and Cs on mass loss measurement performed on cement paste W/C $= 0.525$, cylinders of 3.6×18 cm

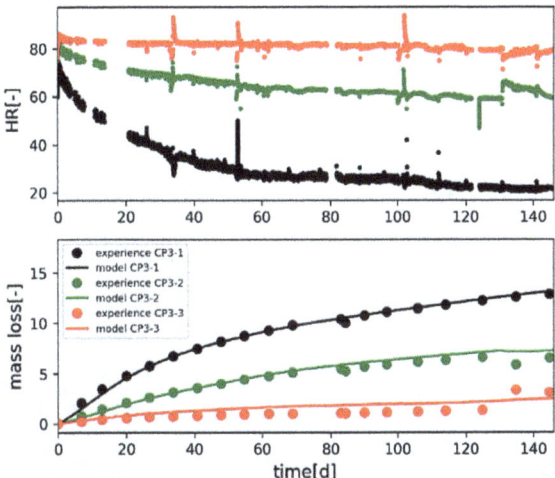

As a further validation of the drying model identification shown on "Fig. 3" following the procedure described above, let us now show that the model is able to reproduce the mass loss evolution of the very thin specimen ($1 \times 5 \times 10$ mm) used in Test CP2, except for the last humidity step of 5% RH. The results are shown on "Fig. 4" where red curve corresponds to simulations results with Robin type boundary condition, (the surface factor used herein is the one identified above) and the black curve is what is obtained using Dirichlet type boundary conditions.

Drying shrinkage: To simulate contraction caused by drying, a full water transport analysis is necessary. A spatio-temporal evolution of water obtained, thanks to the present calibrated Richards-Fick model is used as input for shrinkage model calibration. Then only one parameter has to be identified for all specimen. It is identified

Fig. 4 Prediction of mass loss of small prism (1 × 5 × 10 mm) of 1 mm drying thickness performed by DVS (Test CP2)

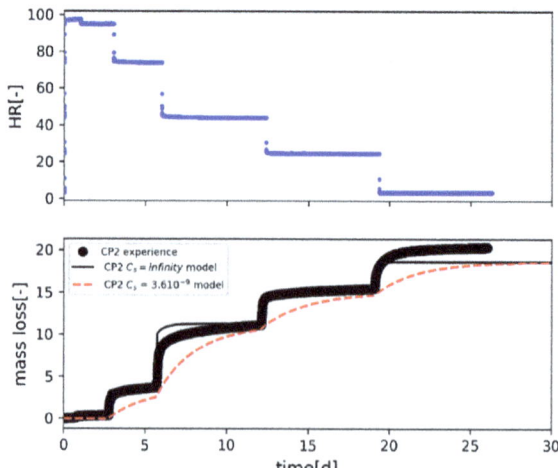

Table 1 Identified parameters of the drying and drying shrinkage models

$K[m^2]$	$n_k[-]$	$a_{mq}[-]$	$C_s[Kg.m^{-2}.s^{-1}]$	$k[-]$
1.0210^{-21}	2.52	5	3.610^{-9}	0.008

Table 2 Fixed parameters of drying model for the studied material

$\varnothing[-]$	$\rho_b[Kg.m^{-3}]^a$	$S_{init[-]}^b$	$b_{mq}[-]$	$P_{vsat}[Pa]$	$\mu_l[Pa.s]$	$\rho_v[Kg.m^{-3}]$	$T[K]$	$R[Kg.m^2s^{-1}]$
0.465	1870	0.99	4.2	3062	0.001	0.018	293	8.314

[a] Bulk density of cement paste
[b] Initial saturation for all studied specimen

on test CP3. All the parameters are displayed on "Table 1" for identified parameters, and "Table 2" for fixed ones. The temperature was kept constant in all experiments, close to 20 °C and the relative humidity condition for each test is displayed on the corresponding figures.

Once the shrinkage parameter identified, the agreement with the experimental results is satisfactory for all drying rates, ref "Fig. 5".

3 Discussion

The Richards-Fick model parameters have been identified successfully on various experimental tests performed on specimens of different geometry, size and level of drying humidity. Usually the model is used for relative humidity above 50%, where the contribution of vapor to drying process might be neglected [5]. For the present study, it is shown that the model can be used to predict mass loss up to 20%

Fig. 5 Identification of drying shrinkage model described above, using Test CP3 1, 2, 3, $k = 0.008$

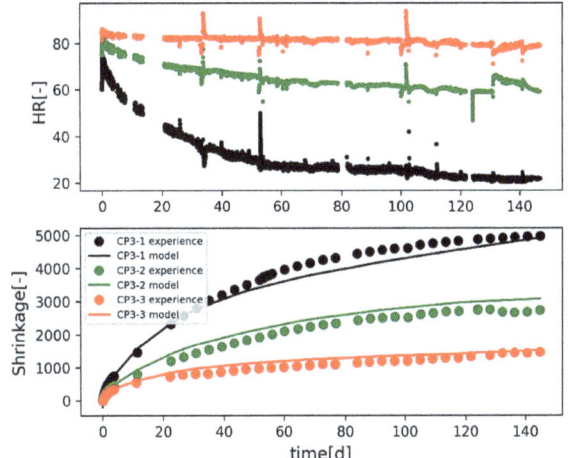

relative humidity tests by taking into account the vapor diffusion contribution in the modeling. A question may raise to know whether the surface factor C_s incorporated for identification cannot imped the identified parameters. To ensure that it is not the case, we identified the contribution of liquid water permeation parameters (K_0, n_k) on Test CP4, performed on vast room regulated on temperature and humidity, enough to allow instantaneous exchange between the exposed drying surface of specimen, with its environment. But in case of Test CP3, the exchange speed at the specimen surface with atmosphere, was very low, because of use of hermetic chamber without any ventilation. It was then necessary to introduce surface factor on boundary conditions, to reproduce the experiments. The prediction made on Test CP2, Fig. 4 demonstrate that C_s is specific to Test CP3 and the drying material parameters identified herein are intrinsic to the material. The calibrated model was able to predict correctly the mass loss evolution of specimen of $36\times$ smaller size, for different steps of relative humidity ranging from 98% up to 20% RH. For the step of 5% RH, the model underestimates the mass loss. Another good surprise, is that we noticed that the simplified drying shrinkage model based on the proportionality between drying contraction and humidity change rate was able to predict the experiment results for different drying rates, and for humidity below 50% relative humidity. We think that is was made possible thanks to use of real sorption isotherm of the material. This point will be investigated further by testing other models.

4 Conclusion

In this study, Richards-Fick model is tested against different sets of experimental data, and the identification of the model is shown trustworthy, since the identified parameters are calibrated and allow to predict mass loss measurement of specimen of different geometry, size and drying conditions. The drying model was able to simulate drying condition down to 20% RH at ambient temperature, and assuming the drying surface exchange factor C_s to be infinite, is quite realistic, if air ventilation at the drying surface is speed enough to allow removing instantaneously the water vapor at the specimen surface. It was also found that if the desorption isotherm of the material is measured and calibrated properly, and if the Richards-Fick drying model to get a realistic moisture distribution inside the specimen, the phenomenological drying shrinkage model $\dot{\varepsilon} = k\dot{h}$, can predict correctly the strain evolution, for different drying rates and for relative humidity up to 20% at ambient temperature. More investigation on drying shrinkage, both for experimental and numerical point of view is undergoing now, and the results will be available soon.

References

1. Mainguy, M., Lassabatere, B.-B.V.T., Coussy, O.: Characterization and identification of equilibrium and transfer moisture properties for ordinary and high-performance cementitious materials. Cement Concr. Res. 1225–1238 (1999)
2. Benboudjema, F., Torrenti, J.-M.: Modelling desiccation shrinkage of large structures. NUCPERF (2012)
3. Acker, P., Ulm, F.-J.: Creep and shrinkage of concrete: physical origins and practical measurements. Nucl. Eng. Design 143–158 (2001)
4. Thiery, M., Baroghel-Bouny, V., Bourneton, N., Vilain, G., Stéfani, C.: Modélisation du séchage des bétons. Analyse des différents modes de transfert hydrique. Revue européenne de génie civil 541–578 (2007)
5. Mainguy, M., Coussy, O., Baroghel-Bouny, V.: Role of air pressure in drying of weakly permeable materials. J. Eng. Mech. **127**, 582–592 (2001)
6. Wittmann, F., Roelfstra, P.: Total deformation of loaded drying concrete. Cement Concr. Res. 601–610 (1980)
7. Rahimi-Aghdam, S., Zdenek, C., Gianluca, P.: Extented Microprestress-Solidification Theory (XMPS) for long-term creep and diffusion size effect in concrete at variable environment. J. Eng. Mech. 145–176 (2019)
8. Mathieu, J.P., Charpin, L., Sémété, P., Toulemonde, C., Boulant, G.H.J., Taillade, F.: Temperature and humidity-driven ageing of the VeRCoRs mock-up. In: Proceedings of the Conference on Computational Modelling of Concrete and Concrete Structures (2018)
9. Charpin, L., Courtois, A., Taillade, F., Martin, B., Masson, B., Haelewyn, J.: Calibration of Mensi/Granger constitutive law: evidences of ill-posedness and practical application to VeRCoRs concrete. In: SMSS, TINCE 2018, 10, Rovinj (2018)
10. Mualem, Y.: A new model for predicting the hydraulic conductivity of unsaturated porous media. Water Resour. Res. **12**, 513–522 (1976)

Self-healing Capacities of Mortars with Crystalline Admixtures

Lina Ammar, Kinda Hannawi, and Aveline Darquennes

Abstract The aim of this research study consists of determining the self-healing capacities of cement-based materials incorporating Crystalline Admixtures (CA) such as permeability and shrinkage reducers. Mortars with three different types of CA were studied. At 28 days old, specimens were cracked by means of a three-point bending test to obtain a single crack characterized by a width varying from between 120 and 200 μm. Thereafter, the specimens were kept under water and the self-healing process was monitored by means of the crack width and area measurements at 35 and 120 days after cracking. From these first experimental results, it appears that specimens without CA and with calcium sulphate are characterized by a higher healing rate. This difference of behavior between the mortar mixtures is probably related to their microstructure. To confirm this hypothesis, their hydration products and their porosity were characterized at 28 days.

Keywords Crystalline admixture · Self-healing · Expansive agent

1 Introduction

Reinforced concrete structures often present cracks due to mechanical loadings or environmental conditions. These cracks are a weakness for the structure's lifespan. Indeed, aggressive agents' penetration is easier and degradation processes are accelerated reducing the service life of concrete structures and increasing their rehabilitation costs [1, 2]. A solution for limiting repair costs and improving the concrete structure's durability consists in the self-healing materials development [3]. A self-healing material can be described as a material having a capacity of repairing itself. This process could take place in concrete structures leading to a partial or complete closing of cracking.

L. Ammar (✉) · K. Hannawi · A. Darquennes
LGCGM, Institut National des Sciences Appliquées de Rennes (INSA Rennes), Rennes, France
e-mail: lina.ammar@insa-rennes.fr

© The Author(s), under exclusive license to Springer Nature Switzerland AG 2022   131
J. Sena-Cruz et al. (eds.), *Proceedings of the 3rd RILEM Spring Convention and Conference (RSCC 2020)*, RILEM Bookseries 34,
https://doi.org/10.1007/978-3-030-76465-4_12

One type of self-healing is "autogenous healing". It takes place in small cracks under moister conditions thanks to the hydration of unhydrated particles and the carbonation of hydration products [4–9]. When anhydrous particles located along the crack come into contact with water, the reaction between them create new hydration products (generally in young concrete). Moreover, the calcium ions released from the portlandite/C–S–H produced by cement hydration, react with carbonate ions in water to form calcium carbonate (generally in mature concrete). The formation of these new products contributes to close partially or totally the cracks [4–6, 10, 11].

Several studies try to enhance the self-healing capacity of concrete using mineral additions (blast-furnace slag and fly ash) thanks to their delayed hydration [12–18]. However, the performance of autogenous healing depends strongly of the cracks dimensions [19], the concrete age [14] and the curing conditions [20, 21]. For these reasons, new concepts of healing have been investigated in the last years like the use of Crystalline Admixtures (CA) [11, 19–22, 25, 28].

As reported in the ACI Comittee 212 [23], CA consist of proprietary active chemicals provided in a carrier of cement and sand. They are generally added to concrete mixture to reduce its permeability and/or shrinkage. The active chemicals react with water to produce supplementary products in the concrete porosity. CA using as permeability or shrinkage reducing admixtures are treated in several studies on the self-healing [10, 15, 19–22, 28]. It appears that several parameters can influence the CA efficiency to seal cracks, i.e. concrete's composition, CA's composition and content. Moreover, it was found that the presence of water is generally critical for the self-healing process of cementitious material with CA [10, 11, 19–22].

The main results of these previous studies are summarized hereafter:

- Cracks smaller than 300 μm can be 90% healed when from 1 to 4% of permeability reducer (by cement weight replacement) is added to mixture [11, 20, 21]. The healing rate evaluated by means of permeability tests is improved in the presence of a permeability reducer [19, 21].
- The shrinkage reducer admixtures especially the expansive agents can reduce effectively the concrete shrinkage. Several types of admixtures can be used: iron powder, alumina powder, calcium oxide, calcium sulfoaluminate, and magnesia [27]. Their chemical reaction with water leads to a macroscopic volume expansion due to the formation of an important content of expansive products at early age. The crystallization and swelling pressures of these products allows compensating/reducing the autogenous shrinkage of cementitious materials [27].

  The cementitious materials with 10% of calcium sulfoaluminate based expansive agent (CSA-type) was able to seal cracks up to 300 μm by the formation of ettringite inside the cracks [16, 21, 28].

  The addition of MgO-based expansive agents leads to the formation of brucite reducing the autogenous shrinkage [24–26]. But, the effect of MgO-based expansive agents on self-healing is not very well investigated.

The main objective of the present study is to analyze the impact of CA on the self-healing of mortar, particularly for the MgO-based expansive agent. Its effect on the healing process is compared to that of two other CA: another shrinkage reducer with calcium sulfoaluminate and a permeability reducer. Image analysis is used to monitor the CA effect on the evolution of crack surface/width. Moreover, the microstructure (porosity, hydration products) of the studied mixtures are investigated in order to understand these experimental results.

2 Experimental Investigations

2.1 Materials and Mixture Proportions

In this study, the healing capacity of four mortars made with a Portland cement (CEM I 52.5 N), standard sand (0/2) and several types of Cristalline Admixtures (CA) is studied. Their water/binder and sand/binder ratios are equal to 0.45 and 2 respectively. Several types of CA are tested: one permeability reducing admixture in the mixture named *Ca1.5* hereafter, as well as two shrinkage reducing admixtures in the mixtures named *Mg5* and *CSA10* hereafter. They differ by their main compound: MgO for *Mg5* and calcium sulphate for *CSA10*. The CA proportions are given in the Table 1. Their proportion is based on the values recommended by the manufacturer, as well as on the results from the literature. To determine the effect of CA, a mortar without CA -named *Ref*- is also tested. To obtain an acceptable and similar workability (Table 1), a superplasticizer (Sika ViscoFlow-800 Power) is added (0.3% of cement mass).

2.2 Experimental Program

After mixing, the prismatic specimens ($28 \times 7 \times 7 \ cm^3$) are demolded at 1 day old and kept at $20 \pm 2 \ °C$ and 95 ± 5 of % Relative Humidity (RH) until 28 days old.

Methodology for the Evaluation of Self-healing. The methodology for monitoring the self-healing process consists of two stages: (1) cracking of specimens, (2) monitoring the evolution of geometrical parameters of cracks (width and total area).

Table 1 CA and cement proportions (%)—Fresh mortars properties

Composition	Ref	Ca1.5	Mg5	CSA10
Cement 52.5 N	100	98.5	95	90
Crystalline additive	–	1.5	5	10
Average slump (cm)	11.5	11.5	11	12
Air content (%)	2.1	1.8	3.5	3.5

Cracking of Specimens. The mortar specimens are cracked at 28 days old by means of three points bending test (Fig. 1a). The loading speed is fixed at 0.03 mm/min, and the crack width is located between 120 and 200 μm. It is controlled by a specific ruler during the cracking test (Fig. 1b). After measuring the geometrical crack parameters (width and total area), all the specimens are kept under water and laid on the no-cracked face. All the samples from the same mortar composition are kept in the same container.

Evaluation of Crack. To monitor the geometrical crack parameters and thus the self-healing process, two zones (height = 3 cm) along the crack located on the lateral and bottom faces of the tested specimens are analyzed (one zone on each face). The crack width is measured on zones representative of the global crack, i.e. without defects, such as missing pieces of the cementitious matrix, parallel cracks, inclined crack path, etc. For each zone, ten points are chosen to measure the crack width during the healing process. The crack evolution is monitored using a reflective microscope. The pictures are analyzed by means of the "Image J" software. Each picture is binarized to separate the solid phase (white pixel) and the void (black pixel) related to the crack (Fig. 2). The pixel size is equal to 14 μm. Crack width and area are measured at the cracking day (28 days old) and 35 and 120 days after cracking.

Microstructure Characterization. Microstructure (hydration products, porosity) of the studied mixtures is characterized on 28 days old specimens kept at $20 \pm 2 \, °C$ and 95 ± 5 of % of RH. The hydration products (portlandite, brucite and calcite) are determined using ThermoGravimetric Analysis (TGA) by means of the Eqs. 1, 2 and 3. During the test, the temperature varies from 20 to 1000 °C with a rate of 10 °C/min. Before the test, samples are crushed and ground, therefore, $\sim 70 - -90$ mg of powdered sample are kept in nitrogen environment during the test.

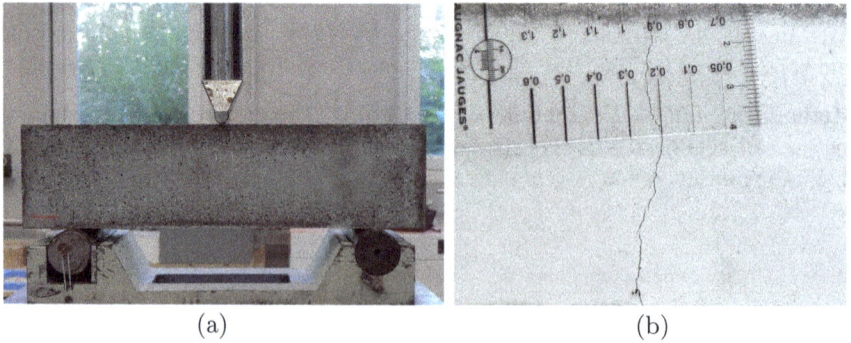

(a)   (b)

Fig. 1 3-Point bending test with a beam span of 21 cm (**a**)—Crack width controlled with a specific ruler (**b**)

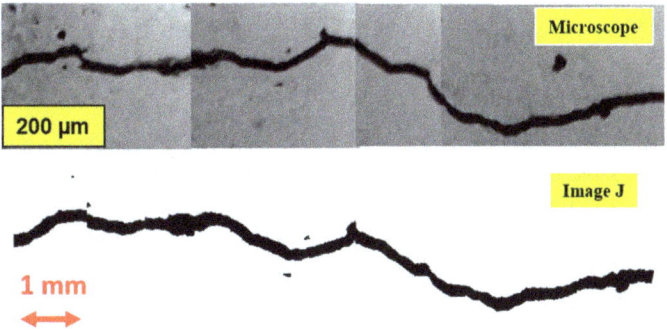

Fig. 2 Panoramic picture of the crack located on the bottom face of specimen just after cracking and its binarized picture

- brucite ~250–400 °C [30];

$$Mg(OH)_2 \longrightarrow MgO + H_2O \tag{1}$$

- portlandite ~400–600 °C [31];

$$Ca(OH)_2 \longrightarrow CaO + H_2O \tag{2}$$

- and calcite ~ 600 − −1000°C [29];

$$CaCO_3 \longrightarrow CaO + CO_2 \tag{3}$$

For porosity, sorption analysis in nitrogen environment is performed on powdered samples (900–1000 mg) dried at 110 °C during 24 h. According to IUPAC [32] , the pore size of cementitious materials can be classified as following: micropores (<2 nm), mesopores (2–50 nm), and macropores (>50 nm). So, the sorption analysis is adapted to identify mesopores. Moreover, to identify the total porosity, three cylindrical specimens (Φ4XH6 cm) of each mixture are tested. At 28 days old, the specimens are saturated in water and under vaccum for 48 h. After that, their air and hydrostatic weighings (M_{air} and M_{hyd} respectively) are measured, and they are placed at 40 °C until constant weight change is achieved (weight change is not greater than 0.05% in 24 h). Then, the dried mass (M_{dry}) is measured and the total porosity is calculated using Eq. 4.

$$Total\ porosity = \frac{M_{air} - M_{dry}}{M_{air} - M_{hyd}} \tag{4}$$

3 Results

3.1 Self-healing Results

Figure 3 shows the evolution of crack on the bottom and lateral faces for the four studied mortar mixtures. On the bottom face, the initial crack width (Day 0) is superior to 120 μm, while the initial crack width on the lateral face varies between 90 μm and 150 μm. The comparison between the pictures shows clearly that the lateral crack area decreases faster. So, the crack size affects the healing rate. Moreover, the healing products content is also more important in the lateral crack, particularly for the mixture without CA (*Ref*).

To quantify the Self-Healing Rate (*SHR*), the Eq. 5 is used and the results are presented on Fig. 4. The crack area is determined using Image J.

$$SHR = 1 - \frac{Crack\ area\ \textbf{at\ time\ t}}{Initial\ crack\ area} \tag{5}$$

Figure 4 presents the Self-Healing Rate for both bottom and lateral cracks. After 35 days of water curing, bottom cracks of *Ref* and *CSA10* are more than 90% healed; while *Ca1.5* and *Mg5* are 40% and ~65% healed respectively. However, lateral cracks are about 92% healed for *Ca1.5* specimens, and completely healed for the other mixtures. These results confirm the previous observations on the crack size effect on the healing kinetics (Fig. 3).

After 120 days of water curing, the bottom crack of *Ref*, *Mg5* and *CSA10* show a slightly better healing results (healing rate is increased around 10%), while the healing rate of *Ca1.5* bottom cracks is increased of 50%. But lateral cracks of all the studied mixtures are completely healed at 120 days.

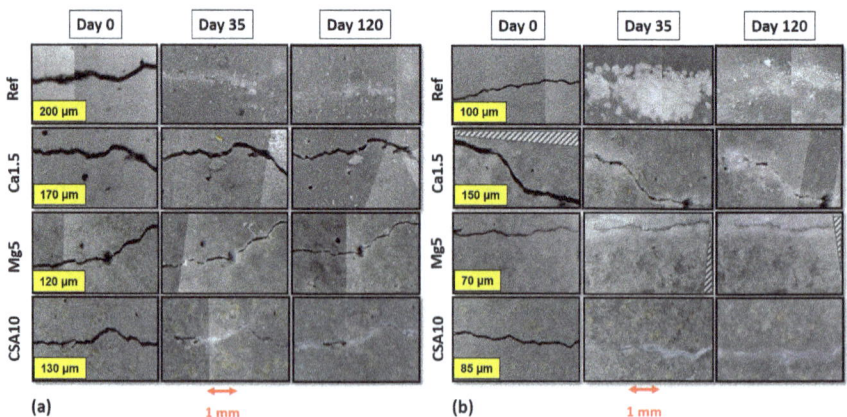

Fig. 3 Cracks on the bottom face (**a**) and on the lateral face (**b**) at the cracking day (Day 0), 35 (Day 35) and 120 days (Day 120) after cracking

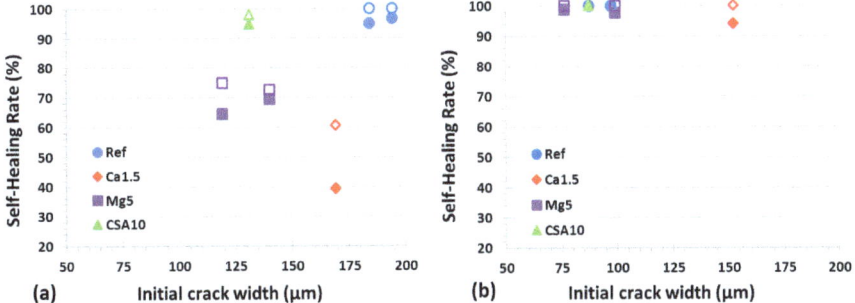

Fig. 4 Self-healing rate of *Ref*, *Ca1.5*, *Mg5* and *CSA10* after 35 days (•) and 120 days (○) of healing for bottom cracks (**a**), lateral cracks (**b**)

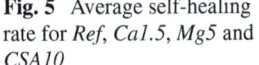

Fig. 5 Average self-healing rate for *Ref*, *Ca1.5*, *Mg5* and *CSA10*

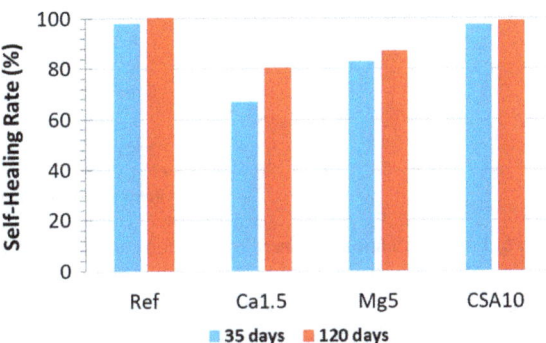

Figure 5 shows the average healing rates after 35 and 120 days of water curing. The mixtures with the highest healing rate are *Ref* and *CSA10*, both with an average *SHR* equal to 95% and 99% respectively after 35 and 120 days. They are followed by *Mg5* with an average *SHR* equal to 83% and 87% after 35 and 120 days respectively; and finally *Ca1.5* shows the lowest *SHR* at 35 days. It increases up to 80% after 120 days.

3.2 Microstructure of Mortar

The microstructure of mortars is studied to understand the influence of crystalline admixtures on the healing rate.

Hydration Products. From TGA measurements, brucite, portlandite and calcite contents are measured using the Eqs. 1, 2 and 3. All mixes contain various contents of classical cement hydration products (portlandite, and calcite) (Table 2). Brucite is only found in *Mg5*, mortar with 5% of MgO-based expansive agent. This confirms

Table 2 Brucite, portlandite, and calcite contents (%) for *Ref*, *Ca1.5*, *Mg5* and *CSA10*

Mass (%)	$Mg(OH)_2$	$Ca(OH)_2$	$CaCO_3$
Ref	0	9.08	2.30
Ca1.5	0	6.62	2.25
Mg5	4.17	9.03	1.90
CSA10	0	7.38	1.14

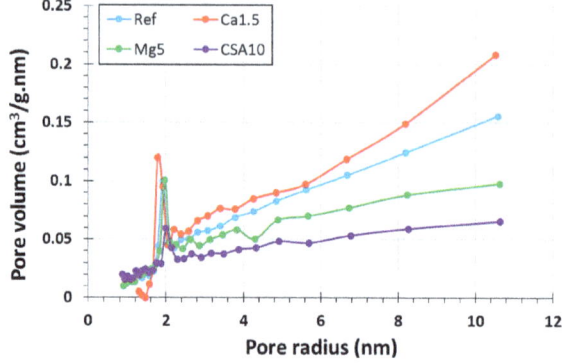

Fig. 6 Pore distribution measured by sorption analysis

Table 3 Total porosity (%) for *Ref*, *Ca1.5*, *Mg5* and *CSA10*

	Ref	Ca1.5	Mg5	CSA10
Total porosity (%)	14.3 ± 0.32	14.0 ± 0.28	13.9 ± 0.14	13.3 ± 0.02

that the growth of brucite is favored in solutions rich in magnesium ions Mg^{2+}, as confirmed by other studies [24, 30].

Porosity of Mortar Mixtures. Figure 6 presents the pore distributions in terms of the pore radius. The results show that *CSA10* and *Mg5* have a lower volume of mesopores (2–50 nm). According to Table 3, it can be seen that all mixtures have a very close total porosity (around 14 ± 1 % of the matrix). The crystalline admixtures (MgO and CSA) added in *Mg5* and *CSA10* mixtures are expansive agents leading to the formation of brucite and ettringite respectively in the porous network. These two hydration products are responsible of the swelling of the matrix when their volumes are greater than that of the pores [24, 27, 30]. It can explain the difference between their mesopore volume and that of *Ref*. Notice that the pore volume of *Ca1.5* is very similar to that of *Ref* (for both mesopores and total porosity). But, *Ca1.5* did not show a remarquable healing rate like *Ref*. To better understand these results, the hydration reaction kinetics of each mixture will be followed by means of calorimetry tests.

4 Discussion

The studied CA affect differently the self-healing rate of mortars. The CA used in *Ca1.5* and *Mg5* seem to reduce the healing capacity of the studied mortar (Fig. 5). These admixtures are generally added to reduce the permeability and the shrinkage respectively. So they affect the microstructure. In our case, the mortar with the permeability reducer *Ca1.5* presents less portlandite free for carbonation and it is more porous at the cracking age. In other studies [10, 21], the mortar with 1.5% of permeability reducer shows a good self-healing results. It is explained by the fact that the hydration of permeability reducer has favored the formation of C–S–H and calcium carbonate inside the crack. So, for this research work, more studies are needed in order to evaluate the hydration process of the permeability reducer by means of different techniques, i.e, calorimetry, XRD and TGA analysis. Another behavior is observed for *Mg5*: formation of a y product like brucite and smaller mesoporosity. It can limit the water flow through the crack, slows down the ionic exchanges and thus, delays self-healing. The addition of a shrinkage reducer containing calcium sulphate (*CSA10*) leads to a self-healing capacity similar to that of the mortar without CA (*Ref*). An explanation of this observation could be related by the fact that the shrinkage reducer content is important, and characterized probably by anhydrous particles at the cracking age—free for crack healing. However, water permability tests will be conducted on a cracked specimen for each mixture in order to better quantify the healing capacity inside the crack. Furthermore, in order to evaluate their mechanical recovery, a three point bending test will be performed on the healed specimens.

5 Conclusions

This paper has presented the results on the self-healing process in the presence of crystalline admixtures (CA) such as permeability (Ca) or shrinkage reducers (MgO and CSA-based expansive agents). Based on the present experimental results, the following conclusions can be drawn:

- The self-healing rate depends strongly on the crack width: smaller cracks can be reduced faster.
- *Ref* shows the highest healing rate-bottom cracks with 200 μm are completely healed after 120 days of healing, followed by *CSA10*-bottom cracks with 130 μm are 99% healed after 120 days of healing.
- The brucite and ettringite formation in *Mgb5* and *CSA10* explain their smaller mesoporal volume.

It is important to keep in mind that the healing process depends on several factors (surrounding ions, saturation in carbonate, pH, temperature, etc.) [1]. So, the difference of mortar's composition can lead to different self-healed products and consequently to different healing performances. However, regarding to the results, more studies are needed to better understand the growth rate of both hydration and healing products of the studied mortars. So, different mortar parameters have to be studied: hydration reaction kinetics, permeability and recovery of mechanical parameters. Different mortars with different CA percentages have to be evaluated as well in order to obtain a better self-healing process.

References

1. Escoffres, P., Desmettre, C., Charron, J.-P.: Effect of a crystalline admixture on the self-healing capability of high-performance fiber reinforced concretes in service conditions. Construct. Build. Mater. **173**, 763–774 (2018)
2. Neville, A.: Chloride attack of reinforced concrete: an overview. Mater. Struct. **28**, 63–70 (1995)
3. Van Breugel, K.: Is there a market for self-healing cement based materials? In: Proceedings of the First International Conference on Self Healing Materials, The Netherlands, 18–20 April 2007
4. Hearn, N.: Self-sealing, autogenous healing and continued hydration: what is the difference? Mater. Struct. **31**, 567–653 (1998)
5. Neville, A.: Autogenous healing, a concrete miracle? Concr. Int. 24 (2002)
6. Edvardsen, C.: Water permeability and autogenous healing of cracks in concrete. ACI Mater. J. 448–454 (1999)
7. Dry, C.: Matrix cracking repair and filling using active and passive modes for smart timed release of chemicals from fibers into cement matrices. Smart Mater. Struct. **3**, 118–123 (1994)
8. Dry, C.M.: Three designs for the internal release of sealants, adhesives, and waterproofing chemicals into concrete to reduce permeability. Cement Concr. Res. **30**, 1969–1977 (2000)
9. Dry, C., McMillan, W.: Three-part methylmethacrylate adhesive system as an internal delivery system for smart responsive concrete. Smart Mater. Struct. **5**, 297–300 (1996)
10. Sisomphon, K., Copuroglu, O., Koenders, E.A.B.: Self-healing of surface cracks in mortars with expansive additive and crystalline additive. Cement Concr. Compos. **34**, 566–574 (2012)
11. Cuenca, E., Tejedor, A., Ferrara, L.: A methodology to assess crack-sealing effectiveness of crystalline admixtures under repeated cracking-healing cycles. Construct. Build. Mater. **179**, 619–632 (2018)
12. Huang, H., Ye, G., Qian, C., Schlangenb, E.: Self-healing in cementitious materials: materials, methods and service conditions. Mater. Design **92**, 499–511 (2016)
13. Van Tittelboom, K., Gruyaert, E., Rahier, H.De, Belie, N.: Influence of mix composition on the extent of autogenous crack healing by continued hydration or calcium carbonate formation. Construct. Build. Mater. **37**, 349–359 (2012)
14. Olivier, K.: Etude expérimentale et modélisation de l'auto-cicatrisation des matériaux cimentaires avec additions minérales. Génie Civil. Université de Sherbrooke et Université de Paris Saclay, Thèse de doctorat (2016)
15. Jaroenratanapirom, D., Sahamitmongkol, R.: Self-crack closing ability of mortar with different additives. J. Metals Mater. Miner. **21**, 9–17 (2011)
16. Termkhajornkit, P., Nawa, T., Yamashiro, Y., Saito, T.: Self-healing ability of fly ash-cement systems. Cement Concr. Compos. **31**, 195–203 (2009)
17. Qian, S., Zhou, J., a, De Rooij, M.R., Schlangen, E., Ye, G., Van Breugel, K.: Self-healing behavior of strain hardening cementitious composites incorporating local waste materials. Cement Concr. Compos. **31**, 613–621 (2009)
18. Hung, C.-C., Su, Y.-F., Su, Y.-M.: Mechanical properties and self-healing evaluation of strain-hardening cementitious composites with high volumes of hybrid pozzolan materials. Compos. Part B **133**, 15–25 (2018)
19. Roig-Flores, M., Moscato, S., Serna, P., Ferrara, L.: Self-healing capability of concrete with crystalline admixtures in different environments. Construct. Build. Mater. **86**, 1–11 (2015)

20. Ferrara, L., Krelani, V., Carsana, M.: A "fracture testing" based approach to assess crack healing of concrete with and without crystalline admixtures. Construct. Build. Mater. **68**, 535–551 (2014)
21. Sisomphon, K., Copuroglu, O., Koenders, E.A.B.: Effect of exposure conditions on self healing behavior of strain hardening cementitious composites incorporating various cementitious materials. Construct. Build. Mater. **42**, 217–224 (2013)
22. Roig-Flores, M., Pirritano, F., Serna, P., Ferrara, L.: Effect of crystalline admixtures on the self-healing capability of early-age concrete studied by means of permeability and crack closing tests. Construct. Build. Mater. **114**, 447–457 (2016)
23. American Concrete Institute ACI Comittee 212. Report on Chemical Admixtures for Concrete (2010)
24. Qureshi, T., Kanellopoulos, A., Al-Tabbaa, A.: Autogenous self-healing of cement with expansive minerals-I: impact in early age crack healing. Construct. Build. Mater. **192**, 768–784 (2018)
25. Sherir, M.A.A., Hossain, K.M.A., Lachemi, M.: Self-healing and expansion characteristics of cementitious composites with high volume fly ash and MgO-type expansive agent. Construct. Build. Mater. **127**, 80–92 (2016)
26. Sherir, M.A.A., Hossain, K.M.A., Lachemi, M.: The influence of MgO-type expansive agent incorporated in self-healing system of engineered cementitious Composites. Construct. Build. Mater. **149**, 164–185 (2017)
27. Yang, L., Shi, C., Wu, Z.: Mitigation techniques for autogenous shrinkage of ultra-high-performance concrete—a review. Compos. Part B **178**, 107456 (2019)
28. Zhang, Y., Teramoto, A., Ohkubo, T.: Effect of addition rate of expansive additive on autogenous shrinkage and delayed expansion of ultra-high strength mortar. J. Adv. Concr. Technol. **16**, 250–261 (2018)
29. Pei, Y.: Effets du chauffage sur les matériaux cimentaires - impact du "self-healing" sur les propriétés de transfert. Génie Civil. Ecole Centrale de Lille, Thèse de doctorat (2016)
30. Wang, J.A., Novaro, O., Bokhimi, X., Lopez, T., Gomez, R., Navarrete, J., Llanos, M.E., Lopez-Salinas, E.: Characterizations of the thermal decomposition of brucite prepared by sol-gel technique for synthesis of nanocrystalline MgO. Mater. Lett. **35**, 317–323 (1998)
31. Alonso, C., Fernandez, L.: Dehydration and rehydration processes of cement paste exposed to high temperature environments. J. Mater. Sci. 3015–3024 (2004)
32. Moretti, J.P., Sales, A., Quarcioni, V.A., Silva, D.C., Oliveira, M.C., Pinto, N.S., Ramos, L.W.: Pore size distribution of mortars produced with agroindustrial waste. J. Clean. Product. **187**, 473–484 (2018)

In-Plane Behavior of Clay Brick Masonry Wallets Strengthened by TRM System

Ali Dalalbashi⊙, Bahman Ghiassi⊙, and Daniel V. Oliveira⊙

Abstract Due to the poor shear capacity of unreinforced masonry (URM) walls, it is necessary to strength these elements in the masonry structures. Textile-reinforced mortar (TRM) composites have received extensive attention as a sustainable solution for seismic strengthening of masonry and historical structures. This new system is composed of textile fibers embedded in an inorganic matrix such as lime-based mortar and is applied on the masonry substrate surface as an externally bonded reinforcement (EBR) system. Lime-based mortars are preferred for application to masonry and historical structures due to compatibility, sustainability issues, breathability and capability of accommodating structural movements.

Owing to the novelty of these materials in application to masonry structures, several aspects related to their performance with URM wallets are still not clear. To that end, a new study has been lunched that looks at the effect of textile-reinforced mortar system on the improvement the shear capacity of clay brick wallets. For this purpose, hydraulic lime-based mortar and the glass fiber are used, as TRM system. The results show TRM composite increases the ultimate load of strengthened wallet, in comparison to the URM wallet. Another key point to remember is that TRM system causes the failure mode of wallet to change and improve its performance under diagonal compression test.

Keywords Textile-Reinforced-Mortar (TRM) system · Diagonal compression test · Shear · Hydraulic lime mortar · Glass fiber

A. Dalalbashi (✉)
PhD student, ISISE, Department of Civil Engineering, University of Minho, Guimarães, Portugal
e-mail: alidalalbashi@gmail.com

B. Ghiassi
Assistant Professor, Centre for Structural Engineering and Informatics, Faculty of Engineering, University of Nottingham, Nottingham, UK
e-mail: bahman.ghiassi@nottingham.ac.uk

D. V. Oliveira
Associate Professor, ISISE, Department of Civil Engineering, University of Minho, Guimarães, Portugal
e-mail: danvco@civil.uminho.pt

© The Author(s), under exclusive license to Springer Nature Switzerland AG 2022   143
J. Sena-Cruz et al. (eds.), *Proceedings of the 3rd RILEM Spring Convention and Conference (RSCC 2020)*, RILEM Bookseries 34,
https://doi.org/10.1007/978-3-030-76465-4_13

1 Introduction

Unreinforced masonry (URM) structures have shown poor seismic performance during the earthquake [1, 2]. Since there are a large number of URM brick walls, the necessity for seismic assessment and considering appropriate strengthening techniques for these structures requires. One possible solution is to use externally bonded reinforcement system (EBRs) such as fiber-reinforced polymers and textile-reinforced mortar [3, 4].

In an attempt to alleviate the drawbacks that arise from the use of fiber-reinforced polymers (FRP), the textile reinforced mortar (TRM) composites have made the former very interesting for externally bonded reinforcement of masonry and concrete structures [5, 6]. The large variety of available fibers and mortar types allow development of TRM composites with a large range of mechanical properties [7, 8]. In addition, TRM system shows a pseudo-ductile response with distributed cracking, which makes them intersting for seismic strengthening apllications [9, 10].

In the literature, some researches can be found in which URM walls have been strengthened by TRM system and different parameters such as effect of the textile and the mortar types, the number of textile layers, and symmetrical or asymmetrical configurations were examined [11, 12]. Although in some studies the debonding between TRM and substrate have not been reported, mechanical anchors cause the performance of the strengthened wallets to increase, especially for the asymmetrical configuration [13–15].

Sandblasting can be performed to increase the TRM-to-substrate bond instead of mechanical anchor systems proposed in the literature. Razavizadeh et al. [16] reported that the sandblasted bricks causes the bond between TRM composite and substrate to increase under single-lap shear test, in comparison with un-sandblasted bricks. The aim of this study is to investigate the effect of glass-based TRM jacketing and the surface treatment of the wallets on the in-plane shear behavior of brick masonry wallets.

2 Experimental Program

The experimental program consists of a series of in-plane shear tests on the wallet specimens. Three control specimens and six strengthened specimens are prepared in which the surface of three strengthened wallets are sandblasted and the others' surface keep as original. In this study, the wallets are nominated as URM, SS, and OS, respectively. The detailed procedure followed for preparation of the specimens and performing the tests are given in this section.

2.1 Material Characterization Tests

TRM Composite. A commercially available hydraulic lime-based mortar (Planitop HDM Restauro) as matrix and a woven biaxial fabric mesh made of an alkali-resistance fiberglass (Mapegride G220) are used.

A high-ductility hydraulic lime mortar, named mortar M1, is prepared by mixing the powder with the liquid provided by the manufacturer in a low-speed mechanical mixer to form a homogenous paste. For mechanical characterization of the mortar, compressive and flexural tests are performed (see Fig. 2a and b), according to ASTM C109 [17] and EN 1015–11 [18]. Five cubic ($50 \times 50 \times 50$ mm^3) and five prismatic ($40 \times 40 \times 160$ mm^3) specimens are used for each test. Specimens are cured for seven days in a damp environment and then stored in the laboratory (20 °C, 67% RH) until the age of 90 days.

The mesh size of reinforcing material is equal to 25×25 mm^2, as shown in Fig. 2a. According to the data technical sheet provided by manufacture, the area per unit of width is 35.27 mm^2/m. The direct tensile test is performed on the dry fibers (5 specimens) to obtain their tensile strength and elastic modulus. A universal testing machine with a maximum load capacity of 10 kN and a rate of 0.3 mm/min is used for these tests.

Masonry Wallets. The solid clay brick and commercial lime-based mortar (Mape-Antique MC) are carried out to construct the masonry wallets. The dimensions of solid clay brick are equal to $200 \times 100 \times 50$ mm^3. The compressive strength and the flexural strength of the brick are characterized according to EN 772–1 [19] and EN 1015–11 [18], respectively, in the flatwise direction. Figure 1c and b show the performed test setup for measuring the compressive and flexural strength of bricks. Six cubic (40 mm) and five prismatic ($40 \times 40 \times 160$ mm^3) specimens are used for each test. In addition, the lime mortar that is used as the bed joint and named mortar M2 is characterized similar to the mortar M1.

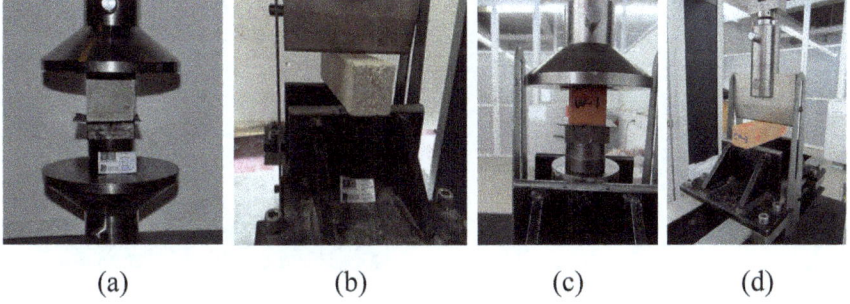

(a)   (b)   (c)   (d)

Fig. 1 Materials mechanical characterization tests: **a** mortar compressive test; **b** mortar flexural test; **c** brick compressive test; **d** brick flexural test

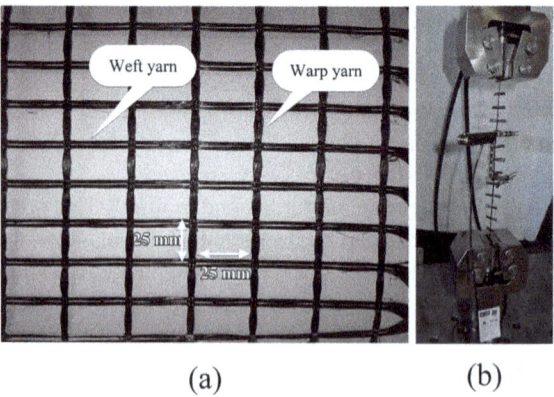

(a)   (b)

Fig. 2 a glass fiber mesh; **b** fiber direct tensile test

2.2 Diagonal Compression Test

The masonry wallet dimensions are equal to $540 \times 540 \times 100$ mm^3, as shown in Fig. 3a. Nine specimens are constructed in which three wallets are left unstrengthened and used as control specimen; whereas the remaining six wallets are reinforced with one layer of glass-based TRM externally bonded on both wallet faces. In order to investigate the effect of sandblasted surface as anchor system, the both surface of three wallets have been sandblasted before strengthening. The preparation of the specimens is as follows: (1) immersing brick in water for 2 h, (2) constructing wallets and curing in the lab environmental condition for 30 days. TRM composite is applying

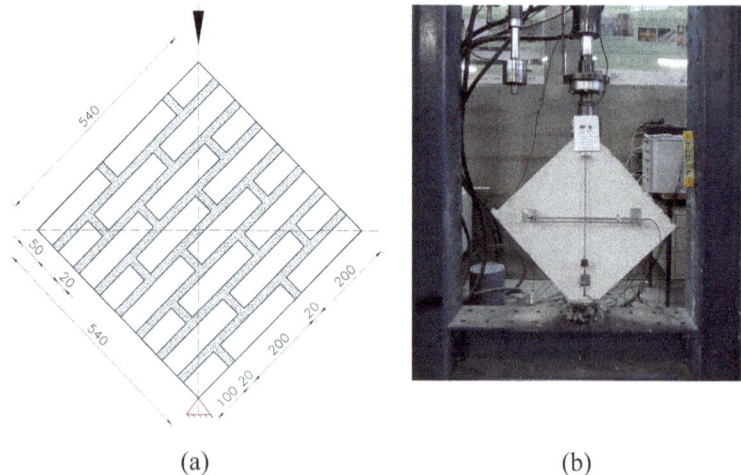

(a)   (b)

Fig. 3 Diagonal compression test: **a** geometry of specimen; **b** detail of test setup

Strength [MPa]	Mortar M1	Mortar M2	Brick
Compressive	16.8 (12)	8.7 (6)	23.5 (6)
Flexural	4.5 (2)	1.7 (10)	4.9 (16)

Table 1 Mechanical properties of the mortars and the brick

as follow: (3) applying the first layer of mortar (5 mm), (4) placing the glass fiber mesh on the surface of the wallet, (5) applying the second layer of mortar (5 mm). Finally, strengthened specimens are cured for seven days in a damp environment and then stored in the laboratory condition (20 °C, 67% RH) until the age of 90 days. All wallets (control and strengthened) are tested at a same age (120 days).

A servo-hydraulic system with a maximum capacity of 300 kN is performed to investigate the in-plane shear behavior of the wallets. The load is applied through steel shoes (115 × 115 × 15 mm³) placed at diagonally opposing bottom and top corners [20]. Figure 3b shows the details of test setup. The compression load is operated under displacement control at a rate of 0.3 mm/min. Two LVDTs with 20 mm range and 2-µm sensibility as well as a measurement base range equal to 473 mm placed in each side of the wallet measure the vertical and horizontal deformation.

3 Experimental Results and Discussion

3.1 Material Properties

The mean compressive and flexural strengths of the mortars and the brick as well as coefficient of variation in percentage (provided inside parentheses) are presented in Table 1. The average tensile strength, Young's modulus, and rapture strain of the warp glass roving are 875 MPa (13%), 65.94 GPa (5%), and 1.77% (10%), respectively, with coefficient of variation provided inside parentheses. These values for the weft glass roving are 676 MPa (13%), 72.91 GPa (3%), and 1.29% (17%), respectively.

3.2 In-Plane Experimental Characterization

The values of the shear stress, shear strain, and modulus of shear stiffness in the center of the panel are also calculated according to ASTM-E519-2 [21]. The shear stress (τ) is obtained by the Eq. (1):

$$\tau = \frac{0.707P}{A_n} \tag{1}$$

In which, P is the applied load and A_n is the net area of the specimen calculated as follows:

$$A_n = \left(\frac{w+h}{2}\right) t.n \tag{2}$$

where, w, h, and t are the width, the height, and the thickness of specimen, respectively. n is the percentage of the gross area of the unit that is solid, expressed as a decimal. The shear strain (γ) is calculated as follows:

$$\gamma = \frac{\Delta_v + \Delta_h}{g} \tag{3}$$

Δ_v, Δ_h, and g are the vertical shortening, the horizontal extension, and the vertical gauge length, respectively.

In order to evaluate the efficiency of both the selected strengthening system and the surface treatment of the wallets, the masonry wallets are tested in diagonal compression. The load increments and corresponding deformations for all specimens are recorded and the results are presented as the stress–strain curves. Figure 4 shows the average curves for each series. The in-plane shear behavior of the control specimens (URM) is brittle and when the maximum stress reaches, the sudden failure occurs with no considerable crack development prior to failure. Figure 5a shows the crack pattern of the control specimens (URM), which is the shear-sliding mechanism along the bed joints.

The in-plane behaviors of the strengthened specimens are completely different with the control specimen. The in-plane behavior of both the SS and OS specimens shows three stages: linear, nonlinear, and post peak stages see Fig. 4. In addition, the obtained failure mode of the strengthened specimens is vertical cracks starting to appear in the central area and involving both the joints and the bricks, as shown in Fig. 5b and c. The un-sandblasted wallets (OS) show the partially debonding between TRM and the substrate so that the out-of-plain failure occurs. While, the sandblasted

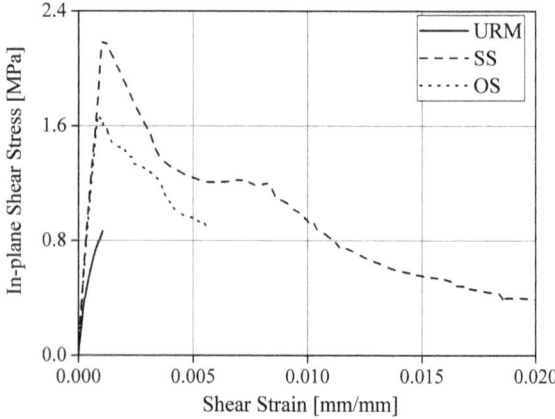

Fig. 4 Shear stress–strain curves of tested wallets

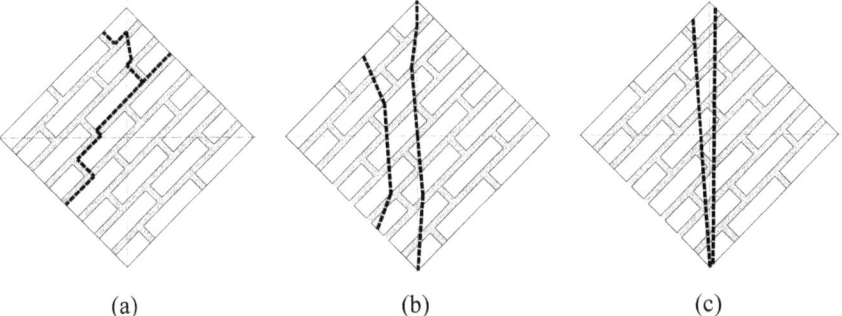

(a)   (b)   (c)

Fig. 5 Crack pattern of the tested wallets: **a** URM; **b** OS; **c** SS

specimens (SS) shows a perfect bond between TRM composite and the wallets due to the high anchor available between TRM and bricks.

Maximum shear stress (τ_{max}), crack shear stress (τ_f), and their corresponding strain (γ) as well as pseudo-ductility (μ) and shear modulus (G) are the parameters characterizing the in-plane shear behavior of masonry wallets [22]. Pseudo-ductility is defined as ultimate shear strain (γ_u) (corresponding to the shear stress, which underwent a 20% reduction) to the crack/yield shear strain (γ_{cr}) (determined as the yield shear stress at 70%) [22]. In addition, G is measured by dividing the crack shear stress to its corresponding strain. Table 2 shows the summary of the in-plane shear behavior of masonry wallets.

Comparison between the control and the strengthened wallets illustrates that TRM composite causes the in-plane shear behavior to increase. For example, the average shear strength of the URM wallets is 0.6 MPa, while the strengthened wallet (SS) resulted in the highest shear strength increase of 367%. At the same time, the strengthened wallets with the un-sandblasted surface (OS) show a considerable shear strength increase of 300%. These ratios are the same for other in-plane shear parameters.

The effect of sandblasted surface on the in-plane shear behavior is dramatic. This anchor system not only improves the failure performance of the wallets under diagonal compression load, but also increases the in-plane behavior, in contrast to the OS specimens. Therefore, the maximum and cracking shear stress and their corresponding strain of the SS wallets increase of 22% and 15%, respectively, compared to un-OS specimens. In addition, thanks to a good bond between glass-based TRM and the sandblasted surface, ultimate shear strain and ductility show a considerable increase of 50 and 30%.

Table 2 Summary of the in-plane shear behavior of masonry wallets

Specimen	τ_{max} [MPa]	γ_{max} [%]	τ_{cr} [MPa]	γ_{cr} [%]	γ_u [%]	μ	G [MPa]
URM	0.6 (31)	0.07 (58)	0.4 (31)	0.04 (50)	0.07 (58)	2.0 (15)	1715 (95)
OS	1.8 (7)	0.09 (2)	1.3 (12)	0.06 (5)	0.16 (42)	2.7 (46)	2189 (7)
SS	2.2 (1)	0.11 (4)	1.5 (1)	0.07 (2)	0.24 (1)	3.5 (3)	2179 (1)

4 Conclusion

A comprehensive experimental study is presented in this paper on the effect of glass-based TRM jacketing as well as the efficiency of the surface treatment of the substrate on the in-plane shear behavior of clay brick masonry wallets. The following conclusions can be drawn from the obtained experimental results:

- The control masonry wallets (URM) fail in the same manner; it means the failure mode is sudden and brittle and the crack pattern is stepwise along the mortar joints.
- The TRM jacketing improves dramatically the behavior of the strengthened wallets under diagonal compression load. The obtained failure mode is TRM composite cracking; by failing the jacket, the cracks appear along the mortar joints and the bricks.
- Sandblasted technique causes the bond of TRM-to-substrate to improve and the in-plane shear behavior to increase, in comparison to the un-sandblasted wallets.

Acknowledgements This work was partly financed by FEDER funds through the Competitivity Factors Operational Programme (COMPETE) and by national funds through the Foundation for Science and Technology (FCT) within the scope of project POCI-01-0145-FEDER-007633. The support to the first author through grant SFRH/BD/131282/2017 is kindly acknowledged.

References

1. Ghiassi, B., Soltani, M., Tasnimi, A.A.: Seismic evaluation of masonry structures strengthened with reinforced concrete layers. ASCE J. Struct. Eng. **138**, 1–18 (2012). https://doi.org/10.1061/(ASCE)ST.1943-541X.0000513
2. Bothara, J.K., Dhakal, R.P., Mander, J.B.: Seismic performance of an unreinforced masonry building: an experimental investigation Jitendra. Earthq. Eng. Struct. Dyn. 1–6 (2010). https://doi.org/10.1002/eqe.932.
3. Basili, M., Vestroni, F., Marcari, G.: Brick masonry panels strengthened with textile reinforced mortar: experimentation and numerical analysis. Constr. Build. Mater. **227**, (2019). https://doi.org/10.1016/j.conbuildmat.2019.117061
4. Valluzzi, M.R., Tinazzi, D., Modena, C.: Shear behavior of masonry panels strengthened by FRP laminates. Constr. Build. Mater. **16**, 409–416 (2002). https://doi.org/10.1016/S0950-0618(02)00043-0
5. Tetta, Z.C., Koutas, L.N., Bournas, D.A.: Textile-reinforced mortar (TRM) versus fiber-reinforced polymers (FRP) in shear strengthening of concrete beams. Compos. Part B Eng. **77**, 338–348 (2015). https://doi.org/10.1016/j.compositesb.2015.03.055
6. Raoof, S.M., Bournas, D.A.: TRM versus FRP in flexural strengthening of RC beams: behaviour at high temperatures. Constr. Build. Mater. **154**, 424–437 (2017). https://doi.org/10.1016/j.conbuildmat.2017.07.195
7. Mazzuca, S., Hadad, H.A., Ombres, L., Nanni, A.: Mechanical characterization of steel-reinforced grout for strengthening of existing masonry and concrete structures. J. Mater. Civ. Eng. **31**, 04019037 (2019). https://doi.org/10.1061/(ASCE)MT.1943-5533.0002669
8. Younis, A., Ebead, U.: Bond characteristics of different FRCM systems. Constr. Build. Mater. (2018). https://doi.org/10.1016/j.conbuildmat.2018.04.216

9. Deng, M., Yang, S.: Cyclic testing of unreinforced masonry walls retrofitted with engineered cementitious composites. Constr. Build. Mater. **177**, 395–408 (2018). https://doi.org/10.1016/j.conbuildmat.2018.05.132

10. Wakjira, T.G., Ebead, U.: FRCM/internal transverse shear reinforcement interaction in shear strengthened RC beams. Compos. Struct. **201**, 326–339 (2018). https://doi.org/10.1016/j.compstruct.2018.06.034

11. Marcari G, Basili M, Vestroni F. Experimental investigation of tuff masonry panels reinforced with surface bonded basalt textile-reinforced mortar. Compos. Part B Eng. 108 (2017). https://doi.org/10.1016/j.compositesb.2016.09.094

12. Wang, X., Lam, C.C., Iu, V.P.: Comparison of different types of TRM composites for strengthening masonry panels. Constr. Build. Mater. **219**, 184–194 (2019). https://doi.org/10.1016/j.conbuildmat.2019.05.179

13. Parisi, F., Iovinella, I., Balsamo, A., Augenti, N., Prota A.: In-plane behaviour of tuff masonry strengthened with inorganic matrix-grid composites. Compos. Part B Eng.45 (2013). https://doi.org/10.1016/j.compositesb.2012.09.068

14. Del Zoppo, M., Maddaloni, G., Balsamo, A., Di Ludovico, M., Prota, A.: Shear capacity of masonry panels reinforced with inorganic strengthening systems. Key Eng. Mater. 817 KEM:486–492 (2019). https://doi.org/10.4028/www.scientific.net/KEM.817.486

15. Yardim, Y., Lalaj, O.: Shear strengthening of unreinforced masonry wall with different fiber reinforced mortar jacketing. Constr. Build. Mater. **102**, 149–154 (2016). https://doi.org/10.1016/j.conbuildmat.2015.10.095

16. Razavizadeh, A., Ghiassi, B., Oliveira, D.V.: Bond behavior of SRG-strengthened masonry units: testing and numerical modeling. Constr. Build. Mater. **64**, 387–397 (2014). https://doi.org/10.1016/j.conbuildmat.2014.04.070

17. ASTM C109/C109M-05, Standard test method for compressive strength of hydraulic cement mortars (Using 2-in. or [50-mm] Cube Specimens). vol. 04 (2005). https://doi.org/10.1520/C0109_C0109M-05

18. BS EN 1015–11, Methods of test for mortar for masonry. Determination of flexural and compressive strength of hardened mortar (1999)

19. EN 772–1. Methods of test for masonry units—Part 1: Determination of compressive strength (2011)

20. Casacci, S., Gentilini, C., Di Tommaso, A., Oliveira, D.V.: Shear strengthening of masonry wallettes resorting to structural repointing and FRCM composites. Constr. Build. Mater. **206**, 19–34 (2019). https://doi.org/10.1016/j.conbuildmat.2019.02.044

21. ASTM E519–02, Standard test method for diagonal tension (shear) in masonry assemblages (2002). https://doi.org/10.1520/E0519-02

22. Wang, X., Lam, C.C., Iu, V.P.: Experimental investigation of in-plane shear behaviour of grey clay brick masonry panels strengthened with SRG. Eng. Struct. **162**, 84–96 (2018). https://doi.org/10.1016/j.engstruct.2018.02.027

Meso-Scale Study of Plain Concrete Beam Under Both Ambient and High Temperature

Biswajit Pal and Ananth Ramaswamy

Abstract In this study, concrete is modelled at meso-level ($10^{-4} - 10$ cm), where concrete is assumed either to be a two-phase or a three-phase composite. First the aggregates are placed in the structural domain randomly following a uniform distribution with the specific size and shape in order to capture actual aggregate gradation and placement process. The remaining space in the domain of concrete structure is then filled with the cement mortar in case of a two-phase assumption. However, in three phase modelling, part of the mortar of finite thickness surrounding the aggregates is assumed as ITZ which happens to be the weakest zone in the concrete structures. In this study, beam specimen in two dimensions (2D) has been considered and its behaviour under 4-point bending before and after it is subjected to a heating–cooling cycle is simulated. A finite element method (FEM) based approach is adopted here for the simulated study. A convergence study has been performed at ambient condition for different number of simulate beam specimens. The simulated load–deflection responses of the beam specimen subjected to different temperatures are then validated with the corresponding available experimental data.

Keywords Meso-scale model · Concrete · 4-point bending · High temperature

1 Introduction

Concrete is one of the commonly used composite materials where the hardened cement paste binds the aggregate and sand particles. Under the event of a fire (i.e., when concrete is subjected to high temperature) reduction of its strength can leads to complete or partial failure of a concrete structures. Simulated studies performed over the last two decades [1–3] in the field of structural fire engineering are mostly confined to the assumption of homogeneity of concrete. In these studies [1–3], concrete with certain mechanical and thermal properties was used for thermo-mechanical stress analysis of concrete structures. However, the behaviour of the constituents of concrete

B. Pal (✉) · A. Ramaswamy
Indian Institute of Science, Bangalore, India

© The Author(s), under exclusive license to Springer Nature Switzerland AG 2022   153
J. Sena-Cruz et al. (eds.), *Proceedings of the 3rd RILEM Spring Convention and Conference (RSCC 2020)*, RILEM Bookseries 34,
https://doi.org/10.1007/978-3-030-76465-4_14

(e.g., cement matrix, aggregates) is substantially different under high temperature. In general, as the temperature in concrete rises, the volume of the aggregates increases while the hardened cement paste surrounding the aggregate shrinks. However, such expansion and contraction of aggregate and cement matrix are strongly dependent upon the maximum temperature reached during the fire event. Due to these contrary expansion/contraction processes, the interface region between the cement matrix and aggregate (known as the interfacial transition zone-ITZ) becomes the weakest region in the concrete and the concrete then suffers damage by cracking at this interface zone. Such differential volume changes of aggregate and cement matrix are the results of various reversible and irreversible physico-chemical changes in these constituents during high temperature [4, 5] and lead to reduction of strength of concrete at high temperature. These processes can also significantly affect the post fire load-carrying capacity of a concrete structure [5]. In terms of loss of post-cooling residual capacity of concrete structures, the following processes lead to such strength reduction. For instance, the Ordinary Portland Cement (OPC) paste specimen disintegrates when it is cooled from 500 °C. This happens due to the rehydration of the Calcium Oxide (CaO) which is formed due to the decomposition of Portlandite at 400 °C. This rehydration is an expansive process and leads to the failure of concrete. On the other hand, at high temperature Calcium-Silicate- Hydrated (C–S–H) gel gets dehydrated and upon cooling it further rehydrated. This rehydration process leads to crack healing of the cement matrix and associated strength recovery. Further, calcium enriched aggregates decomposes above 750 °C and upon cooling it rehydrates and form $Ca(OH)_2$ by absorbing moisture. This leads to 44% increase in volume causing severe cracking of the concrete specimen and subsequent reduction in strength [4]. In contrast, at around 570 °C, a phase change occurs in siliceous aggregate. This causes large volume expansion and significant post-cooling residual strain. Cracks which formed in the aggregates at those high temperatures never be healed as contrary to cement matrix. Therefore, in order to capture realistic cracking and associated spalling, strength degradation phenomenon in concrete during and after fire, it becomes necessary to consider such differential thermal behaviour of concrete constituent in the numerical modelling. A recent study by Bošnjak [6] showed that a numerical study of concrete without explicit modelling of aggregate can well predict the realistic temperature and pore pressure distribution. However, spalling initiation and associated localised failure pattern cannot be captured when concrete was considered as a homogeneous material. Therefore, in this study aggregates are explicitly modelled and thermo-mechanical stress analysis of concrete has been done to realistically capture the initiation and propagation of crack in concrete. It is aimed to incorporate the differential thermal properties of the concrete constituents in the numerical meso-scale study. This study is a one step towards the direction through which the behaviour of concrete during and after exposure to high temperature can be captured more accurately.

2 Meso-Scale Model for Concrete

At meso-scale, concrete is assumed to be as either a two phase or a three phase composite which consist of aggregate, mortar matrix and ITZ. In the numerical model for concrete with two phases, the aggregates (which follow some standard aggregate gradation) are placed in the structural domain randomly following a uniform distribution with the specific size and shape in order to capture actual aggregate gradation and placement process. The remaining space in the domain of concrete structure is then filled with the cement mortar in case of a two-phase assumption. However, in three phase modelling, part of the mortar of finite thickness surrounding the aggregates is assumed as ITZ which happens to be the weakest zone in the concrete structures. This approach of modelling concrete numerically is known as Random Aggregate Model (RAM) [7].

2.1 Generations of Aggregate Particles

In this study, shape of the aggregate is assumed as circular. In order to generate the aggregates of circular shape in 2D based on an aggregate gradation [8], below are the steps being followed (which can be extended for other shape particles):

Step-1: Calculate the area of aggregate in each size range $[d_i, d_{i+1}]$ such that in each grading segment, area of aggregate is

$$A_{agg}\left[d_{i+1} - d_i\right] = \frac{P(d_{i+1}) - P(d_i)}{P(d_{max}) - P(d_{min})} \times P_{agg} \times A \qquad (1)$$

where, $P(d_i)$ is the percentage of aggregate smaller than diameter d_i; d_{min} and d_{max} represent the minimum and maximum diameter of the aggregate in the gradation, P_{agg} represents the total volume percentage of aggregate in concrete, A denotes the total concrete area.

Step-2: Starting with the range of aggregates containing largest size particles, an aggregate of diameter d is generated, such that:

$$d = d_i + \eta(d_{i+1} - d_i) \qquad (2)$$

where, η is a uniform random number lies between 0 and 1.

Step-3: The generated aggregate is placed in the given concrete domain, such that

$$x_i = x_{i,min} + \eta(x_{i,max} - x_{i,min}) \qquad (3)$$

where, x_i is the ith coordinates of the newly generated particle in the e_i Cartesian basis. It has to be ensured that there is no overlapping with the previously generated particles as well as with the domain boundary.

Step-4: Repeat steps 2–3 till the remaining area/volume of aggregate in that particular grading segment is not sufficient for the generation of another particle,

Step-5: Repeat steps 2–4 for the next smaller grading segment in the aggregate gradation.

2.2   Constitutive Models of the Constituents and Material Parameters Choice

In this study, coupled plasticity and damage based model [7] is used as constitutive law for the cement mortar and ITZ. In this model, the stress–strain relationship is expressed as

$$\sigma = \left(1 - d'\right) D_0^{el} : \left(\varepsilon - \varepsilon^{pl}\right) = \left(1 - d'\right)\overline{\sigma} \qquad (3)$$

where, σ and $\overline{\sigma}$ represent Cauchy stress and effective stress tensor, $d' = d'_{c/t}(\overline{\sigma}, \varepsilon^{pl}_{c/t}, T)$ is the isotropic scalar damage variable under compressive or tensile loading, $\varepsilon^{pl}_{c/t}$ is the plastic strain tensor and D_0^{el} denotes the initial constitutive tensors. On the other hand, aggregates are considered to be as a liner elastic brittle material at ambient condition due to its high stiffness and strength as compared to mortar.

3   Results and Discussion

In this study, behaviour of plain concrete beam at both ambient condition and at high temperature is simulated through the above discussed meso-scale model of concrete. The beam of dimensions 350 mm × 100 mm × 100 mm is taken for which experimental load–deflection data are available in [9]. Details of the mix-proportion of concrete can be found in [9]. Based on the volume fraction (which is approximately 0.40) and gradation of coarse aggregate (maximum size of coarse aggregate is 19 mm), aggregates are generated within the given beam domain. As a simplification, the shape of the aggregate is considered as circular. The one such possible configuration of the generated aggregate particles within the given beam domain is shown in Fig. 1 in case of a two phase assumption (i.e., aggregate and mortar). However, in case of a three phase composite, a part of the mortar around the aggregate is considered as ITZ. Such a configuration is shown in Fig. 2a, b. In

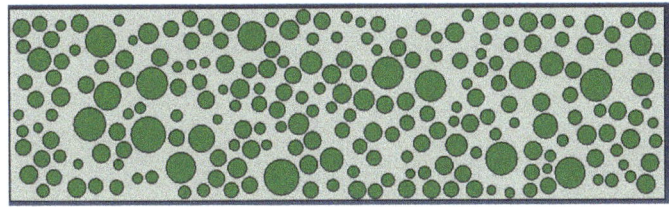

Fig. 1 Simulated random distribution of aggregates in 2D in a beam of size 350 mm × 100 mm × 100 mm

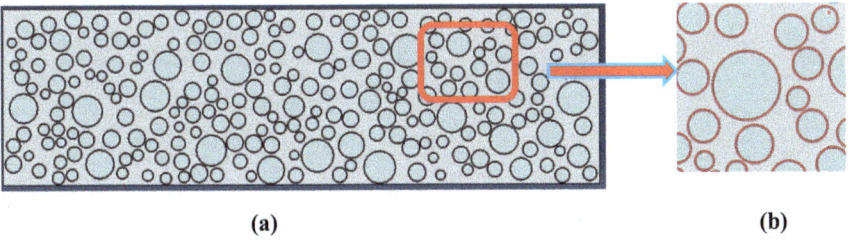

(a)   (b)

Fig. 2 a Simulated random distribution of aggregates and ITZ in 2D in a beam of size 350 mm x 100 mm x 100 mm, **b** Enlarged view

literature [7, 10], the thickness of ITZ was considered in the range between 0.05 and 1 mm based on either nano-indentation test [10] or numerical study [7]. Here, the thickness of the ITZ is considered as 0.5 mm.

In order to perform the numerical study, the exact material data of the concrete constituents are not available. Therefore, the material properties of cement mortar are chosen based on the experimental study performed in [11] for different composition mortar. These data are chosen so as to get close fit with the experimental load–deflection results [12]. The ratio of ITZ to mortar material properties were taken in the range of 0.5–0.8 [7, 10] based on nano-indentation test performed on ITZ [10] or from numerical simulation study [7]. So, in this study a factor of 0.8 is chosen as the ratio of material properties of ITZ to mortar. The properties of mortar during high temperature and in post-cooling period are taken based on the experimental study performed in [4, 13, 14]. On the other hand, material properties of aggregates at high temperature are taken from [15]. Some of the properties of mortar and aggregate at ambient conditions are shown in Table 1.

Table 1 Material properties of mortar and aggregate at ambient condition

Material properties	Mortar	Aggregate
Modulus of Elasticity (GPa)	20	38.37
Poisson's ratio	0.22	0.3
Compressive strength (MPa)	30	–
Tensile strength (MPa)	3.6	–

Fig. 3 Load–deflection of
plain concrete beam at
ambient temperature

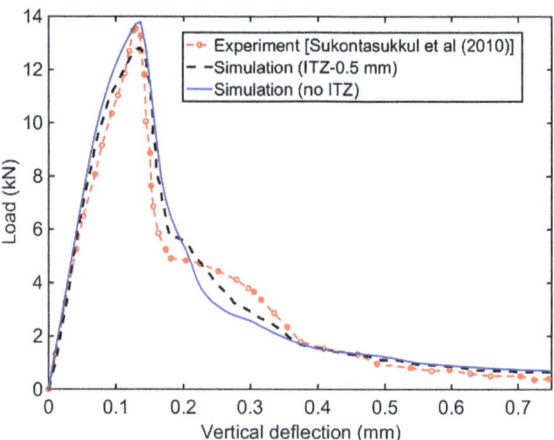

At ambient condition, the beam is subjected to two equal displacements at 1/3rd points [9]. The simulated load–deflection response of a specimen is shown in Fig. 3 with and without ITZ. Simulated responses match quite well with that of experiments. However, incorporation of ITZ make the responses little softer. The formation and propagation of micro-cracks as has been simulated at different stages of loading are shown in Figs. 4a–d. Here, the possible formation of cracks is shown in terms of the tensile damage variable (d_t). Since, the aggregate is assumed to have high strength than the mortar at ambient condition, formation of cracks follows a path which do not goes through an aggregate. Moreover, due to heterogeneity of concrete in terms of the random distribution of aggregates, the final crack does not necessary will be formed at the centroid of the beam.

In Figs. 3 and 4, simulated load–deflection and consecutive failure pattern of a concrete sample has been shown. One such specimen is nothing but the one possible random distribution of aggregates in the concrete domain. However, in reality it could be any distribution of aggregates within the concrete domain. Therefore, the simulation has been done for different number of samples. The load–deflection responses generated for 100 such samples are shown in Fig. 5. It can be seen from Fig. 5 that the variation in load–deflection response of the beam due to random size (with specific size range) and distribution occurs during and after the post-peak region. Figure 6a, b show the mean and standard deviation of the peak load with different number of concrete samples. These figures suggest that at least fifty number of such concrete samples need to simulate in order to get a converged mean peak load. However, one such sample does not give a response which has a significant variation from the mean curve.

On the other hand, in case of high temperature simulation, thermo-mechanical stress analysis is performed. In the simulation, the beam is placed over a steel plate and is heated up to a specified temperature following [9]. After cooling, the specimen is tested under four-point bending. Details of the test can be found in [9]. The simulated crack pattern for the specimen which is heated to 400 °C and then cooled

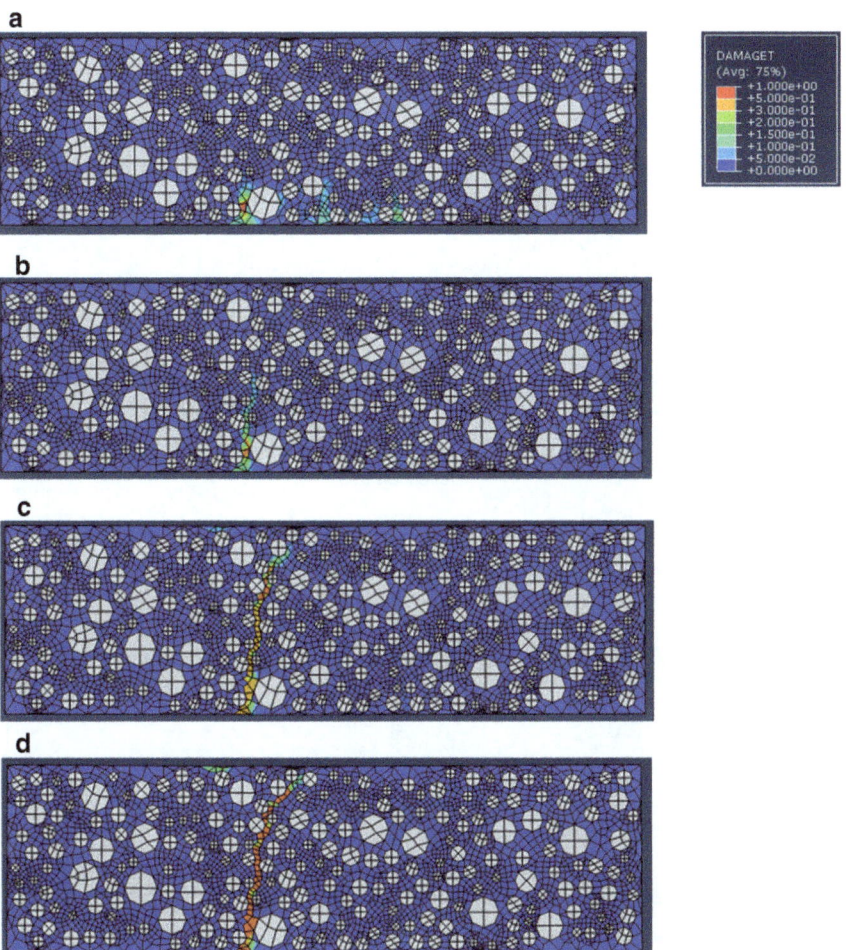

Fig. 4 **a** Possible failure pattern under mechanical load of plain concrete beam at ambient temperature at central vertical displacement of 0.08 mm and tensile damage value, **b** Possible failure pattern under mechanical load of plain concrete beam at ambient temperature at central vertical displacement of 0.14 mm, **c** Possible failure pattern under mechanical load of plain concrete beam at ambient temperature at central vertical displacement of 0.34 mm, **d** Possible failure pattern under mechanical load of plain concrete beam at ambient temperature at central vertical displacement of 0.68 mm

is shown in Fig. 7. Several micro-cracks can exist in a specimen which is cooled down from a temperature of 400 °C. This post-cooled specimen is then tested under four-point bending. The failure pattern of this post-cooled specimen after a 0.5 mm of vertical displacement of centre is shown in Fig. 8. The associated load–deflection response of this specimen is shown in Fig. 9 along with the experimental response

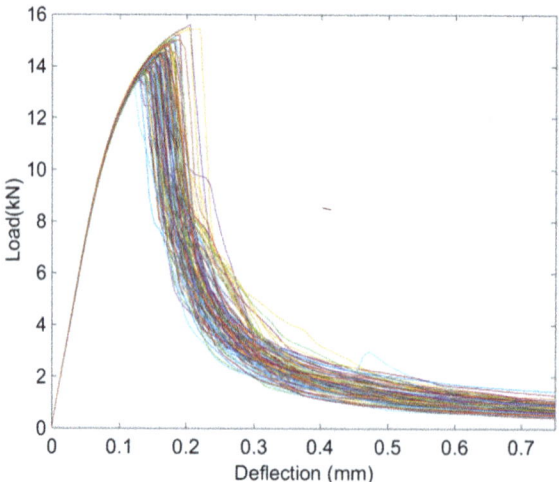

Fig. 5 Simulated load–deflection of plain concrete beam at ambient temperature for 100 different samples

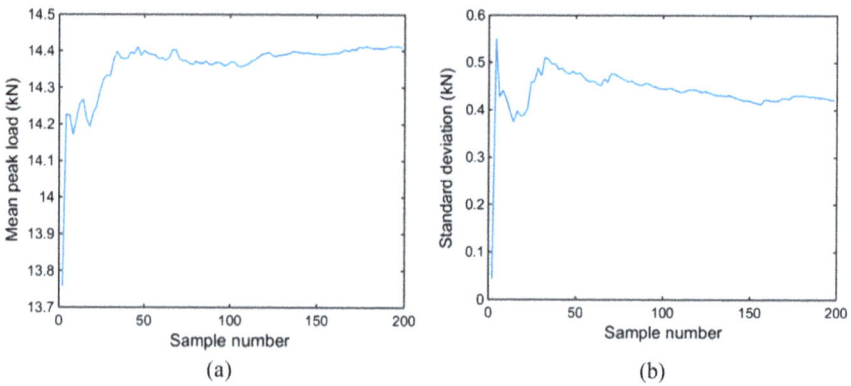

(a)   (b)

Fig. 6 **a** Mean and **b** standard deviation of average peak load with number of samples

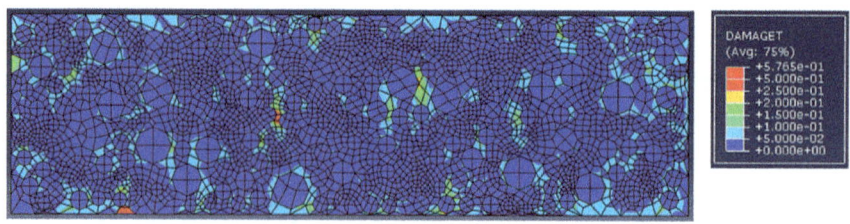

Fig. 7 Possible failure pattern of plain concrete beam after heating (400 °C)-cooling cycle

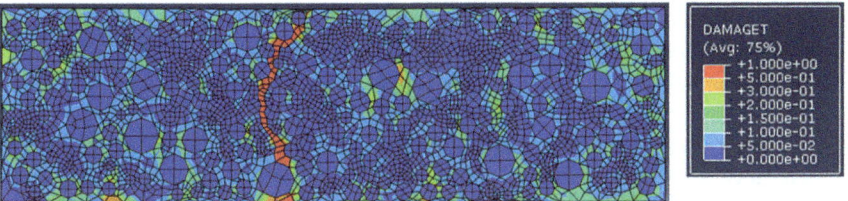

Fig. 8 Possible failure pattern under mechanical load applied after heating (400 °C)-cooling cycle at a central vertical displacement of 0.50 mm

Fig. 9 Load–deflection of plain concrete beam under mechanical load after heating to specified temperature and then cooled

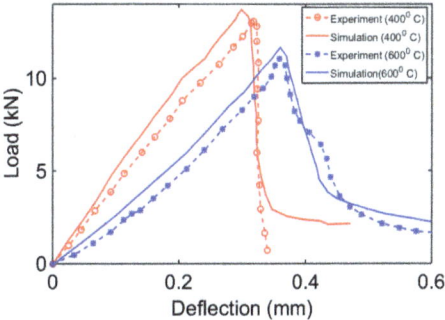

[9]. The simulated load–deflection for this specimen agrees quite well with the available experimental data. Some discrepancies could occur due to the assumption of circular shape of aggregates, 2D modelling assumptions etc. In Fig. 9, load–deflection response of the specimen which is heated to 600 °C is also shown. The failure pattern of this specimen after the heating–cooling cycle and after the mechanical load is not shown here since it is quite similar with that of the specimen which is heated to 400 °C. However, the simulation is not tried for the specimen which is heated to 800 °C due to non-availability of several thermo-mechanical properties of mortar.

4 Conclusions

In this study, beam specimens are modeled in 2D and their behavior under both ambient and high temperature is simulated. Under ambient condition, the beam is subjected to two applied displacements at 1/3rd points. On the other hand, in case of high temperature simulation, the beam is heated up to specified temperature and then it is cooled and tested under four-point bending loading conditions. Based on the study performed at meso-scale for these concrete beam specimens, the following conclusions can be drawn:

- Considering ITZ will soften the response and gives a conservative failure load,

- At least 50 monte carlo samples in terms of randomness of aggregate size gradation and distribution would be sufficient to get a converged mean peak load. However, the variation of response of one such sample from the mean peak load is less than 5%,
- The major crack in case of a normal strength concrete beam under 4-point bending need not always be at the center of the beam. This asymmetry of crack development could be due to the random distribution of comparatively stronger aggregates at both ambient and post-cool specimen
- Depending on the exposed surface and rate of heating, there could be several micro-cracks exist in beam specimen which is cooled down from a specified maximum temperature.

References

1. Bratina, S., Cas, B., Saje, M., Planinc, I.: Numerical modelling of behaviour of reinforced concrete columns in fire and comparison with Eurocode 2. Int. J. Solids Struct. **42**, 5715–5733 (2005)
2. Gao, W.Y., Dai, J.G., Teng, J.G., Chen, G.M.: Finite element modeling of reinforced concrete beams exposed to fire. Eng. Struct. **52**, 488–501 (2013)
3. Kodur, V.K.R., Agrawal, A.: An approach for evaluating residual capacity of reinforced concrete beams exposed to fire. Eng. Struct. **110**, 293–306 (2016)
4. Fu, Y.F., Wong, Y.L., Poon, C.S., Tang, C.A., Lin, P.: Experimental study of micro/macro crack development and stress–strain relations of cement-based composite materials at elevated temperatures. Cem. Concr. Res. **34**, 789–797 (2004)
5. Hager, I.: Behaviour of cement concrete at high temperature. Bulletin Polish Acad. Sci. Tech. Sci. **61**(1), 1–10 (2013)
6. Bošnjak, J.: Explosive spalling and permeability of high performance concrete under fire– numerical and experimental investigations. Doctoral thesis, Institut für Werkstoffe im Bauwesen der Universität Stuttgart (2014)
7. Chen, H., Xu, B., Mo, Y.L., Zhou, T.: Behavior of meso-scale heterogeneous concrete under uniaxial tensile and compressive loadings. Constr. Build. Mater. **178**, 418–431 (2018)
8. Xu, Y., Chen, S.: A method for modeling the damage behavior of concrete with a three-phase mesostructure. Constr. Build. Mater. **102**, 26–38 (2016)
9. Sukontasukkul, P., Pomchiengpin, W., Songpiriyakij, S.: Post-crack (or post-peak) flexural response and toughness of fiber reinforced concrete after exposure to high temperature. Constr. Build. Mater. **24**, 1967–1974 (2010)
10. Xiao, J., Li, W., Corr, D.J, Shah, S.P.: Effects of interfacial transition zones on the stress–strain behavior of modeled recycled aggregate concrete. Cement Concr. Res. **52**, 82–99 (2013)
11. ElNemr, A.: Role of water/binder ratio on strength development of cement mortar. Am. J. Eng. Res. **8**(1), 172–183 (2019)
12. Grassl, P., Gregoire, D., Solano, L.R., Cabot, G.P.: Meso-scale modelling of the size effect on the fracture process zone of concrete. Int. J. Solids Struct. **49**, 1818–1827 (2012)
13. Cülfik, M.S., Özturan, T.: Effect of elevated temperatures on the residual mechanical properties of high-performance mortar. Cement Concr. Res. **32**, 809–816 (2002)
14. Vieira, J.P.B., Correia, J.R., Brito, J.: D: Post-fire residual mechanical properties of concrete made with recycled concrete coarse aggregates. Cem. Concr. Res. **41**, 533–541 (2011)
15. Li, X.X., Xin, K.Z., Ming, J., Xuan, G.W., Jing, C.: Research of microcosmic mechanism of brittle-plastic transition for granite under high temperature. Proced. Earth Planet. Sci. **1**, 432–437 (2009)

Electrochemical Realkalisation of Carbonated "Dalle de Verre" Windows

Shishir Mundra⬡, Götz Hüsken⬡, and Hans-Carsten Kühne⬡

Abstract Concrete-glass windows or commonly referred to as "Dalle de Verre" windows are prevalent in historical monuments across Europe. "Dalle de Verre" windows were made by placing cut stained glasses, of various shapes and sizes, into a steel reinforced concrete frame. In addition to the steel reinforcement, a few steel wires run across the concrete elements to hold the glass windows into place. Much of these structures have been damaged due to carbonation of the thin concrete covers and consequent corrosion of the steel reinforcement; and exhibit several cracks due to the expansive nature of corrosion products formed at the steel–concrete interface. This study focussed on understanding the efficacy of electrochemical realkalisation in reinstating the passive state of the steel reinforcement embedded in such concrete-glass windows. Simulant "Dalle de Verre" windows (representative of windows at Kaiser-Wilhelm Gedächtniskirche) were produced using different cements with a water/cement ratio of 0.6 and carbonated under accelerated conditions. Post carbonation, the concrete glass windows were electrochemically realkalised using a sacrificial anode. The influence of highly alkaline conditions due to electrochemical realkalisation on the glass-concrete interface has also been investigated. This study shows that the passive state of the steel reinforcement in "Dalle de Verre" windows, particularly at the Kaiser-Wilhelm Gedächtniskirche (memorial church) in Berlin, can be reinstated using the electrochemical realkalisation method.

Keywords Electrochemical realkalisation · Carbonation · Concrete-glass windows · Passivation · Dalle-de-verre

S. Mundra (✉) · G. Hüsken · H.-C. Kühne
Bundesanstalt für Materialforschung und -prüfung (BAM), Unter den Eichen 87, 12205 Berlin, Germany
e-mail: smundra@ethz.ch

163

1   Introduction

"Dalle de Verre", a French artform to make concrete-stained glass windows of the 1920s, are a common occurrence (Fig. 1) in many monuments across western Europe and the United States of America, such as the Buckfast Abbey, the New York Hall of Science, the Church of St. Odile, Church of St. Theodore and the Kaiser-Wilhelm-Gedächtniskirche [1].

These installations have been around for more than 5 decades now and have shown significant amounts of degradation. The windows at the Kaiser-Wilhelm-Gedächtniskirche in Berlin, in particular, have developed cracks and are nearing the end of their service life. One of the major causes for the degradation mechanism is the corrosion of the steel-reinforcement frame within these windows, particularly due to the carbonation of the thin concrete cover (approximately 5 mm). Corrosion of the steel reinforcement in concrete structures occurs either due to the presence of chloride above a threshold concentration or due a reduction in the pH caused by leaching or carbonation of the concrete cover. Carbonation of concrete occurs via the chemical reaction between atmospheric carbon dioxide and the cement matrix, adversely impacting the long-term durability of reinforced concrete structures [4, 5]. The carbonation of Portland cement (PC) has been thoroughly studied in the literature. Gaseous carbon dioxide from the surrounding environment penetrates

Fig. 1 "Dalle de Verre" used in various churches across the world. The Buckfast Abbey (top left) [2], the New York Hall of Science (top right) [3], the Church of St. Odile (bottom left) [1], Church of St. Theodore (bottom middle) [1] and the Kaiser-Wilhelm-Gedächtniskirche (bottom right)

through the porous concrete structure and dissolves in the pore solution forming weak acidic ionic species of HCO_3^- and CO_3^{2-} [4, 5]. The dissolution of gaseous carbon dioxide and formation of weak acids can be represented by the following reactions (Eq. 1–3) [6]:

$$CO_2(g) \rightarrow CO_2(aq.) \tag{1}$$

$$CO_2(aq.) + H_2O \rightarrow H^+ + HCO_3^-(aq.) \tag{2}$$

$$HCO_3^-(aq.) \rightarrow H^+ + CO_3^{2-}(aq.) \tag{3}$$

The release of acidic species such as H^+ leads to a reduction in the alkalinity of the pore solution which causes the dissolution of Ca-rich hydration products such as portlandite, calcium silicate hydrate gel, AFm and ettringite [7–9]. The dissolved calcium and carbonate ions react to precipitate various polymorphs of $CaCO_3$—vaterite, aragonite, amorphous $CaCO_3$, or calcite (Eq. 4) [7].

$$Ca^{2+}(aq.) + CO_3^{2-}(aq.) \rightarrow CaCO_3(s) \tag{4}$$

The reduction of alkalinity due to carbonation in reinforced structures can result in the breakdown of the passive film formed around steel. One possible technique used to increase the alkalinity at the steel–concrete interface is electrochemical realkalisation [10].

Electrochemical realkalisation works on the principle of applying a current (as shown in Fig. 2) w.r.t. the surface area of the embedded steel-reinforcement (acting as the cathode). An electrochemical cell is established by using an auxiliary anode (steel, TiO_2 mesh) immersed in an aqueous alkali-carbonate electrolyte (usually a Na_2CO_3 solution). Upon the application of a current between 0.8 and 2 A/m^2 w.r.t the steel-reinforcement, the realkalisation of the steel–concrete interface is achieved by formation of OH^- due to the electrolysis of water [10]. At the cathode (or the steel reinforcement), the reduction of water to OH^- is controlled by the availability of oxygen, and once the oxygen is consumed, H_2 (g) is released (Eq. 5). The anodic reaction occurring at the external anode is dominated by the oxidation of OH^- to oxygen and water (Eq. 6), which can further be oxidised to release H^+ ions [10].

$$Cathode : 2H_2O(aq.) + 2e^- \rightarrow H_{-2}(g) + 2OH^-(aq.) \tag{5}$$

$$Anode : 2OH^-(aq.) \rightarrow {}^1\!/_2O_2(g) + 2e^- + 2H_2O(aq.) \tag{6}$$

In addition to the redox reactions occurring at the anode and cathode, other processes like electromigration (movement of ions due to the applied current, based on their polarity), electroosmosis, diffusion, and capillary absorption also influence the process of realkalisation [10]. However, due to the high electrical current applied during the realkalisation process, electromigration of ions is the dominant mechanism controlling the efficacy of electrochemical realkalisation. Negatively charged ions (such OH^-, Cl^-, CO_3^{2-}) from the pore solution at steel–concrete interface migrate towards the anode or the surface of the concrete, whereas the positively charged ions such as Na^+ from the electrolyte migrate inwards towards the steel–concrete interface to maintain electrical neutrality [10]. Under equilibrium conditions post electrochemical realkalisation, the pH at the steel–concrete interface is >13, enabling the re-establishment of passivity [10].

2 Materials and Methods

2.1 Materials

Three different cements (CEM I, CEM III/A and CEM II/B-S) and stained glasses of different sizes and colours were obtained from Heidelberg Cement and Glasmalarai Peters, respectively, to simulate "Dalle de Verre" windows (330 mm × 330 mm × 25 mm) present at Kaiser-Wilhelm Gedächtniskirche. A mild steel reinforcement frame ($\varphi = 10$ mm) was embedded in the simulant samples, ensuring a cover depth of 5 mm. The mix designs used in this study are listed in Table 1.

2.2 Sample Preparation

"Dalle de Verre" windows were simulated (Fig. 3) in the laboratory by placing the steel reinforcement frame and the cut stained glasses in predetermined positions within the mould. The stained glasses were placed firmly within the mould using a silicone gel, to ensure no movement while casting. As shown in Fig. 3, stainless steel wires of diameter 2 mm, were welded onto the mild steel reinforcement cage at the centre of each edge to ensure electrical connection for the application of electro-chemical methods. Each of the simulated "Dalle de Verre" windows were demoulded after 24 h of casting and then cured under water for 14 days and conditioned for an additional 14 days at a temperature of 20 ± 2 °C and a relative humidity of 65%.

Table 1 Mix designs used in this study

Cement type	Cement content (kg/m^3)	w/c	Aggregate content (kg/m^3)
CEM I	450	0.6	1350
CEM III/A	450	0.6	1350
CEM II/B-S	450	0.6	1350

2.3 Exposure Duration

The conditioned samples were then exposed to an accelerated carbonation environment (1% CO_2, 65% RH, 20 ± 2 °C) in a carbonation chamber. The duration of exposure was determined based on whether the steel was in the active state or passive state, according to the half-cell potential (<-250 mV versus Ag/AgCl) of the embedded steel-reinforcement. Once the steel reinforcement was in the active state, the samples were kept in the carbonation chamber for roughly 3 weeks in the carbonation chamber prior to the application of electrochemical realkalisation.

2.4 Electrochemical Realkalisation

Post accelerated carbonation, the steel samples were exposed to a 1 M Na_2CO_3 solution with a TiO_2 mesh as the sacrificial anode and connected to a DC power source. A current of 1 and 2 A/m^2 was applied for a duration of 14 days and the half-cell potential of the steel reinforcement was continuously monitored prior to and post realkalisation [10]. The duration of electrochemical realkalisation was selected based on earlier works in the literature [11–13]. The half-cell potential was measured using an Ag/AgCl reference electrode (sat. KCl solution) and a wet sponge at least on 15 different points on the surface of the sample.

3 Results and Discussion

Figure 4 shows the measured half-cell potentials for simulated Dalle de Verre windows from different cements, and the efficacy of electrochemical realkalisation treatment with a constant applied current density of 1 A/m^2. The efficacy of realkalisation in this study is monitored by the rate at which passivity of the steel reinforcement is re-established, post realkalisation treatment. Figure 4 clearly indicates that the realkalisation treatment works most effectively in carbonated CEM I mortars, and the half-cell potential almost moves back to vales close -0.20 V versus Ag/AgCl in a span of approximately 40 days. In the case of carbonated CEM II/B-S and CEM III/A mortars, the half-cell potential also recovers after approximately 50–60 days and rests at around -0.25 V versus Ag/AgCl.

The half-cell potential measured after electrochemical realkalisation of all carbonated binders are slightly lower than those generally expected for the steel to be in the passive state. Recording more negative potentials after electrochemical realkalisation is not uncommon and has been reported by several researchers [14–17], because the resistivity of the concrete cover is much lower than untreated carbonated binders. The lower resistivity of the concrete cover after electrochemical realkalisation primarily stems from the uptake of Na_2CO_3 electrolyte from the exterior [10]. Additionally,

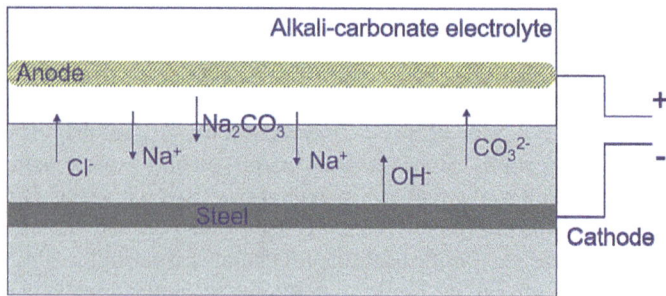

Fig. 2 Schematic describing the electrochemical realkalisation method [10]

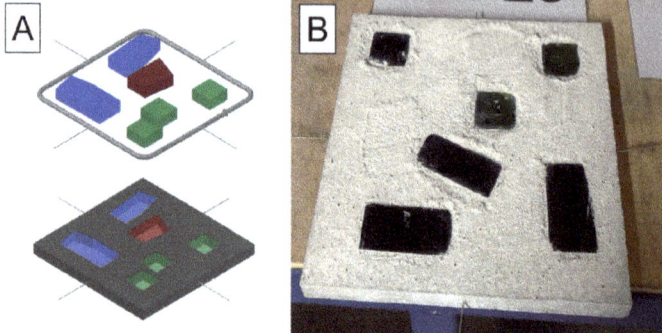

Fig. 3 **a** Schematic of the simulated "Dalle de Verre" windows produced at BAM and **b** simulated "Dalle de Verre" window made from CEM I and cut stained glasses. A mild steel reinforcement frame is embedded in the mortar with a cover depth of 5 mm

the pH of the pore solution at the steel–concrete interface of realkalised structures has been reported to be >13.5 and the passive steel surface acts as a pH electrode and in this case the potential decreases upon increasing the pH [10]. As highlighted in [10], the passive state of the steel-reinforcement can be considered to be reinstated if the measured potentials were found to be homogenous across the element being considered. As mentioned earlier, the half-cell potentials measured in this study post the electrochemical realkalisation treatment were found to be stable after 40 days for carbonated CEM I Dalle de Verre windows, and 50–60 days for carbonated CEM II/B-S and CEM III/A Dalle de Verre windows. Therefore, stable half-cell potentials are −0.20 and −0.25 V versus Ag/AgCl after the electrochemical realkalisation (Fig. 4) are indicative of repassivation of the steel-reinforcement.

Figure 4 also shows the efficacy of electrochemical realkalisation as a function of the binder type. As mentioned above, the rate of repassivation of the embedded steel reinforcement was found to be the fastest for carbonated Dalle de Verre windows made from CEM I, when compared to CEM II/B-S and CEM III/A. Similar observations were made by Ribeiro et al. [16], where the authors related the longer time for repassivation to the lower amounts of portlandite in concretes based on Portland

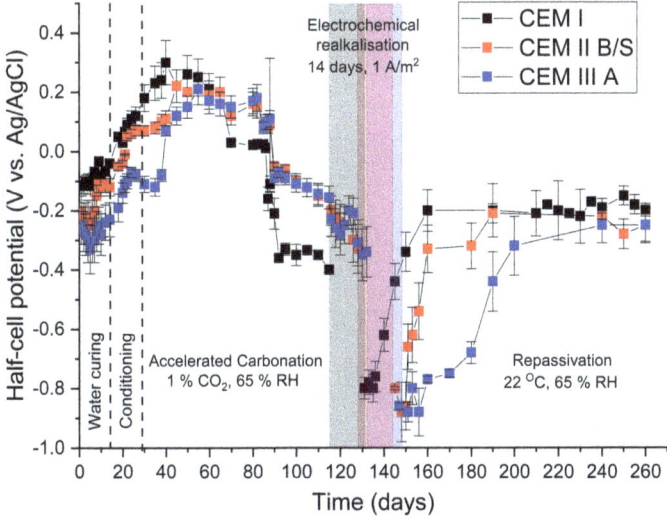

Fig. 4 Electrochemical realkalisation of simulant "Dalle de Verre" windows made from CEM I, CEM II/B-S and CEM III/A using 1 M Na_2CO_3 solution, a current density of 1 A/m^2 and a TiO_2 anode mesh

pozzolana cement (CEM II), and higher charge required to achieve repassivation, when compared to CEM I. Aguirre-Guerrero and Mejía de Gutiérrez [17] measured the E_{corr} and i_{corr} for electrochemically realkalised steel-reinforced concretes made out of Portland cements, and blended cements with 10% silica fume and metakaolin, and observed the re-establishment of the passive film to be quicker in PC based concretes when compared those from blended cements. As electrical migration of ions is the dominant mechanism determining the efficacy of electrochemical realkalisation, the slower rate of repassivation of the embedded reinforcement could be linked to the higher resistivities of mortars made from CEM II/B-S and CEM III/A when compared to CEM I.

Figure 5 shows the influence of current used w.r.t. surface area of the steel reinforcement, during electrochemical realkalisation, on the rate at which the steel-reinforcement repassivates. Current densities of 1 and 2 A/m^2 were applied on the embedded reinforcement post carbonation in different specimens. The half-cell potential was found to be stable at around −0.20 V versus Ag/AgCl after very similar times for both the current densities used for electrochemical realkalisation. Mietz [14] measured the realkalisation depth from the outside of cylindrical samples as a function of the current density used and observed similar realkalisation depths of 5–6 mm for current densities of 1 and 5 A/m^2 after a 14 day realkalisation treatment. The realkalisation front from the steel–concrete interface in the study by Mietz [14] was found to be directly proportional to the current density. For Dalle de Verre windows with a cover depth of 5 mm studied here, it could be credibly asserted that the current density used during electrochemical realkalisation does not play a major role in the rate at which the passivity of the steel reinforcement is reinstated.

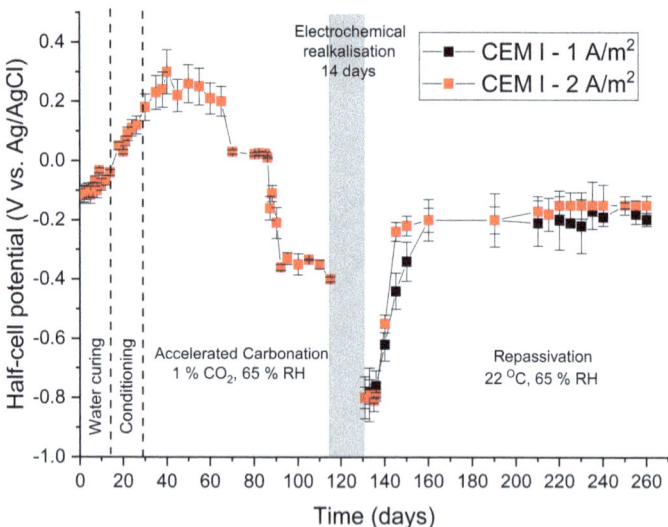

Fig. 5 Influence of current density used on the efficacy of electrochemical realkalisation of "Dalle de Verre" windows made from CEM I. Electrochemical realkalisation was carried out using 1 M Na_2CO_3, TiO_2 anode mesh, and current densities of 1 and 2 A/m^2

4   Conclusion and Future Work

This study showcases the successful application of electrochemical realkalisation treatment to reinstate the passive state of the steel reinforcement in carbonated Dalle de Verre windows at the Kaiser-Wilhelm Gedächtniskirche, using simulated samples. As several windows at the Kaiser-Wilhelm Gedächtniskirche are damaged significantly, on the verge of spalling and need to be completely replaced with newer materials, the applicability of electrochemical realkalisation on windows made from blended cements such as CEM II/B-S and CEM III/A were assessed. The rate of realkalisation was found to be the quickest for Dalle de Verre windows made from CEM I followed by CEM II/B-S and CEM III/A, and the rate of repassivation can be directly related to the resistivity of the binders to ionic diffusion. Additionally, the current density used during the electrochemical realkalisation treatment was found to not significantly alter the rate of repassivation of the embedded reinforcement.

To ensure electrochemical realkalisation is a suitable way for restoring the alkalinity at the steel–concrete interface of degraded cultural heritage structures, the following work is currently being conducted:

1. The change in microstructure at the steel–concrete interface due to carbonation and electrochemical realkalisation treatment.
2. The efficacy of using cellulose pulp fiber as a substrate to hold the electrolyte (1 M Na_2CO_3) on the surface of vertical specimens to be electrochemically realkalised, and possible staining/etching of glass due to the release of H^+ at the anode/electrolyte.

3. The successive applicability of electrochemical realkalisation on already treated windows.
4. Repair protocol for Dalle de Verre windows at the Kaiser-Wilhelm Gedächt-niskirche.

References

1. De Vis, K.: The Consolidation of Architectural Glass and Dalle de Verre; Assessment of Selected Adhesives—Thesis. Universiteit Antwerpen (2014)
2. Kennedy, N.: Stained glass window at Buckfast Abbey (2004). https://www.geograph.org.uk/photo/207061. Accessed 2 Dec 2019
3. Goldberg, J.: New York Hall of Science Reopens Great Hall with Renovations from Todd Schliemann (2015) . https://www.archdaily.com/769609/new-york-hall-of-science-reopens-great-hall-with-renovations-from-todd-schliemann/5595ead5e58ece2c83000558-new-york-hall-of-science-reopens-great-hall-with-renovations-from-todd-schliemann-photo?next_project=no. Accessed 2 Dec 2019
4. Morandeau, A., Thiéry, M., Dangla, P.: Investigation of the carbonation mechanism of CH and C–S–H in terms of kinetics, microstructure changes and moisture properties. Cem. Concr. Res. **56**, 153–170 (2014)
5. Morandeau, A., Thiéry, M., Dangla, P.: Impact of accelerated carbonation on OPC cement paste blended with fly ash. Cem. Concr. Res. **67**, 226–236 (2015)
6. Adamczyk, K., Prémont-Schwarz, M., Pines, D., et al.: Real-time observation of carbonic acid formation in aqueous solutions. Science **326**, 1690–1694 (2009)
7. Black, L., Breen, C., Yarwood, J., et al.: Structural features of C–S–H(I) and its carbonation in air-A Raman spectroscopic study. Part II: carbonated phases. J. Am. Ceram. Soc. **90**, 908–917 (2007)
8. Rodway, D.I., Groves, G.W., Richardson, I.G.: The carbonation of hardened cement pastes. Adv. Cem. Res. **3**, 117–125 (1990)
9. Anstice, D.J., Page, C.L., Page, M.M.: The pore solution phase of carbonated cement pastes. Cem. Concr. Res. **35**, 377–383 (2005)
10. EFC Working Party 11.: Electrochemical rehabilitation methods for reinforced concrete structures: a state of the art report. IOM Communications Ltd., Institute of Materials (1998)
11. Bertolini, L., Carsana, M., Redaelli, E.: Conservation of historical reinforced concrete structures damaged by carbonation induced corrosion by means of electrochemical realkalisation. J. Cult. Herit. **9**, 376–385 (2008)
12. Bertolini, L., Bolzoni, F., Elsener, B., Pedeferri, P.: La realcalinización y la extracción electro-química de los cloruros en las construcciones de hormigón armado. Mater. Construcción **46**, 45–55 (1996)
13. Pedeferri, P., Bertolini, L., Elsener, B., et al.: Corrosion of Steel in Concrete—Prevention. John Wiley & Sons, Diagnosis, Repair (2013)
14. Mietz, J.: Electrochemical realkalisation for rehabilitation of reinforced concrete structures. Mater. Corr. **46**, 527–533 (1995)
15. Tong, Y., Bouteiller, V., Marie-Victoire, E., Joiret, S.: Efficiency investigations of electrochemical realkalisation treatment applied to carbonated reinforced concrete. Cem. Concr. Res. **42**, 84–94 (2012)
16. Ribeiro, P.H.L.C., Meira, G.R., Ferreira, P.R.R., Perazzo, N.: Electrochemical realkalisation of carbonated concretes-Influence of material characteristics and thickness of concrete reinforcement cover. Constr. Build. Mater. **40**, 280–290 (2013)
17. Aguirre-Guerrero, A.M., Mejía de Gutiérrez, R.: Efficiency of electrochemical realkalisation treatment on reinforced blended concrete using FTIR and TGA. Constr. Build. Mater. **193**, 518–528 (2018)

Micro-To-Meso Scale Mechanisms for Modelling the Fatigue Response of Cohesive Frictional Materials

Antonio Caggiano, Diego Said Schicchi, Swati Maitra, Sha Yang, and Eddie A. B. Koenders

Abstract This work deals with a research study on the fatigue behavior of cohesive frictional materials like concrete or mortars. A two-scale approach will be proposed for analyzing concrete specimens subjected to either low- or high-cycle fatigue actions. The multiscale technique is based on a combination of a micro-scale procedure, to describe the microstructural defects affecting the cyclic response, which is subsequently homogenized to the upper meso- and macroscopic failure. More specifically, projected microscopic representative volume elements, equipped with a fracture-based model and combined with a continuous inelastic constitutive law that accumulates damages induced by cycle behaviors, represents the lower scale. A plastic-damage based model, for concrete subjected to cyclic loading, is developed combining the concept of fracture-energy theories (within the family of fictitious crack approaches) with stiffness degradation, the latter representing the key phenomenon occurring in concrete under cyclic responses. The work will numerically explore the potential of the various techniques for assessing the fatigue formation and growth of (micro-, meso- and macro-) cracks, and their influence on the macroscale behavior.

Keywords Fatigue · Fracture · Cement-based materials · Micro · Meso · Fictitious crack approaches

A. Caggiano (✉) · D. Said Schicchi · S. Yang · E. A. B. Koenders
Institute of Construction and Building Materials, TU-Darmstadt, Germany
e-mail: caggiano@wib.tu-darmstadt.de

S. Yang
e-mail: yang@wib.tu-darmstadt.de

E. A. B. Koenders
e-mail: koenders@wib.tu-darmstadt.de

A. Caggiano
Universidad de Buenos Aires, LMNI, INTECIN, CONICET, Buenos Aires, Argentina

S. Maitra
Indian Institute of Technology Kharagpur, Kharagpur, India
e-mail: swati@iitkgp.ac.in

© The Author(s), under exclusive license to Springer Nature Switzerland AG 2022
J. Sena-Cruz et al. (eds.), *Proceedings of the 3rd RILEM Spring Convention and Conference (RSCC 2020)*, RILEM Bookseries 34,
https://doi.org/10.1007/978-3-030-76465-4_16

173

1 Introduction

Concrete structures such as bridges, roads, towers, modern building, and runways, among others, suffer gradual deterioration when subjected to fatigue loading, usually caused by mesoscopic inhomogeneities and defects present in the material. Fatigue failure can be defined as a progressive and permanent change in the internal structure of the material in the form of microcracks which coalesce into a macrocrack [1].

Plenty of experimental studies are available in the scientific literature dealing with the fatigue behavior of cement-based composites, such as unreinforced, steel bar reinforced and fiber-reinforced concretes [2–4]. Many of these studies exhibit a significant scatter in the fatigue performance which can be attributed to the variability of the materials nature, inner meso-structures, fibers distributions, hardening/softening mechanisms and testing conditions [5, 6].

Codes and design guidelines for concrete structures propose empirical-based rules for concrete-fatigue calculations. These laws are in most cases based on the well-known Wöhler concepts [7], where material input in the analyses are the so-called stress-life (S–N) curves. Research activities have mainly been addressed to the determination of these S–N curves. These approaches have been very valuable from a practical point of view, but completely avoid deeper explanations of the mechanisms driving the damage initiation and triggering of concrete-fatigue behavior [8].

In classical Wöhler curves, three different fatigue types can be differentiated: (i) Oligocyclic (O), (ii) Low (L) and (iii) High (H) Cycle Fatigue (CF). OCF is mainly due to very high stress levels. The material is hugely fractured and after very few number of cycles the crack propagation becomes unstable and failure is quickly reached. On the other hand, LCF and HCF are characterized by stable crack propagation processes. In LCF, the nucleation and failure processes are driven by the increase of meso-to-macro cracks evolving with every cycle. The number cycles of failure is commonly lower than 10^4. Constitutive models based on fatigue damage-based description mechanisms are commonly available in literature for this kind of fatigue description [9]. In HCF, the stress level to reach the failure is much lower than LCF and the accumulation of damage is driven by the generation of micro-cracks, at lower scales and under very high number of cycles, and their coalescence into macro-cracks up to failure. Main differences between these two is that HCF is characterized by remaining on the elastic range while LCF by the repeated plastic accumulations and inelastic work-spent cycle by cycle.

Constitutive models classically based on plasticity theory and damage mechanics, often suitable for capturing the nonlinear response of specimens under LCF, need to be extended for HCF [10]. The "number of cycles" is usually taken as a variable capable of modifying the internal plastic evolution, necessary to capture the strength loss occurring under HCF tests [11].

The present research is focused on a unified coupled damage-elastoplastic constitutive model for the numerical analysis of concrete specimens subjected to both LCF and HCF actions. A fracture-based damage model will be combined with microscopic mechanics for capturing damage accumulations in HCF.

2 Unified Fatigue Damage-Mechanisms Model Formulation

2.1 Main Assumptions

The theoretical model proposed to simulate the cyclic response of cement-based materials is based upon the following key assumptions:

- A unified coupled damage-elastoplastic constitutive model will account for both LCF and HCF;
- LCF: unloading/reloading stiffness (damage) degradation depends on the actual value of the "fracture work", spent in each interface point;
- HCF: a multiscale approach, and microcracking mechanisms, describes the HCF phenomena;
- Small displacements are assumed;
- Quasi-static loads (inertial forces can be ignored) are assumed; and lastly,
- Strain-rate effects are omitted.

2.2 Constitutive Response

A discontinuous-based approach, based on the use of zero-thickness interface elements (0IEs) model, for the numerical analysis of concrete specimens subjected to (both LCF and HCF) fatigue action (Fig. 1), is presented in a unified way.

A coupled damage-elasto-plastic constitutive equation is proposed as follows

$$\dot{\mathbf{t}} = \mathbf{C_d} \cdot \left(\dot{\mathbf{u}} - \dot{\mathbf{u}}^{cr} \right) \tag{1a}$$

$$\dot{\mathbf{u}} = \dot{\mathbf{u}}^{el} + \dot{\mathbf{u}}^{cr} \tag{1b}$$

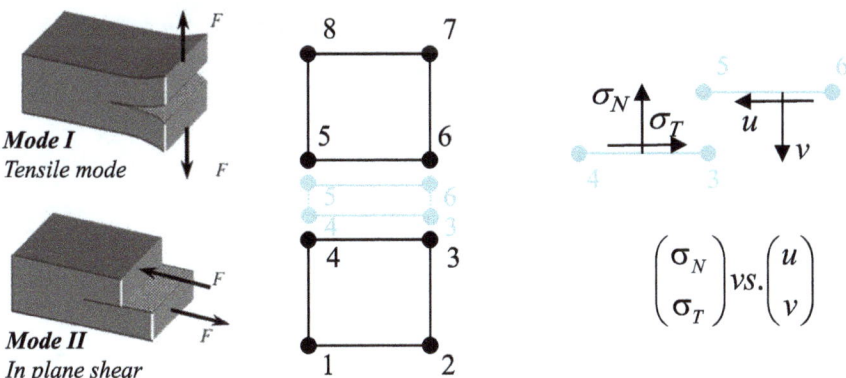

Fig. 1 Discontinuous-based approach, based on the use of zero-thickness interface elements

where $\dot{\mathbf{t}} = (\dot{\sigma}_N, \dot{\sigma}_T)^T$ is the interface stress rate vector, $\mathbf{C_d}$ the interface elastic-damage stiffness matrix, $\dot{\mathbf{u}} = (\dot{u}, \dot{v})^T$ is the interface relative displacement rate vector, with $\dot{\mathbf{u}}^{el}$ and $\dot{\mathbf{u}}^{cr}$ being their elastic and cracking (inelastic) components, respectively.

The effective interface stress, $\tilde{\mathbf{t}}$, is defined by [12]

$$\dot{\mathbf{t}} = (\mathbf{I} - \mathbf{D}) \cdot \dot{\tilde{\mathbf{t}}} \tag{2}$$

being $\mathbf{I}$ the second order identity matrix and $\mathbf{D}$ the damage matrix. This allows to express $\mathbf{C_d} = (\mathbf{I} - \mathbf{D}) \cdot \mathbf{C}$ in terms of the undamaged elastic matrix $\mathbf{C}$.

Particularly, the following global relationship is proposed to integrate the damage behavior under general cyclic events (i.e., either L- or H-CF)

$$\dot{\mathbf{D}} = \dot{\mathbf{D}}^{LCF} + \dot{\mathbf{D}}^{HCF} \tag{3}$$

where the timely accumulated damage can be evaluated as

$$\mathbf{D} = \int_{t_0}^{t_f} \dot{\mathbf{D}} \, dt \quad \in \quad [0, 1] \tag{4}$$

Low cycle fatigue

A fracture-based model combined with a continuous damage theory extension equipped the 0 IE in order to account for the LCF effects [13].

A cracking surface (Fig. 2) defines the (effective) stress level at which the post-elastic displacements starts

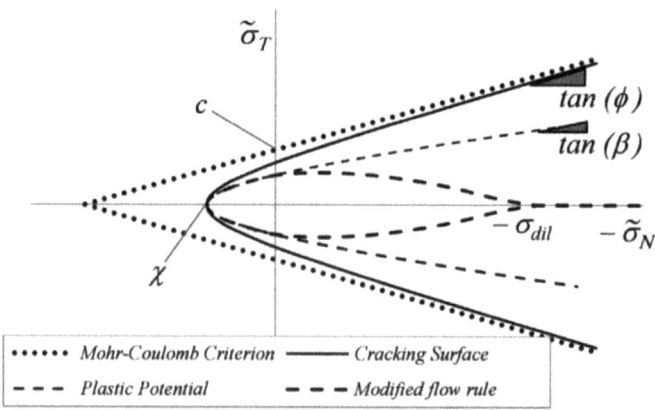

Fig. 2 Failure hyperbola [13], Mohr–Coulomb surface, plastic potential and modified flow rule

$$f = \tilde{\sigma}_T^2 - (c - \tilde{\sigma}_N \tan \phi)^2 + (c - \chi \tan \phi)^2 \qquad (5)$$

where χ is the tensile strength, c the cohesion and ϕ the frictional angle.

The vector of crack displacement rates $\dot{\mathbf{u}}^{cr}$ can be defined following a non-associated flow rule as

$$\dot{\mathbf{u}}^{cr} = \dot{\lambda}\mathbf{m} \qquad (6a)$$

$$\mathbf{m} = \mathbf{A} \cdot \mathbf{n} \qquad (6b)$$

where $\dot{\lambda}$ is the non-negative plastic multiplier, deriving from the classical Kuhn-Tucker ($\dot{\lambda} \geq 0$, $f \leq 0$, $\dot{\lambda}f = 0$) and consistency ($\dot{f} = 0$) conditions, $f = f[\tilde{\sigma}_N, \tilde{\sigma}_T]$ is the yield function, $\mathbf{m}$ is the vector controlling the direction of the fracture displacements by means of a non-associated flow rule, and finally $\mathbf{n} = \frac{\partial f}{\partial \mathbf{t}} = \left(\frac{\partial f}{\partial \tilde{\sigma}_N}, \frac{\partial f}{\partial \tilde{\sigma}_T}\right)^T$.

The softening post-cracking behavior is driven by scaling functions based on the ratio between the plastic work spent W_{cr} and the available fracture energies (G_f^I or G_f^{IIa})

$$\xi_{p_i} = \begin{cases} \frac{1}{2}\left[1 - \cos\left(\pi \frac{W_{cr}}{G_f^{\#}}\right)\right] & \text{if } W_{cr} \leq G_f^{\#}, \ \# = I \text{ or } IIa \\ 1 & \text{otherwise} \end{cases} \qquad (7)$$

where p_i alternatively represents χ, c and $\tan \phi$.

Finally, the following scalars describe the damage in LCF

$$d^{\#} = e^{-\beta_d}\left(\frac{W_{cr}}{G_f^{\#}}\right)^{\alpha_d}\left[1 + (e^{-\beta_d} - 1)\left(\frac{W_{cr}}{G_f^{\#}}\right)^{\alpha_d}\right]^{-1}, \quad \# = I, IIa \qquad (8a)$$

$$\mathbf{D}^{LCF} = \begin{pmatrix} d^I & 0 \\ 0 & d^{IIa} \end{pmatrix} \qquad (8b)$$

with $\alpha_d \geq 0$ and β_d representing the parameters controlling the decay shape of the damage induced by fracture.

High cycle fatigue

Plastic strains and/or cracks are almost negligible in HCF failures when compared against LCF results. Therefore, classical approaches based on fracture energy, plasticity theory and damage cannot be used in this case. Instead, microscopic

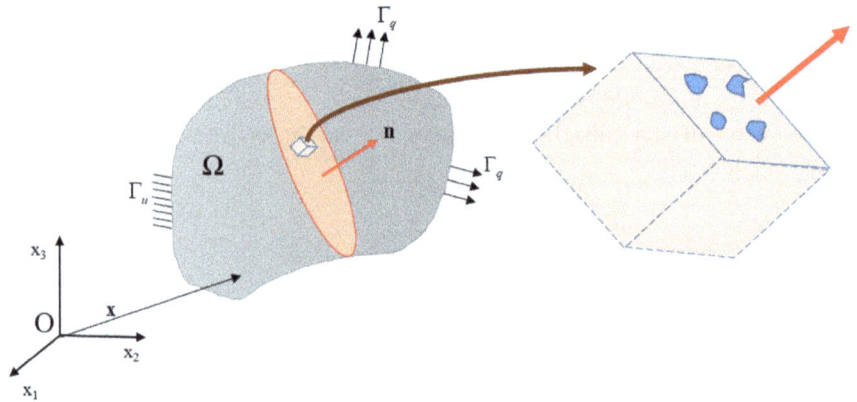

Fig. 3 Two scale approach for HCF

plastic strains and cracks are the main responsible of fatigue damage accumulation under HCF. After huge cycle reversals they mainly increase and coalesce into few macro-cracks [14].

Based on this idea, a two scale approach (Fig. 3) is adopted where no plastic strains take place at the macro-scale while microplastic strains (cracks) are modelled at the micro-level.

Hollomon´s hardening power law is assumed at the microscale

$$\tilde{\sigma}_{eq} = K\delta_{micro}^{M} \tag{9a}$$

$$\tilde{\sigma}_{eq} = \begin{cases} \sqrt{(\tilde{\sigma}_N)^2 + (\tilde{\sigma}_T)^2} & \tilde{\sigma}_N \geq 0 \\ \tilde{\sigma}_T - |\tilde{\sigma}_N|\tan(\phi) & \tilde{\sigma}_N < 0 \end{cases} \quad \wedge \quad \delta_{micro} = \begin{cases} \sqrt{(u_{micro}^{cr})^2 + (v_{micro}^{cr})^2} & \tilde{\sigma}_N \geq 0 \\ v_{micro}^{cr} & \tilde{\sigma}_N < 0 \end{cases} \tag{9b}$$

being K the strength index (or strength coefficient) and M a crack-hardening exponent, which in the most general case will be frequency-dependent: $M = M[f]$; u_{micro}^{cr} and v_{micro}^{cr} are the cracking microscale relative displacements, in the normal and tangential direction, respectively.

Following an isotropic damage approach, it can be stated that

$$\frac{\sigma_{eq}}{1-D} = K\delta_{micro}^{M} \quad \Rightarrow \quad \delta_{micro} = \frac{\sigma_{eq}^{1/M}}{(1-D)^{1/M}K^{1/M}} \quad \Rightarrow \dot{\delta} = \frac{\sigma_{eq}^{1/M-1}}{M(1-D)^{1/M}K^{1/M}}\dot{\sigma}_{eq} \tag{10}$$

Then, the microscopic damage description can be done by means of the following rule

$$\dot{D} = \begin{cases} 0 & \text{if } \frac{\sigma_{eq,m}-\sigma_{eq,m}}{2} \leq \sigma_f \\ L_{cl^{-1}}\frac{Q[f]\sigma_{eq}^{q[f]}}{(1-D)^{q[f]+1}}\dot{\sigma}_{eq} & \text{if } \frac{\sigma_{eq,m}-\sigma_{eq,m}}{2} > \sigma_f \end{cases} \tag{11a}$$

Fig. 4 Representation of the fatigue stress limit

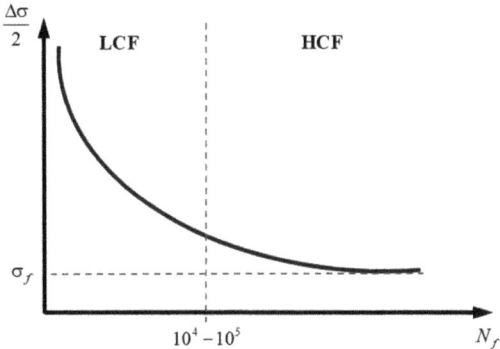

$$\mathbf{D}^{\mathbf{HCF}} = \begin{pmatrix} \dot{D} & 0 \\ 0 & \dot{D} \end{pmatrix} \qquad (11b)$$

being $q[f] = 1/M[f] - 1$, $Q[f] = K^{-1/M[f]}/M[f]$ and $L_{cl^{-1}}$ (keeping in mind that $\dot{D} = L_{cl^{-1}}\dot{\delta}_{micro}$) a characteristic length value.

The latter three variables are mainly aiming at representing the link between the microscale and the macroscale; $\sigma_{eq,M}$ and $\sigma_{eq,m}$ are the "*sup*" and "*inf*" values of σ_{eq} under cyclic protocols, while σ_f is a fatigue stress limit to be calibrated and shown in Fig. 4.

3   Numerical Example

This section is aimed at validating the proposed model by performing and comparing numerical examples against experimental data available in literature on concrete specimens submitted to fatigue loads.

The numerical results deal with the simulation of concrete flexural fatigue tests, with a target strength of approximately 40 MPa. The experimental data are available in Oh [15] who tested un-notched concrete beams (100 × 100 × 500 mm) under four-point bending. The fatigue protocol was designed considering that the bottom fiber stress varies from zero to a predetermined maximum stress. The maximum cyclic stress ratio, namely f_R^{MAX}/f_{mon}, was analyzed in each series to reach the fatigue failure after a certain number of reversals, namely $\mathbf{N_{FAILURE}}$.

For the calibration purpose of the numerical examples, key geometric and material properties were chosen according to the experimental evidences [15], while only two model parameters were fitted:

$$q[f] = 24.30; \quad \frac{q[f]+1}{2(q[f]+2)Q[f]L_{cl^{-1}}} = 2.1\,\text{MPa} \qquad (12)$$

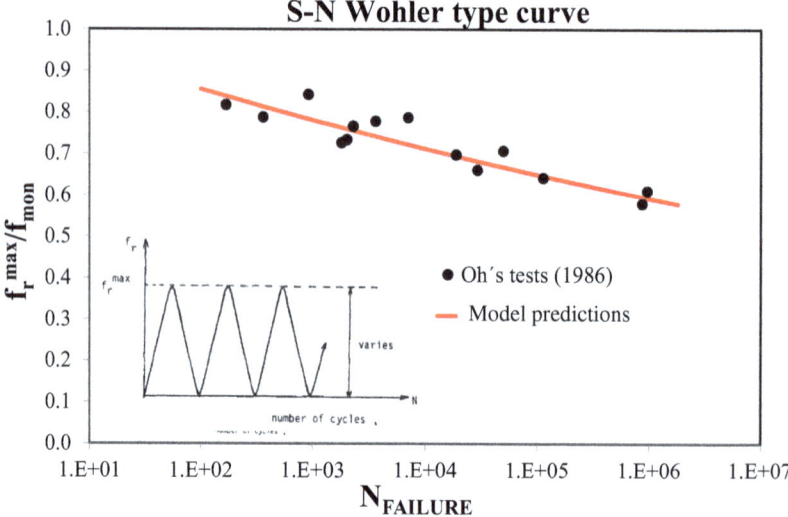

Fig. 5 Numerical examples versus the experimental tests by Oh [15]

For the model integration, the following assumptions were considered: $D = 0$ (when the number of cycles is null) and $D = 1$ reaching the fatigue failure (i.e. $N = N_{FAILURE}$).

Numerical S–N curves can be predicted from this approach and compared with experimental data as shown in Fig. 5. Results are promising in predicting the experimental data and actually follow the well-known Wöhler expression for describing the S–N curve:

$$f_R^{MAX}/f_{mon} = a + b \log_{10} N_{FAILURE} \qquad (13)$$

The capability of the proposed formulation to reproduce the strong sensitivity of concrete mechanical behavior on the stress level affecting the HCF is hereby demonstrated.

4 Conclusions

A unified numerical approach was proposed for analyzing low- and high-cycle fatigue actions in concrete. The discontinuous-based model is aimed at describing the evolving micro- and meso-structural changes caused by cyclic protocols by employing a damage accumulation rule. Concepts of fracture-energy theory, damage stiffness degradations and microscopic inelasticity were combined for predicting failure mechanisms occurring in concrete under cyclic responses.

Acknowledgements The DFG Priority Program SPP 2020 Project "Cyclic Damage Processes in High-Performance Concretes in the Experimental Virtual Lab" is gratefully acknowledged. The first author wishes also to acknowledge the AvH Foundation (www.humboldt-foundation.de/) for funding his position at the TU-Darmstadt under the grant ITA-1185040-HFST-P (2CENERGY project).

References

1. Simon, K.M., Chandra Kishen, J.M.: A multiscale approach for modeling fatigue crack growth in concrete. Int. J. Fatigue **98**, 1–13 (2017)
2. Makita, T., Brühwiler, E.: Tensile fatigue behaviour of ultra-high performance fibre reinforced concrete (UHPFRC). Mater. Struct. **47**(3), 475–491 (2014)
3. Makita, T., Brühwiler, E.: Tensile fatigue behaviour of ultra-high performance fibre reinforced concrete combined with steel rebars (R-UHPFRC). Int. J. Fatigue **59**, 145–152 (2014)
4. Wille, K., El-Tawil, S., Naaman, A.E.: Properties of strain hardening ultra-high performance fiber reinforced concrete (UHP-FRC) under direct tensile loading. Cement Concr. Compos. **48**, 53–66 (2014)
5. Abbas, S., Nehdi, M.L., Saleem, M.A.: Ultra-high performance concrete: mechanical performance, durability, sustainability and implementation challenges. Int. J. Concr. Struct. Mater. **10**(3), 271–295 (2016)
6. Afroughsabet, V., Biolzi, L., Ozbakkaloglu, T.: High-performance fiber-reinforced concrete: a review. J. Mater. Sci. **51**(14), 6517–6551 (2016)
7. Wöhler, A.: Über die Festigkeitsversuche mit Eisen und Stahl. Zeitschrift für Bauwesen XX, pp. 73–106 (1870)
8. Hordijk, D.A., Reinhardt, H.W.: Numerical and experimental investigation into the fatigue behavior of plain concrete. Exp. Mech. **33**, 278–285 (1993)
9. Darabi, M.K., Al-Rub, R.K.A., Masad, E.A., Little, D.N.: Constitutive modeling of fatigue damage response of asphalt concrete materials with consideration of micro-damage healing. Int. J. Solids Struct. **50**(19), 2901–2913 (2013)
10. Lubliner, J.: Plasticity theory. Dover, Mineola, NY (2008)
11. Turon, A., Costa, J., Camanho, P.P., Dávila, C.G.: Simulation of delamination in composites under high-cycle fatigue. Compos. A **38**, 2270–2282 (2007)
12. Lemaitre, J.: How to use damage mechanics. Nucl. Eng. Des. **80**(2), 233–245 (1984)
13. Caggiano, A., Etse, G., Ferrara, L., Krelani, V.: Zero-thickness interface constitutive theory for concrete self-healing effects. Comput. Struct. **186**, 22–34 (2017)
14. Harenberg, S., Caggiano, A., Koenig, A., Said, D., Gilka-Bötzow, A., Koenders, E.: Micromechanical behavior of UHPC under cyclic bending-tensile loading in consideration of the influence of the concrete edge zone. PAMM **18**, e201800363 (2018)
15. Oh, B.H.: Fatigue analysis of plain concrete in flexure. J. Struct. Eng. **112**, 273–288 (1986)

Deployment of a High Sensor-Count SHM of a Prestressed Concrete Bridge Using Fibre Optic Sensors

F. I. H. Sakiyama, F. Lehmann, and H. Garrecht

Abstract To deploy an accurate safety-relevant structural health monitoring, one must assure the utmost care and in-depth knowledge of the monitored structure, which may present challenges such as the reliability to detect unexpected anomalies due to the failure of a component, the correct setting of thresholds and triggers to discern changes due to environmental conditions from critical events, and the high expense in terms of hardware and personnel availability. The fibre optic (FO) technology can provide integrated sensing along with extensive measurements lengths with high sensitivity, durability, and stability, which makes them ideal for SHM of concrete structures. Therefore, an SHM system using quasi-distributed FO FBG sensors is proposed to continuously monitor the strain changes of a 57 m long prestressed concrete bridge due to traffic loads and environmental changes. A total of 89 long-gauge strain sensors were installed to monitor the strain distribution in two lines along the complete length and five arrays in the shear direction. Additionally, 2 FO acceleration sensors and 94 FO FBG temperature sensors were installed for correct and precise temperature compensation of the strain sensors and to correctly detect the strain changes due to temperature variation on the bridge. In this work, the installation processes of the FO sensors and the operational hardware is shown. Furthermore, initial measurement values are presented to demonstrate the potential of FO to provide a reliable SHM system to monitor large concrete structures.

Keywords Structural health monitoring · Fibre optic sensors · Fibre bragg gratings · Prestressed concrete maintenance

F. I. H. Sakiyama (✉) · F. Lehmann · H. Garrecht
Materials Testing Institute, University of Stuttgart, Stuttgart, Germany
e-mail: felipe.sakiyama@ufvjm.edu.br

F. I. H. Sakiyama
Federal University of the Jeq. and Muc. Valleys, Teófilo Otoni, Brazil

© The Author(s), under exclusive license to Springer Nature Switzerland AG 2022
J. Sena-Cruz et al. (eds.), *Proceedings of the 3rd RILEM Spring Convention and Conference (RSCC 2020)*, RILEM Bookseries 34,
https://doi.org/10.1007/978-3-030-76465-4_17

1   Introduction

1.1   *Overview*

The early detection of cracks in concrete structures is a common practice within building inspection procedures, as it can be a good indicator for necessary maintenance or reinforcement measures of damaged structures. Likewise, documenting any progress of existing cracks is essential to differentiate between static and dynamic damage states, and thus to enable the assessment of the cracks relevance concerning durability and stability; therefore, existing and new cracks and their lengths are recorded during periodic building inspections.

The interval of such on-site inspections is usually every six years, but it can vary between different standards and according to the current structure condition. Although this has worked well in the past, every year the traffic load increases on old bridges that have been reaching its lifespan or which are known not to meet today's demands both on the load capacity and in terms of the construction method and the materials used [1].

Under these circumstances, significant damage can go unnoticed between inspections, with an increasing probability every year. Also, the periodic inspection depicts the state of the building only at the time of inspection and is influenced by the inspector's subjectivity [9]; hence, particularly high loads in the period between, e.g. from heavy traffic events or temperature, are not taken into account unless this has already caused visible damage. Periodic inspections alone are insufficient for maintaining the health of structures and assuring the users' safety [2].

Structural health monitoring (SHM) offers a way to supplement regular building inspections [10]. For this purpose, sensors are installed at suitable locations on the building, which continuously record, save and transmit information about the structure's behaviour. Ideally, this data is automatically evaluated, and an alarm is triggered when specified limit values are exceeded. However, this often proves to be difficult in practice, since the real structural behaviour, particularly in the case of load redistribution and crack formation, deviates from the underlying idealised model, and therefore cannot be evaluated solely by software. Thus, limit values are often set based on the usual fluctuation range of the measured parameters based on the previous continuous monitoring, and the measurement data is interpreted manually.

The selection of suitable sensors takes place in the balance between spatial resolution, number of sensors and sensitivity. For example, the natural frequencies of a building can be determined with just a few vibration velocity sensors, and global changes in load behaviour can be inferred from this; however, damage to the structure must already be considerable. In contrast, strain gauges, e.g., offer the possibility of detecting the smallest changes in crack widths or local strains; however, an extraordinarily large number of sensors would be necessary to be able to monitor the entire structure.

Although the current approaches of SHM systems using traditional single-point sensors—such as electric strain sensors, accelerometers, and GPS-based sensors—have appropriate measurement precision for SHM purposes [8], they present challenges when deployed in real scale applications [6], given the limited number of possible points to assess the structural behaviour and the harsh environmental conditions during operation [4]. When it comes to reinforced concrete structures, the development of health monitoring and damage identification presents further challenges, since this type of structure is affected by a variety of chemical, physical and mechanical degradation processes, and has a heterogeneous composition and nonlinear behaviour [8]. On the other hand, the fibre optic (FO) technology can provide integrated sensing along with extensive measurements lengths with high sensitivity, durability, and stability, which makes them ideal for SHM of concrete structures [7].

The Fibre Bragg Grates (FBG) based FO sensors, for instance, offer new approaches, in particular for the detection of new cracks, since it enables high-resolution (~1 μm) strain monitoring over discrete segments. Within these segments, crack openings, e.g., can be detected without the previous knowledge of their locations.

Many researches have already been done on the topic of crack detection in concrete, but the application on real structures is still missing. This project aims to investigate the possibilities and limits of fibre-optic building monitoring, especially concerning the detection of cracks. Therefore, an SHM system using quasi-distributed FO FBG sensors is proposed to continuously monitor the strain changes of a 57 m long prestressed concrete bridge due to traffic loads and environmental changes. In this work, the installation processes of the FO sensors and the operational hardware is shown. Furthermore, initial measurement values are presented to demonstrate the potential of FO to provide a reliable SHM system to monitor large concrete structures.

1.2 FBG Sensing Technology

An FBG sensor is a microstructure with 5 to 10 mm in length [11] and consists of a series of evenly spaced etchings at a tuneable distance Λ from each other, written using a UV laser in the core of a standard telecom FO [5]. When the light passes through the FO, a portion of the light is reflected at each etching, where the spacing of the etchings causes the reflected light to have a different phase for most wavelengths; thus, destructive interference causes the individual reflections to cancel each other out. However, the wavelength harmonic to the etching spacing will be reflected in phase and experience constructive interference. Hence, the reflected spectrum λ_B (or Bragg wavelength) will contain essentially one wavelength, which can be directly related to the etching spacing [3], as illustrated in Fig. 1. If the FO is subjected to external loads, e.g., strain or temperature variation, the fibre's length and consequently the spacing between the gratings will change, causing a shift in the Bragg wavelength. With the proper calibration, the external load can be quantified from the Bragg wavelength shift.

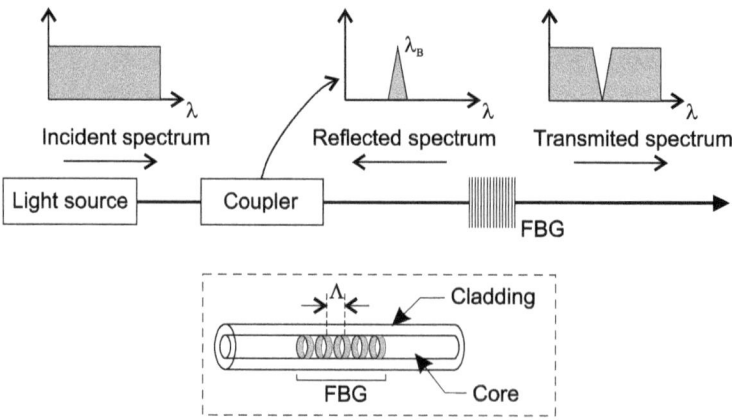

Fig. 1 Measurement principle of the FBG sensor [13]

The long-gauge FBG sensors (LGFBG), also known as quasi-distributed FO sensors, are characterised by their in-lie multiplexing feature, wherewith a series of individual sensors can be connected and measured in a single fibre-array using techniques such as wavelength-division multiplexing (WDM). When combined with the long-gauge attribute, the integral length of a structure can be monitored with a distributed, yet discrete, array of sensors; thus the name "quasi-distributed". The discretisation resolution, i.e. the smallest distance wherein changes can be measured, is determined by the gauge's length of the sensors, ranging from millimetres up to 10 m long [12].

More details about FBG sensors and their application can be found in [10].

2 SHM System

2.1 Monitored Structure Characterisation

The monitored structure is a prestressed hollow-core concrete bridge built in 1964. The design load class is 60/30, according to DIN 1072. With a width of 11.08 m, it has three continuously spans with a total length of 57.00 m (17.00–23.00–17.00 m) without coupling joints (Figs. 2 and 3). The two centre columns are designed as individual supports with pot bearing. On the southern abutment, the superstructure is supported by two linear rocker bearings, and on the northern abutment, by two roller bearings.

Like most of the prestressed concrete structures designed until the '70s in Germany, the bridge was built with prestressing steel types St 145/60 Sigma, KA 141/40 and KA35/10, which are known for their high vulnerability to corrosion-induced cracking.

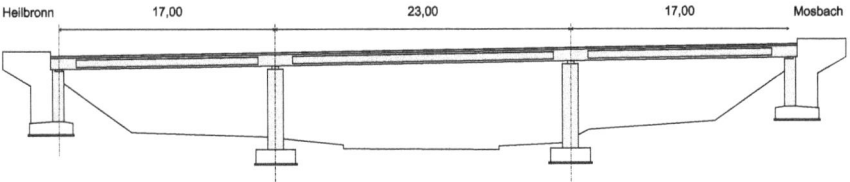

Fig. 2 Longitudinal view of the bridge (dimensions in meter)

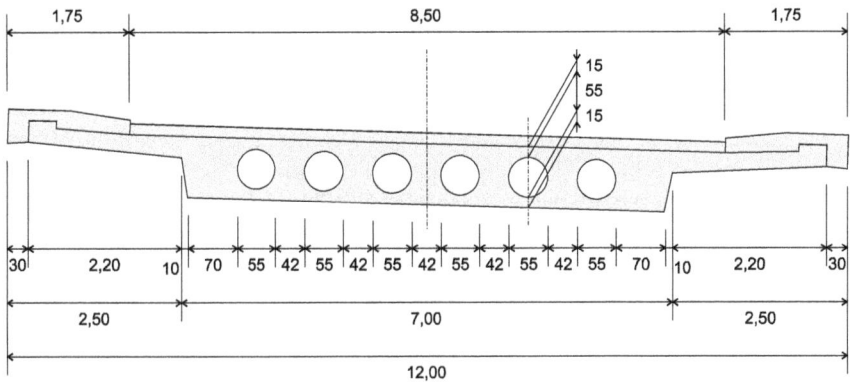

Fig. 3 Bridge's cross-section (dimensions in centimetre)

In addition to the high increase in traffic loads compared to the year of construction in 1964, and the corrosion-induced cracking risk, other critical problems may arise due to construction methods and the design standards adopted back then. Construction failures can already appear during construction caused by misplacement of the hollow-core bodies, and difficulties in compacting the surrounding concrete. From the structural point-of-view, the hollow-core bodies prevent two-axis load transfer and thus in the redistribution of forces in the transversal direction. Likewise, shear forces and temperature loads were not taken into account to the extent that it is considered necessary from today's standards at the time the building was planned. Finally, the hollow-core cannot be examined as part of the building inspection, which means that any damage inside them may not be early detected.

2.2 Monitoring System

A fibre-optic monitoring system based on long-gauge FBG (LGFBG) sensors were installed to continuously monitor strain and temperature changes, and vibration of the bridge superstructure.

The strain monitoring consists of two parallel measuring lines, each with 27 LGFBG sensors along the complete longitudinal direction, and five measuring lines in the shear direction, with seven LGFBG sensors each. For every LGFBG strain sensor, an embedded temperature sensor is present for temperature compensation on the FO. Besides, the concrete temperature is monitored at five different points in the middle bridge field, and the acceleration is measured at 2 locations. The sensors were divided into eight quasi-distributed arrays equipped with redundancy connection fibres, which enable measurements to be continued if a primary connection cable fails. A schema of the sensors is given in Fig. 4 and overview photos are shown in Fig. 5. The following sensors are installed on the structure:

– Strain in the longitudinal direction - sensors S01-S54 and S80-89: For monitoring in the longitudinal direction, two distributed strain sensor-measuring lines are attached to the bottom of the bridge, located in the area of the prestressed cables. The sensors S80 to S89 were installed on the side of the structurer at about 50 cm above the lower surface. The sensors S01 to S54 have a gauge length of 2.05 m, and the sensors S80 to S89 a gauge length of 0.50 m.
– Strain in the transverse direction – sensors S55-S79: Five strain sensor-measuring lines across the cross-section were installed on the underside of the bridge in the area of the maximum bending moments and on the two supports. These sensors have a gauge length of 1.35 m.
– Temperature sensors T01-T05: temperature sensors in the middle of the bridge, transverse to the direction of travel.

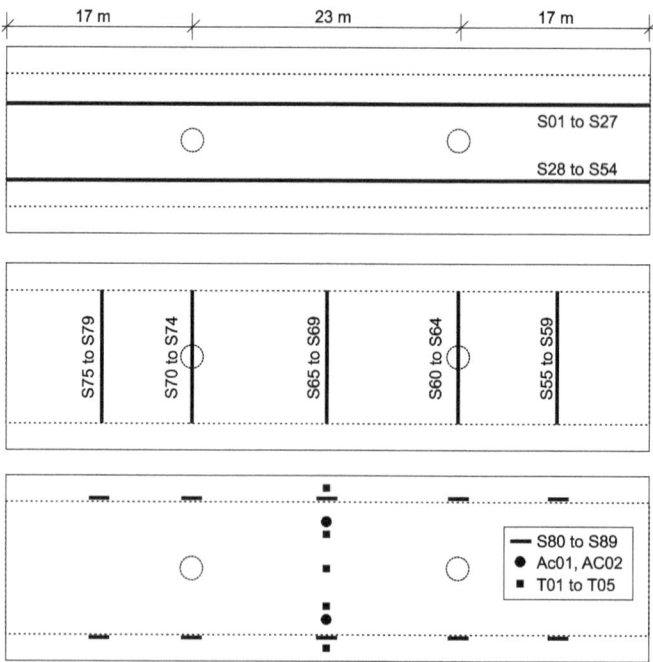

Fig. 4 Schema of the sensors' configuration (bottom view of the superstructure)

a. Overview of the bridge.

b. View of the configuration of the sensor

c. View of the configuration of the sensor

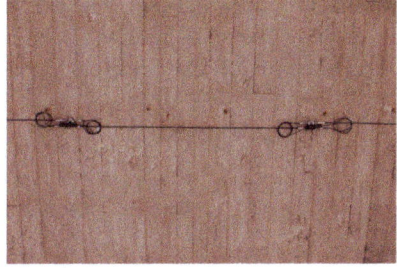

d. An LGFBG sensor with 2.05 m gauge.

Fig. 5 Overview of the bridge and the monitoring system

– Acceleration - sensors AC01-AC02: Two accelerometers with a vertical measuring direction in the middle of the main field each centred between the main axis and the edges of the support plate.

The optical interrogator was installed together with an industrial computer with an embedded LTE modem, a power bank supply to sustain power cuts up to 30 min, and additional components in a temperature-controlled control cabinet on the southern abutment. The measuring system can be entirely operated via remote access, which also enables the transfer of all stored files.

The measurement has been running continuously since 8th October 2019, with a measurement frequency of 200 Hz. All measurement data are available at any time for live visualisation. The data saving is only carried out if the measured change in strain is higher than a specified trigger. Also, a report is automatically generated every 15 min, in which the maximum, minimum and average values of all sensors are documented.

The monitoring system generates a considerable amount of data, given its high-frequency rate and the high sensor-count. To improve data management and the analysis of the results, the measured data is automatically stored in a MySQL database. The SQL database is accessed using customised MATLAB scripts written specifically for this application, where all the analysis and results, including graphics, are automatically processed. After that, older measured data can be deleted to free up memory space.

3 First Measurement Data

The strain-variation measured by the longitudinal sensors during the crossing of a truck over the bridge is shown in Fig. 6. The graphics Fig. 6a–c represent the line of sensors at one side of the superstructure, while the graphic Fig. 6d–f the line of sensors on the opposite side (refer to). The vehicle was travelling in the direction from S01 to S27. The total duration of the measurement is 4 s, and it was acquired with a frequency of 200 Hz. Each curve presented in the graphics can be understood as

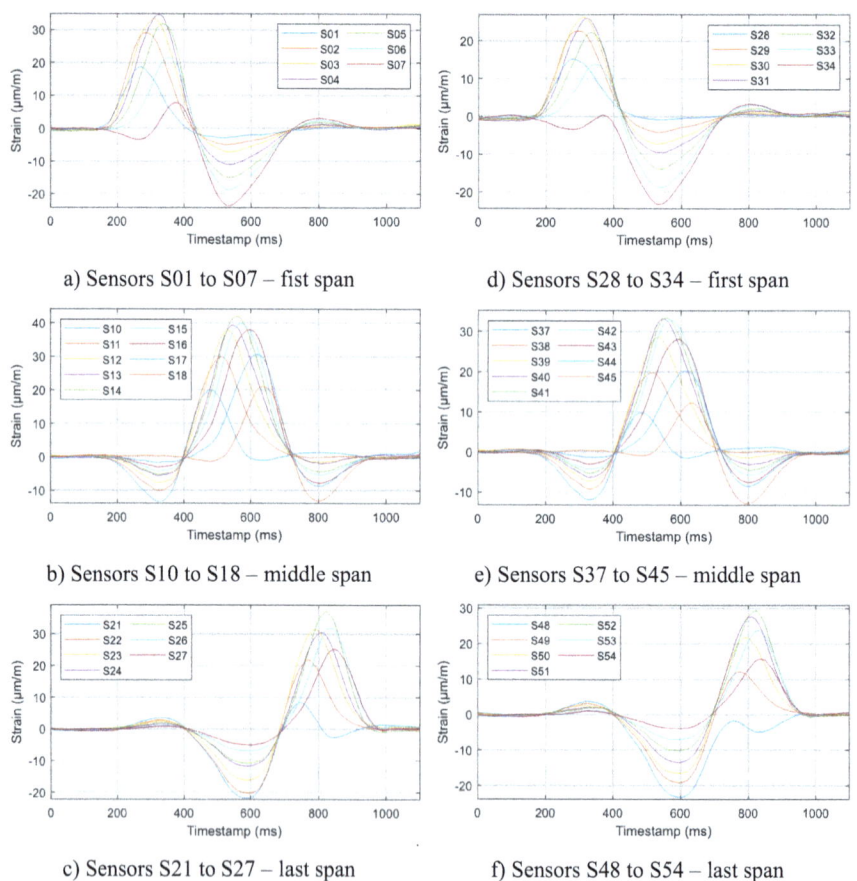

a) Sensors S01 to S07 – fist span

d) Sensors S28 to S34 – first span

b) Sensors S10 to S18 – middle span

e) Sensors S37 to S45 – middle span

c) Sensors S21 to S27 – last span

f) Sensors S48 to S54 – last span

Fig. 6 Dynamic strain measurement of the LGFBG sensors located under the spans in the longitudinal direction during the crossing of a truck over the bridge

the strain influence line of the bridge sections corresponding to each of the installed sensors.

When the vehicle enters the bridge, the sensors S01 to S07 and S28 to S34, located under the first span, displayed peaks of positive strain, while the sensors S10 to S18 and S37 to S45, located in the middle span, measured negative strain, and, finally, all the sensors under the last span displayed values close to zero. When the vehicle passes over to middle span, it can be noted an inversion of the values' signal for all sensors, where now the sensors in the middle span display positive peaks and the sensors on both side-spans negative peaks.

The sensors also tend to converge to zero at the same timestamp when the vehicle passes above the two columns, around timestamps 400 and 700 ms, which is expected since the columns absorb the vehicle's load when it is standing over the column. At last, it is observed symmetric behaviour between the sensors in Fig. 6a–d on the first span and the sensors in Fig. 6c–f on the last span, as the vehicle crosses it and leaves the bridge superstructure.

In Fig. 7 is shown the maximum and minimum strain diagram for the longitudinal sensors. The two lines of LGFBG longitudinal sensors are represented separately in Fig. 7a for sensors S01 to SS27, and in Fig. 7b for sensors S28 to S54. The amplitude, i.e. the difference between the maximum and the minimum strains, is plotted vertically for each sensor. Three positive strain peaks occur in the central region on each of the three spans, whereas the two negative peaks arise where the two centre columns are located. The maximum measured strain is 44.11 μm/m on sensor S14, the minimum -24.36 μm/m on sensors S34, and the maximum strain amplitude is 50.5 μm/m on sensor S25.

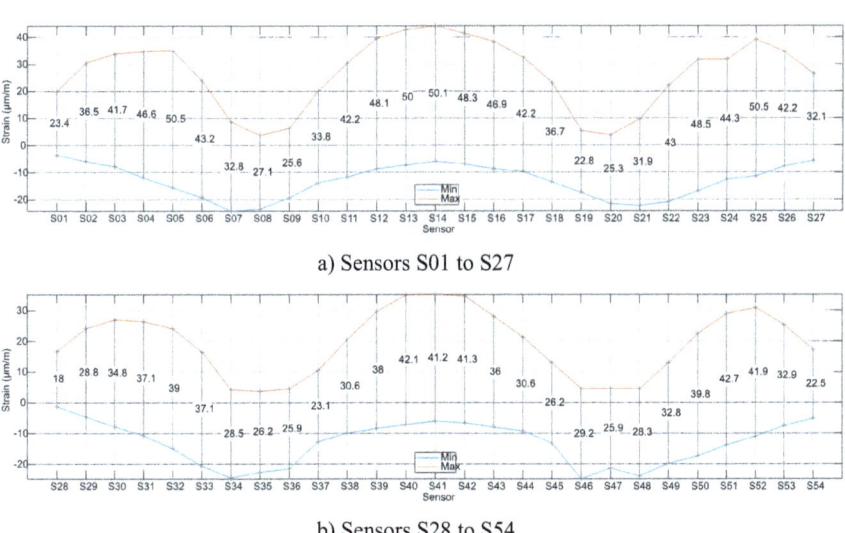

a) Sensors S01 to S27

b) Sensors S28 to S54

Fig. 7 Maximum and minimum strain diagram for the longitudinal LGFBG sensors during the crossing of a truck over the bridge

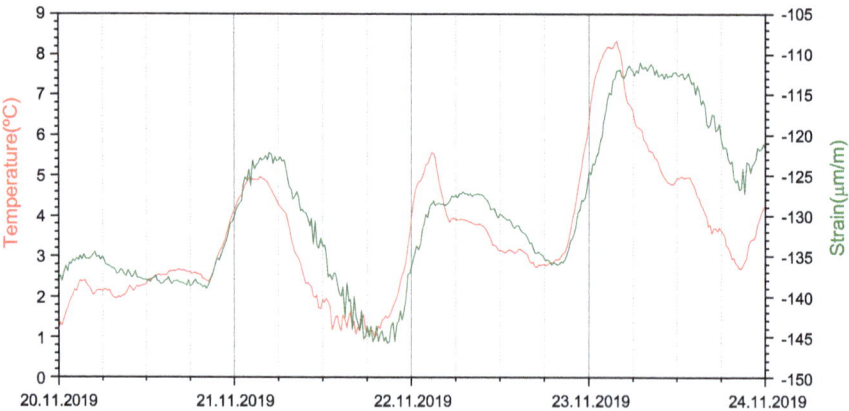

Fig. 8 Influence of the temperature variation on the measured strain during four days

Figure 8 shows an example of the measured strains and temperatures on sensor S41 over four days. It can be seen that the strain variation follows the temperature curve. The phase shift and the damping of individual measurement peaks from solar radiation are a consequence of the temperature inertia of the reinforced concrete. So far, it has been found that the changes in elongation at all sensors due to the temperature fluctuations, with a maximum of 180 μm/m so far, are significantly bigger than the strains from traffic loads with a maximum of 40 μm/m so far. For each Celsius temperature increase, the bridge stretches approx. 8 μm/m in the longitudinal direction, meaning a coefficient of thermal expansion of 8×10^{-6} °C^{-1}, which is in the range of the literature value for concrete according to DIN 1045–1 (between 5 and 14×10^{-6} °C^{-1}).

Figure 9 shows the single-sided amplitude spectrums for the sensors AC01 on two different events where a large truck drove over the bridge. The FFT was done using 1024 samples taken from a measurement at 200 Hz and smoothed using a Hanning window. It can be seen two prominent peaks at 45.3 Hz in the first event and 43.4 Hz in the second event.

The authors believe that the reliability and precision of the measure data combined with the continuously high-frequency measurements will allow an in-depth analysis of the structure's actual condition and remaining lifetime. With algorithms such as the rain flow analysis it is possible to access the more repetitive, with smaller amplitude load cycles due to the traffic and the more spaced, with greater amplitude load cycles due to the temperature shift along the year. The comprehensive strain and temperature distribution in both the longitudinal and the transversal direction enable the calibration of numerical models to identify local degradation of the structure's stiffness, and thermos-energetic models to a better understanding of the temperature influence within the bridge.

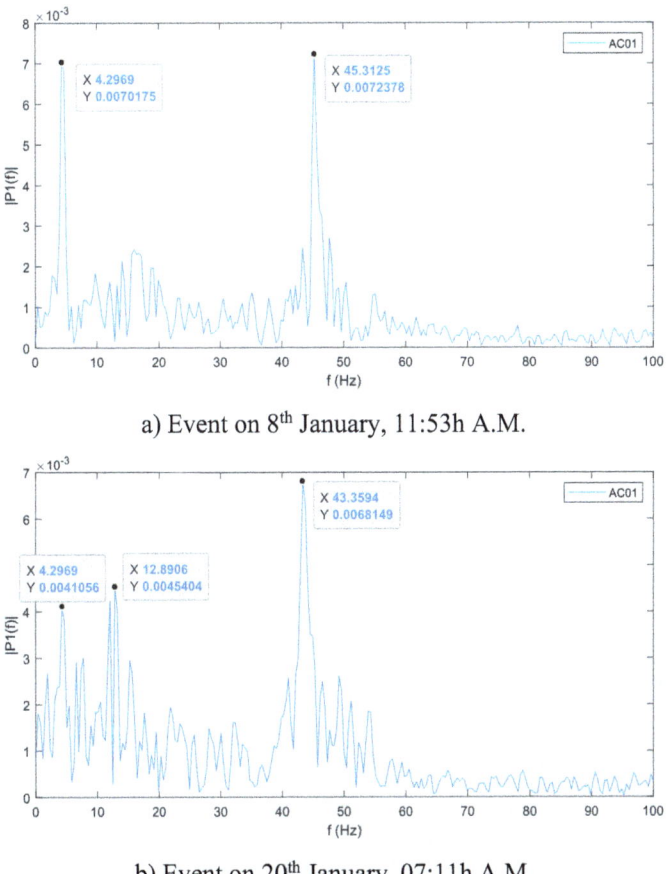

a) Event on 8th January, 11:53h A.M.

b) Event on 20th January, 07:11h A.M.

Fig. 9 Single-Sided Amplitude Spectrum for sensor AC01 during the crossing of a truck over the bridge in two different instances

4 Conclusion

This work presented the deployment of a large-scale application of fibre optic-based sensors for structural health monitoring of a prestressed concrete bridge, and its initial results obtained during operation.

The use of long-gauge FBG sensors has the potential to delivery in-depth knowledge of the strain distribution and the structural behaviour of large civil infrastructures. The dynamic measurements allow the representation of the strain influence line in each section of the bridge covered by a sensor and can be used for the detection of local damages, such as crack openings and the rupture of prestressed cables, as well as the influence of temperature variation loads and the structure's remaining lifetime.

Due to the continuous monitoring, the system can be set-up to notify extreme events, such as an abrupt shift of the strain measurement, increasing public safety.

The high resolution of the measurement data along with numerical analysis would allow an estimation of the actual traffic loads and damage state on the bridge after calibration.

Acknowledgements The authors would like to acknowledge the Brazilian Federal Agency for Support and Evaluation of Graduate Education (CAPES, finance code 001), the Federal University of the Jequitinhonha and Mucuri Valleys (UFVJM, Brazil), the Materials Testing Institute University of Stuttgart (MPA, Germany), and the Regional Council of Stuttgart (RPS, Germany), which collectively funded this work.

References

1. An, Y., Spencer, B.F., Ou, J.: A test method for damage diagnosis of suspension bridge suspender cables. Computer Aided Civil Infrast. Eng. **30**(10), 771–784 (2015). https://doi.org/10.1111/mice.12144
2. Cho, S., Giles, R.K., Spencer, B.F.: System identification of a historic swing truss bridge using a wireless sensor network employing orientation correction. Struct. Control Health Monit. **22**(2), 255–272 (2015). https://doi.org/10.1002/stc.1672
3. Giles, R.K., Spencer Jr, B.F.: Development of a long-term, multimetric structural health monitoring system for a historic steel truss swing bridge (2015)
4. Glisic, B., Inaudi, D.: Development of method for in-service crack detection based on distributed fiber optic sensors. Struct. Health Monit. **11**(2), 161–171 (2012). https://doi.org/10.1177/1475921711414233
5. Huang, J., Zhou, Z., Zhang, D., Yao, X., Li, L.: Online monitoring of wire breaks in prestressed concrete cylinder pipe utilising fibre Bragg grating sensors. Measurement **79**, 112–118 (2016). https://doi.org/10.1016/j.measurement.2015.10.033
6. Kudva, J.N., Marantidis, C., Gentry, J.D., Blazic, E.: Smart structures concepts for aircraft structural health monitoring. In: SPIE Proceedings. SPIE, pp. 964–971 (1993). https://doi.org/10.1117/12.152828
7. Lopez-Higuera, J.M., Cobo, L.R., Incera, A.Q., Cobo, A.: Fiber optic sensors in structural health monitoring. J. Lightwave Technol. **29**(4), 587–608 (2011). https://doi.org/10.1109/JLT.2011.2106479
8. Mahadevan, S., Adams, D., Kosson, D.: Challenges in concrete structures health monitoring. In: Daigle, M.J., Bregon, A. (eds.) Annual Conference o the Porgnostics and Health Management Society 2014. Fort Worth, TX, USA, pp. 561–567 (2014)
9. Phares, B.M., Washer, G.A., Rolander, D.D., Graybeal, B.A., Moore, M.: Routine highway bridge inspection condition documentation accuracy and reliability. J. Bridg. Eng. **9**(4), 403–413 (2004). https://doi.org/10.1061/(ASCE)1084-0702
10. Sakiyama, F.I.H., Lehmann, F.A., Garrecht, H.: Structural health monitoring of concrete structures using fibre optic based sensors: a review. Mag. Concr. Res. 1–45, (2019). https://doi.org/10.1680/jmacr.19.00185
11. Udd, E., Spillman, W.B.: Fiber optic sensors. An introduction for engineers and scientists (2nd ed.). John Wiley & Sons, Hoboken N.J. (2011)

12. Wu, B., Wu, G., Yang, C., He, Y.: Damage identification and bearing capacity evaluation of bridges based on distributed long-gauge strain envelope line under moving vehicle loads. J. Intell. Mater. Syst. Struct. **27**(17), 2344–2358 (2016). https://doi.org/10.1177/1045389X16629571

13. Ye, X.W., Su, Y.H., Han, J.P.: Structural health monitoring of civil infrastructure using optical fiber sensing technology: a comprehensive review. Scientif. World J. **652329**, (2014). https://doi.org/10.1155/2014/652329

Enhanced Fatigue Life of Old Metallic Bridges—Application of Preloaded Injection Bolts

Bruno A. S. Pedrosa, Carlos A. S. Rebelo, José A. F. O. Correia, Milan Veljkovic, and Luís A. P. S. Silva

Abstract There is a significant number of old metallic bridges with high levels of structural degradation due to their long service period. Fatigue problems are especially important in these structures since the majority of them were not designed taking into account this phenomenon. Several investigations showed that riveted joints are critical details since several fatigue cracks were found in these joints. In this sense, strengthening methodologies need to be studied. The strategy that has been considered a good solution is the implementation of injection bolts to replace faulty rivets. The structural performance of injection bolts has been demonstrated essentially under quasi-static conditions presenting good results. This paper intends to contribute to the scientific knowledge regarding the fatigue behavior of connections with preloaded injection bolts in the context of a bridge strengthening scenario. An experimental investigation was conducted to compare the fatigue performance of connections with preloaded injection bolts and preloaded standard bolts. Single and double shear connections were tested. New S–N design curves were proposed based on a statistical analysis of the results and compared with the S–N curves proposed in EC3-1–9. The obtained results showed that the use of injection bolts lead to lower scatter and improvement of fatigue life. It was verified that the Eurocode 3 is not able to represent the fatigue strength of connections whose performance is influenced by old metallic materials. Additionally, the fatigue behavior of these connections was assessed by numerical analysis. The relevance of the fatigue crack initiation was evident.

Keywords Fatigue · Injection bolts · Old metallic bridges · Strengthening

B. A. S. Pedrosa (✉) · C. A. S. Rebelo · L. A. P. S. Silva
University of Coimbra, ISISE, Department of Civil Engineering, Rua Luís Reis Santos, Pólo II, 3030-788 Coimbra, Portugal
e-mail: bruno.pedrosa@uc.pt

J. A. F. O. Correia
Faculty of Engineering, University of Porto, Rua Dr. Roberto Frias, 4200-465 Porto, Portugal

M. Veljkovic
Faculty of Civil Engineering and Geosciences, Delft University of Technology, Delft, Netherlands

© The Author(s), under exclusive license to Springer Nature Switzerland AG 2022
J. Sena-Cruz et al. (eds.), *Proceedings of the 3rd RILEM Spring Convention and Conference (RSCC 2020)*, RILEM Bookseries 34,
https://doi.org/10.1007/978-3-030-76465-4_18

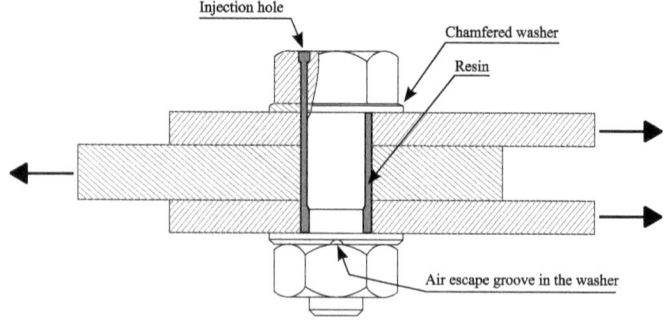

Fig. 1 Schematic representation of an injection bolt

1 Introduction

Maintenance and safety assurance of old riveted bridges designed in the period which steel has become an economically viable material for bridge construction, i.e., between the end of the 19th and the beginning of the 20th century, deserve special attention. These structures are prone to present high levels of structural degradation due to their long service life [1]. Repairing and strengthening operations of old riveted steel bridges may use alternative fastening techniques, such as rivets, welding, high strength friction grip bolts, fitted bolts and resin-injected bolts. The mechanical performance of resin-injected bolts have been demonstrated essentially based on quasi-static or creep tests [2–4], which strengthen the importance to assess its fatigue strength [5]. Injection bolts, represented in Fig. 1, can be produced from standard bolts adapting them for the resin injection process as mentioned in Annex K of EN 1090-2 [6].

The fatigue behavior of details composed with injection bolts is defined in Eurocode 3 part 1–9 [7] through the proposal of S–N curves. These standard states the same fatigue design curve for bolted connections with preloaded injection bolts than for bolted ´connections with standard preloaded bolts.

2 Experimental Campaign

Experimental fatigue tests were performed to assess the fatigue strength of bolted connections with preloaded injection bolts in comparison to bolted connections with preloaded high strength bolts in the context of a structural rehabilitation of an old metallic bridge.

A total of 40 experimental tests were performed, half of them using bolted connections composed with standard bolts and the other half using bolted connections with injection bolts. Specimens are bolted connections in which the connecting element (bolt) is under shear load and both typologies of single shear connections and double

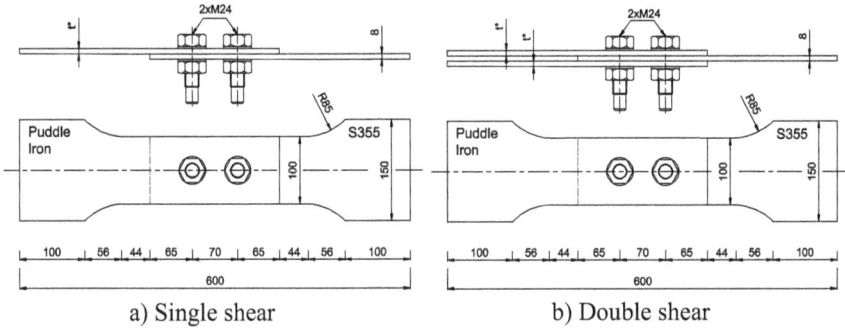

a) Single shear   b) Double shear

Fig. 2 Geometry of specimens (dimensions in mm) *thickness varies between 6 and 8 mm

shear connections were studied. Since one of the main goals was to reproduce as close as possible a structural strengthening scenario of an old metallic bridge, specimens are composed by metallic plates extracted from structural elements of the Eiffel bridge located in Viana do Castelo and by metallic plates produced with S355. The geometry of the specimens is presented in Fig. 2. Each of the specimens is composed by two M24 bolts with high strength class 10.9. These elements were preloaded following the recommendations on EN 1090–2 [6]. The assembly and preparation of specimens with injection bolts was conducted with the recommendations on the Annex K of EN 1090–2 [6]. Within the experimental campaign, different levels of applied load were tested aiming to allow the determination of reliable S–N curves for the studied structural details.

The adhesive used for injection bolts was the epoxy based resin Sikadur®-52. This material is characterized by its low viscosity (good for the injection process), good adhesion to steel and adequate mechanical properties. It is used in rehabilitation of civil engineering structures, such as bridges, to fill voids and cracks and assure that the two connecting elements are structurally bonded.

Specimens were tested on a WALTERBAI Universal Testing Machine rated to 600 kN. All fatigue tests were carried out under constant amplitude loading with a stress R-ratio equal to 0.1. The test frequency was set to 5 Hz for all tests except for one high-cycle fatigue tests where test frequency was defined as 10 Hz. If no rupture is found before 5 million cycles, the test is stopped and it is considered a run-out. This value corresponds to the fatigue limit defined in EN 1993–1–9 [7] for constant amplitude loading scenario.

Fatigue results are normally presented using logarithmic scales in both applied stress range, $\Delta\sigma$, and number of cycles, N, axes. This methodology allows to define a linear relation between those parameters as described in the following equation:

$$\log N_i = \log a + m \cdot \log \Delta\sigma \tag{1}$$

where m is the slope and $\log a$ is the intersection with the axis $\log\Delta\sigma$. Thereby, a mean S–N curve can be defined for the obtained results using a linear regression base on the least squares estimation method.

Moreover, a characteristic S–N curve can be establish using the requirements prescribed in EN 1993–1–9 [7] which are 75% confidence level and 95% probability of $\log N_i$ be exceeded. As stated by Drebenstedt and Euler [8], this curve can be described as:

$$\log N_i = \log a + m \cdot \log\Delta\sigma - k_{1-\alpha,p,n} \cdot s \tag{2}$$

where $k_{1-\alpha,p,n}$ is the statistical factor which depends on the confidence level, $1 - \alpha$, probability of failure, p and the sample size, n, and s is the standard deviation.

In the present study, the statistical analysis was performed aiming to define characteristic curves using a slope equal to 5, since this is the value that has been proposed by different authors [9] for structural details composed by old metallic materials.

2.1 Single Shear

Concerning the experimental tests with single shear specimens, it was verified that the failure was obtained in the puddle iron plate. In Fig. 3 are presented the obtained fatigue data from the experimental campaign referred above together with additional

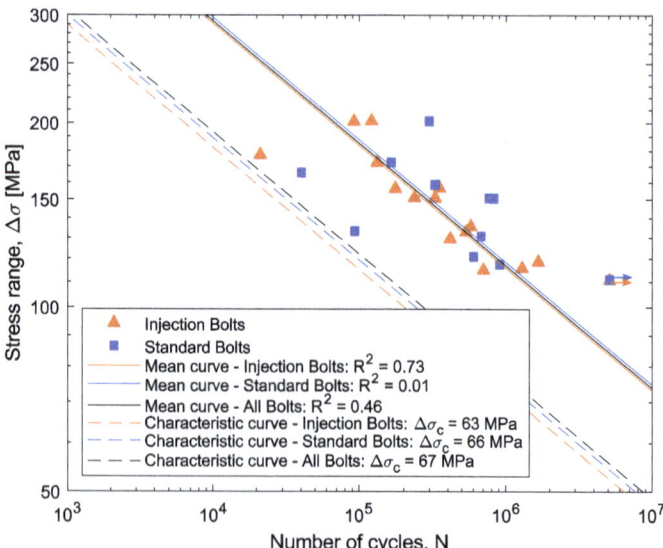

Fig. 3 Experimental data for single shear specimens; mean and characteristic curves

data from similar tests made by Jesus et al. [10]. The statistical analysis allowed to define both mean and characteristic curves. It is observed from the graphic that mean curves are nearly coincided but the scatter for standard bolts is significantly higher as is proved by the value of the coefficient of determination. For the characteristic curves the analysis is similar, the use of injection bolts does not lead to significant changes in the fatigue strength of single shear bolted connections. This fact led to the definition of a mean and characteristic curves for all the obtained fatigue data. The characteristic curve obtained for all fatigue data was found with a detail category of 67 MPa.

In order to understand the feasibility of the design curves proposed in the Eurocode 3 part 1–9 [7], in Fig. 4 is defined the detail 90 proposed in the standard for single shear bolted connections. It is clear that this curve does not represent a design safe criterion for the presented fatigue data. In fact, as previously discussed by several authors [11, 12], S–N curves proposed in EC3–1–9 are not able to represent the fatigue behaviour of structural details composed by old metallic materials such as puddle iron. An additional S–N curve is presented in Fig. 4, which is the design curve defined by the detail category 71 and slope equal to 5 as proposed by Taras and Greiner [9]. They conducted and analysed several fatigue tests on components from old metallic bridges. It can be observed that this design approach is very well suitable to the presented fatigue experimental data.

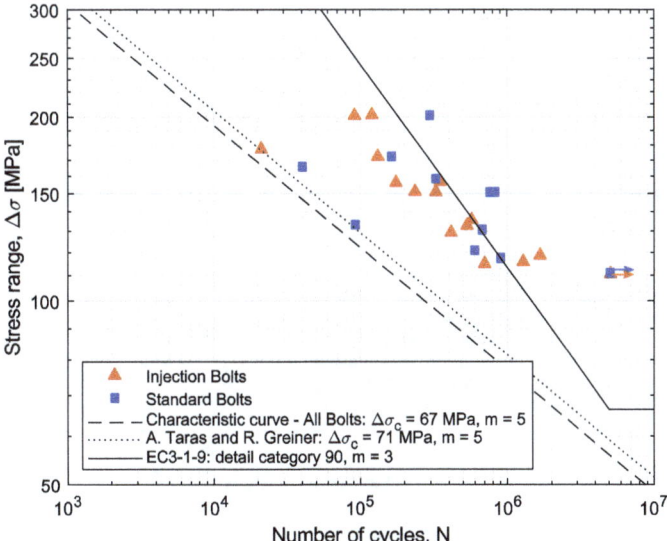

Fig. 4 Experimental data for single shear specimens; design curves

2.2 Double Shear

For double shear specimens, a total of 21 fatigue tests were conducted, 9 using injection bolts and 12 using standard bolts. These specimens were assembled with one S355 steel plate in one side and two puddle iron plates in the other side leading to higher stresses on the plate made with current steel, therefore fatigue failure was found on this component. In Fig. 5 are presented the obtained data and both mean and characteristic curves defined by the statistical analysis.

The comparison of the obtained data with the design curve proposed in Eurocode 3 part 1–9 [7] defined by the detail category 112 shows that it is not completely reliable, however there are only two data points with lower fatigue strengths. Additionally, the curve proposed by Taras and Greiner [9] represents a very conservative approach. In this case, the scenario of scatter reduction by using injection bolts is also present as is verified by the values of the coefficient of determination. Despite the similarity of mean curves, the difference in the standard deviation (0.23 for injection bolts and 0.43 for standard bolts) conducted to non-similar characteristic curves. In fact, the characteristic curve related to injection bolts represents higher fatigue strengths and its detail category is 22% higher than the opponent.

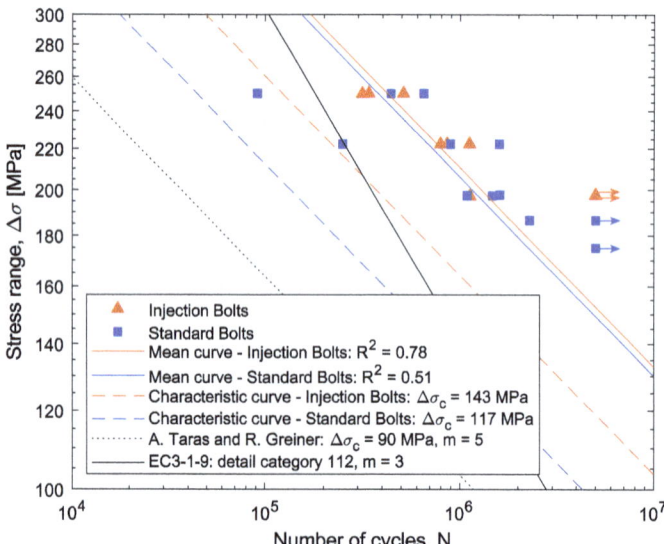

Fig. 5 Experimental data for double shear specimens; design curves

3   Numerical Analysis

3.1   Fatigue Model

Fatigue life predictions of bolted connections under shear loading has been studied by several authors [13, 14]. The most commonly used fatigue models assume that fatigue life can be divided in two stages: fatigue crack initiation and fatigue crack propagation. The first stage can be modelled using strain-life relations while the second stage can be computed using the concept of linear elastic fracture mechanics (LEFM). The crack length which is considered as the beginning of the second stage varies within the range of 0.1 and 1 mm [14]. The strain-life relation used in the present study was developed by Morrow [15]:

$$\frac{\Delta \varepsilon_{loc}}{2} = \frac{\sigma'_f}{E}(2N_i)^b + \varepsilon'_f(2N_i)^c \tag{3}$$

Where $\Delta \varepsilon_{loc}$ is the local elastoplastic strain, σ'_f and b are the fatigue strength coefficient and exponent, respectively, ε'_f, and c are the fatigue ductility coefficient and exponent, respectively, E is the Young modulus and N_i is the number of cycles in the initiation phase. The values for σ'_f, b, ε'_f, c e E were determined by Correia [16] for puddle iron from the Eiffel bridge and S355. The computation of the local elastoplastic strain was made with the analytical approach developed by Neuber [17] together with cyclic curve defined by the Ramberg–Osgood [18] formulation:

$$\begin{cases} \frac{\Delta \sigma_{loc}^2}{E} + 2\Delta \sigma_{loc}\left(\frac{\Delta \sigma_{loc}}{2K'}\right)^{1/n'} = \frac{K_t^2 \Delta \sigma_{nom}^2}{E} \\ \Delta \varepsilon_{loc} = \frac{\Delta \sigma_{loc}}{E} + \left(\frac{\Delta \sigma_{loc}}{2K'}\right)^{1/n'} \end{cases} \tag{4}$$

Where $\Delta \sigma_{loc}$ is the local stresses, $\Delta \sigma_{nom}$ is the nominal stress, K' and n' are the cyclic strain hardening coefficient and exponent, respectively (whose values where obtained from Correia [16]) and K_t is the stress concentration factor (computed by finite element modelling).

The number of cycles in the fatigue crack propagation phase was calculated through the integration of the Paris law [19]:

$$\Delta N = \frac{1}{C} \frac{1}{\Delta K^m} \Delta a \tag{5}$$

Where C and m are material constants whose values were extracted from Jesus et al.[20] and ΔK is the stress intensity factor (computed by finite element modelling).

3.2   *Finite Element Modelling*

A set of finite element models were developed based on specimens used in the experimental campaign aiming to compute both stress concentration and intensity factors. The geometry, boundary conditions and applied load on numerical models follows the experimental campaign. The symmetry of specimens along the plane XZ was used to reduce the computational cost, as presented in Fig. 6.

These models were created using solid finite elements with 8 nodes, C3D8 and the mesh size was calibrated as a function of the stress concentration value, K_t. The vicinity of hole in the plates is the critical area for the stress concentration factor and, therefore, the mesh size was reduced in this zone, as presented in Fig. 7.

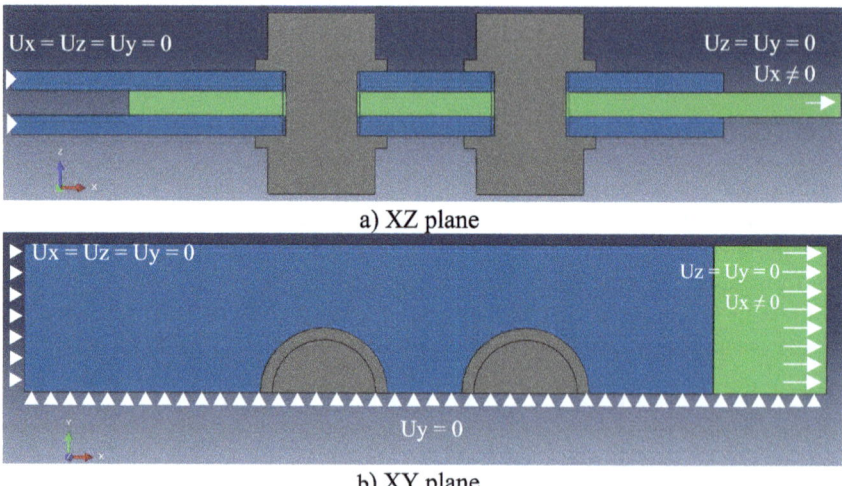

Fig. 6 Schematic representation of numerical models; boundary conditions

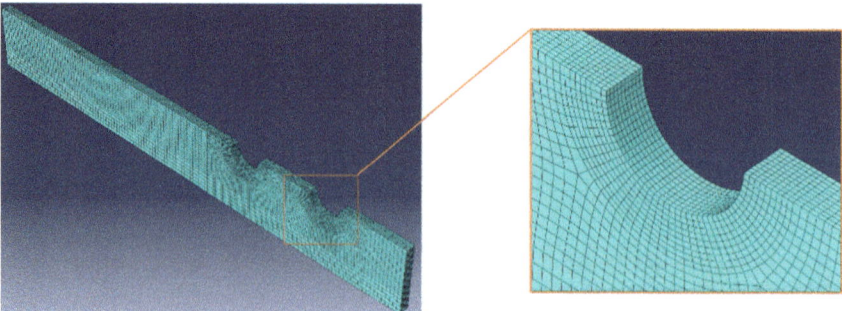

Fig. 7 Finite element mesh in the zone of stress concentration

Contact surfaces between plates and between plates and bolts were defined for its normal and tangential behaviour. For the first scenario, "har contact" formulation [21] was chosen and for the second scenario the "penalty" formulation [21].

Bolt preload forces were applied before the external load and its value was set to 185 kN. The main goal of these numerical analysis was to compute stress concentration factor and stress intensity factor which are elastic parameter, therefore, material models used in the analysis correspond to its linear elastic material properties. Thus, puddle iron material was defined with $E = 193.1$ GPa [16], for S355 steel the value $E = 211.6$ GPa [16] was used and $E = 210$ GPa for bolts. The Poisson coefficient was set to 0.3 for all materials.

The applied load was defined as the maximum value used in the experimental campaign.

From the experimental tests was possible to identify the origin of the fatigue crack in the vicinity of the hole of the plates. Thus, a set of numerical models were created to compute the stress intensity factor with different values for the crack length which starts in the hole and growths perpendicular to the applied load through the net area of the plate. It was modelled with constant length through the thickness of the plate.

The computation of stress intensity factor was made with the modified Virtual Crack Closure Technique (VCCT) developed by Krueger [22]. Then, it is possible to adjust a six-degree polynomial law between the factor $Y = K/\sigma$ and the crack length, a, and obtain the number of cycles in the propagation phase by integration of this polynomial law.

3.3 Results

The fatigue model and the numerical analysis previously discussed allowed to compute S–N prediction curves for the studied details.

In Fig. 8 are presented S–N prediction curves for the crack initiation and propagation curves for single shear specimens, as well as total S–N prediction curve compute as the sum of the S–N prediction curves in initiation and propagation phases. It is observed that numerical S–N prediction curves are very well adjusted to the experimental results. The influence of the crack initiation phase is most evident for high cycles fatigue regime and the influence of the crack propagation phase is evident for low cycles fatigue regime.

In Fig. 9 are presented the numerical S–N prediction curves for double shear specimens. In this case, the S–N curve for the initiation phase is dominant for all fatigue regimes since it is nearly coincident with the S–N for the total fatigue life. The crack propagation phase for double shear specimens is not relevant.

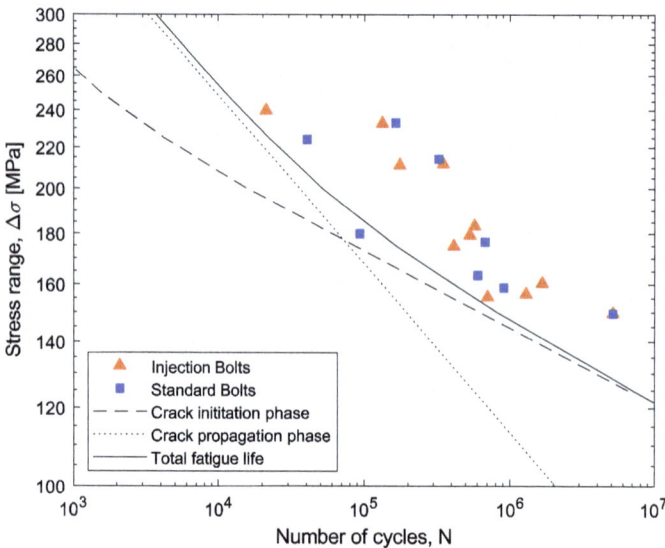

Fig. 8 Fatigue life prediction using numerical and analytical approaches for single shear specimens

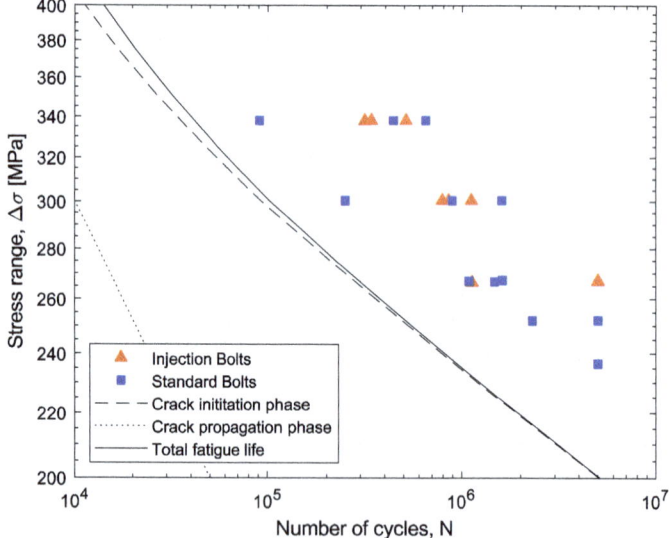

Fig. 9 Fatigue life prediction using numerical and analytical approaches for double shear specimens

4 Conclusions

The main conclusions from this study are the following:

1. The use of injection bolts contributes to reduce the random uncertainty of fatigue resistance in comparison with bolted connections with standard bolts.
2. Fatigue failure on single shear specimen was found on puddle iron plates which led to low fatigue resistance values especially compared to the fatigue strength prediction from the Eurocode 3 part 1 – 9.
3. A statistical analysis was implemented and characteristic S–N curves were defined with the statistical parameters defined in the Eurocode 3 part 1–9.
4. It was verified that the influence of the adhesive around the bolt does not produce significant changes in the fatigue strength of single shear connections, however, it contributed to increase the fatigue strength in 22% in the case of double shear connections.
5. Numerical S–N predictions were computed based on elastic parameters. It was possible to observe the relevance of the initiation phase on the total S–N prediction curve, especially for double shear specimens.

Acknowledgements The authors would like to acknowledge the Fundação para a Ciência e Tecnologia (FCT) for funding the scholarships SFRH/BPD/107825/2015 and SFRH/BD/145037/2019. This work was also financed by FEDER funds through the Competitivity Factors Operational Programme–COMPETE and by national funds through FCT within the scope of the project POCI-01-0145-FEDER-007633 and through the Regional Operational Programme CENTRO2020 within the scope of the project CENTRO-01-0145-FEDER-000006.

References

1. Akesson, B.: Fatigue life of riveted steel bridges (1st ed). Taylor & Francis Group, Ed., CRC Press (2010)
2. ECCS European recommendations for bolted connections with injection bolts (1st ed). Publication No 79 (1994)
3. Mattes, J.: Substituição de rebites por parafusos injectados com resina. MSc Thesis, IST, University of Lisbon, Portugal, 2007 (in Portuguese) (2007)
4. Gresnigt, A., Sedlacek, G., Paschen, M.: Injection bolts to repair old bridges, pp. 349–360 (2000). https://citeseerx.ist.psu.edu/viewdoc/download?doi=10.1.1.552.4959& rep=rep1&type=pdf. Accessed 5 May 2016
5. Correia, J., Pedrosa, B., Raposo, P., Jesus, A., Rebelo, C., Gervásio, H., Calçada, R., Silva, L.: Fatigue strength evaluation of resin-injected bolted connections using statistical analysis. Engineering 3(6), 795–805 (2017)
6. CEN EN 1090–2, Execution of Steel Structures and Aluminium Structures-Part 2: Technical Requirements for Steel Structures, European Committee for Standardization (2008)
7. CEN EN 1993–1–9: Eurocode 3, Design of Steel Structures–Part 1–9: Fatigue., European Committee for Standardization, Brussels (2005)
8. Drebenstedt, K., Euler, M.: Statistical analysis of fatigue test data according to eurocode 3. In: Powers, N., Frangopol, D., Al-Mahaidi, R., Caprani, C. (eds.) Maintenance, Safety, Risk,

Management and Life-Cycle Performance of Bridges : Proceedings of the Ninth International Conference on Bridge Maintenance, Safety and Management. CRC Press, Melbourne, p. 2945 (2018)

9. Taras, A., Greiner, R.: Development and application of a fatigue class catalogue for riveted bridge components. Struct. Eng. Int. J. Int. Assoc. Bridg. Struct. Eng. **20**(1), 91–103 (2010)
10. De Jesus, A., Silva, J., Figueiredo, M., Ribeiro, A., Fernandes, A., Correia, J., Silva, A., Maeiro, J.: Fatigue behaviour of resin-injected bolts: an experimental approach. In: Iberian Conference on Fracture and Structural Integrity. Porto (2010)
11. De Jesus, A., Silva, A., Correia, J.: Fatigue of riveted and bolted joints made of puddle iron-an experimental approach. J. Constr. Steel Res. **104**, 81–90 (2015)
12. Pedrosa, B., Correia, J., Rebelo, C., Lesiuk, G., De Jesus, A., Fernandes, A., Duda, M., Calçada, R., Veljkovic, M.: Fatigue resistance curves for single and double shear riveted joints from old portuguese metallic bridges. Eng. Fail. Anal. **96**, 255–273 (2019)
13. Correia, J., De Jesus, A., Pinto, J., Calçada, R., Pedrosa, B., Rebelo, C., Gervásio, H., Da Silva, L.: Fatigue behaviour of single and double shear connections with resin-injected preloaded bolts. In: IABSE Congress Stockholm 2016 Challenges in Design and Construction of an Innovative and Sustainable Built Environment. Stockholm (2016)
14. De Jesus, A., Da Silva, A., Correia, J.: Fatigue of riveted and bolted joints made of puddle iron—a numerical approach. J. Constr. Steel Res. **102**, 164–177 (2014)
15. Morrow, J.: Cyclic plastic strain energy and fatigue of metals. Intern. Frict. Damping Cycl. Plast. 45–48 (1965)
16. Correia, J.: An integral probabilistic approach for fatigue lifetime prediction of mechanical and structural components. Ph.D. Thesis, Faculty of Engineering, University of Porto, Porto (2014)
17. Neuber, H.: Theory of stress concentration for shear-strain prismatic bodies with arbitrary nonlinear stress-strain law. J. Appl. Mech. **28**, 544–550 (1961)
18. Ramberg, W., Osgood, W.: Description of stress-strain curves by three parameters, NACA Technical Note 902 (1943)
19. Paris, P., Erdogan, F.: A critical analysis of crack propagation laws. J. Basic Eng. **85**(4), 528–533 (1963)
20. De Jesus, A., Matos, R., Fontoura, B., Rebelo, C., Da Silva, L., Veljkovic, M.: A Comparison of the fatigue behavior between S355 and S690 Steel Grades. J. Constr. Steel Res. **79**, 140–150 (2012)
21. Hibbit, D., Karlsson, B., Sorensen, P.: ABAQUS/Standard User's Manual, Ver. 6.10. Pawtucket, Rhode Island (2004)
22. Krueger, R.: Virtual crack closure technique: history, approach, and applications. Appl. Mech. Rev. 109–143 (2004)

Efficient Labelling of Air Voids in Hardened Concrete for Neural Network Applications Using Fused Image Data

Fabian Diewald⑩, Nicolai Klein⑩, Maximilian Hechtl⑩, Thomas Kraenkel⑩, and Christoph Gehlen⑩

Abstract Every concrete structure incorporates a system of air pores of different sizes which is characterized by design parameters according to the respective building codes. The system's characteristics affect properties of the material, specifically the durability of concrete exposed to freeze-thaw action. Testing techniques like the traverse line method, the point count method or the enhanced contrast method have been established and standardized to determine the relevant parameters according to construction material standards in order to empower the future life cycle analysis of the structure. These procedures are typically governed by a major effort in terms of specimen preparation and microscopic examination. We present an approach for a time-efficient procedure to analyze the air void system in optical microscopic images of hardened concrete by means of a neural network that is reliable and highly reproducible at the same time. Height-based segmentation of images using confocal laser scanning microscopy data allowed the composition of an image dataset with automatically labelled air voids. The set consists of 32,750 images of air voids with diameters between $25\,\mu m$ and $1.1 \times 10^3\,\mu m$. By translating the gained information about the size and the location of air voids via a binary mask, we created a corresponding dataset generated by means of an optical microscope. The dense one-stage object detector *RetinaNet* with a *Resnet* backbone, fed with the optical microscopic image dataset, demonstrates the effectiveness of the method referring to localization and characterization of air voids in images of hardened concrete. The presented approach supports the successive characterization of the standardized parameters of the air void system and advances the modelling and prediction of structural durability with regard to freeze-thaw resistance.

Keywords Air void system · Hardened concrete · Confocal laser scanning microscopy · Automation · Neural network

F. Diewald (✉) · N. Klein · M. Hechtl · T. Kraenkel · C. Gehlen
Chair of Materials Science and Testing, Centre for Building Materials,
Technical University of Munich, Franz -Langinger -Straße 10, 81245 Munich, Germany
e-mail: fabian.diewald@tum.de

J. Sena-Cruz et al. (eds.), *Proceedings of the 3rd RILEM Spring Convention and Conference (RSCC 2020)*, RILEM Bookseries 34,
https://doi.org/10.1007/978-3-030-76465-4_19

209

1 Introduction

Functionality of our built environment, at present, is based on reinforced concrete (RC) structures. The service life of these structures is affected by physical and chemical loading depending on their environmental exposure. Although, the required material properties are generally adjusted with a safety margin to resist the specific impacts, unintended deviations from the design state may occur. To quantify these deviations and to ensure structural safety and durability, efficient and less resource-intense maintenance and monitoring technologies are necessary.

A major cause of deterioration of RC structures is the combined effect of freeze-thaw exposure and chloride ingress. On a microstructural level, every concrete structure incorporates a pore system with its characteristic production mechanisms and stability [1, 2]. The composition of the pore system in hardened concrete primarily affects the freeze-thaw resistance, whereas the content of pores impacts the diffusivity and the transport potential of corrosive agents [3]. In general, for a concrete mix which provides more water than necessary for the hydration process, a system of pores develops that accelerates the process of capillary suction [4, 5]. The damage evolution can be characterized by ultrasonic pulse velocity measurements or resonance frequency analysis [6]. Both methods describe the damage evolution and its effect on the service life, e.g. crack development affecting the elastic properties and the characteristic strength. Procedures and standards exist to quantify the relevant parameters of the pore system which, in many cases, are expensive and time-consuming in order to create statistically relevant information. The construction material standards *EN 480-11:2005-12* [7] and *ASTM C457/C457M* [8] govern the determination of the total air content, the specific surface, the spacing factor, the pore size distribution and the content of micro air pores. Additionally, the standards include the characterization procedures like the traverse line method, the point-count method and the contrast enhanced method. Contrast-based automatic systems, first discussed in 1977 [9], have been proposed to analyze the parameters of the air void system and establish a testing procedure with high rates of reproducibility and repeatability [10–12].

Artificial neural networks were proposed in the context of characterization of the pore system, e.g. to predict the effect of asphalt components on the air void content in asphalt mixtures [13] or to quantify the effect of various mixture parameters on the durability of high performance concrete [14]. The large number of pore examples empowers the approach to automate the identification of pores within a defined range of diameters by means of a neural network.

2 Materials and Methods

2.1 Specimen

The sample mortar specimen consisted of natural sand with a maximum grain size of 4 mm and a total sand content of 57 % by vol. It was subjected to the Austrian Standard *ONR 23303:2010-09* [15] to evaluate of the freeze-thaw-resistance of fine aggregates . Therefore, a paste content of 43% by vol. made of 525 kg/m^3 cement with a w/b ratio of 0.42 was used. To ensure the required air void content of 9–15% by vol. in the fresh mortar, an air entraining agent was added. The air void characteristics of the hardened mortar are determined with the traverse line method according to *EN 480-11:2005-12* [7]. The air content was 8.85% by vol., the content of micro air pores was 5.44% by vol. and the spacing factor was 0.12 mm. The specimen is shown together with the three examined strip-shaped sections on its surface in Fig. 1a.

2.2 Imaging Methods

The data acquisition device used for this study was a confocal laser scanning microscope (CLSM) which simultaneously produces light images by optical microscopy (OM). The data obtained by the measurements are light intensity values, which represents the height for every data point within the measured area. The heights of the data points with the highest intensity represents the surface distance of the specimen to the microscope. The resulting geometrical 3D information is shown in Fig. 1b for a typical section. We used a magnification of 5× in order to generate images with a pixel width of 2.84 μm. Additionally to the height measurement, an optical digital image of the measured surface was taken.

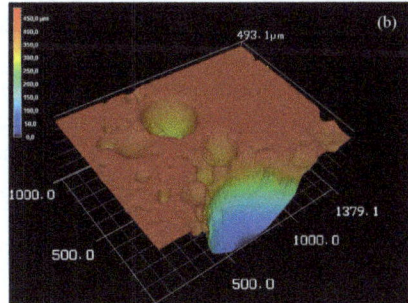

Fig. 1 Air-entrained concrete specimen with an area of 10 × 15 cm^2 limited by the blue lines and the specimen edges. **a** digital image, **b** 3d image of the surface measured with a confocal laser scanning microscope with a magnification of 20×

For automation of the image acquisition, an object table controlled by the microscope's software was used. This facilitates the recording of an area up to 100 x
100 mm^2. Additionally, an automated focus was used to scan the height area for
every single picture, thereby setting the measurement height for the software, minimizing the effort regarding manual specimen evaluation.

2.3   Object Detector

The *RetinaNet* network architecture works as a dense one-stage object detector. It
uses a Feature Pyramid Network (FPN) backbone on top of a *ResNet* feedforward
architecture, a classification subnet, and a box regression subnet [16]. We employed
a one-stage detector which runs at faster speed compared to two-stage detectors like
Faster R-CNNs with a FPN [17].As the network backbone, we tested *Resnet-50* and
Resnet-101 [18] with variable pixel input and an adaptive learning rate.

3   Data

The collected data is the result from the surface examination of a concrete specimen.
We created two corresponding image datasets by means of confocal laser scanning
microscopy and optical microscopy, by examining the same section of the specimen
surface for both sets. Direct labelling of air voids in optical microscopic images is
dependent on the examiner and is usually not straightforward. Thus, we employed
the information about the surface height profile to perform a segmentation operation
on every confocal microscopic image.

The segmentation operation consists of morphological closing with a (15, 15)
kernel to improve the signal-to-noise ratio in the digital image by means of a threshold
condition in the grayscale image, followed by a Canny edge detection operation
using median filtering. These steps form the preprocessing procedure that allows
the localization of the segment contours, i.e. the edges of the air voids (see Fig. 2a).
The height-based information is furthermore transferred to the corresponding optical
microscopic image via a binary mask that represents the computed segments as the
inner part of the contours. For every segment, the consecutive list of air voids expands
by adding a label and the location. Figure 2b presents the result of the preprocessing
algorithm for an exemplary image.

By applying the preprocessing segmentation algorithm to a set of images and
marking rectangular sections, we created an image stack of air void examples which
serves as an input for the training of the object detector *RetinaNet*. Figure 3 shows
three examples of labelled air voids with various diameters as a subset of the data
stack along with their binary masks. We labelled air voids with diameters between
5 μm and 1.1×10^3 μm. The optical lens with a magnification of 5× limits the
resolution of the optical image and the detectable minimum air void diameter. As

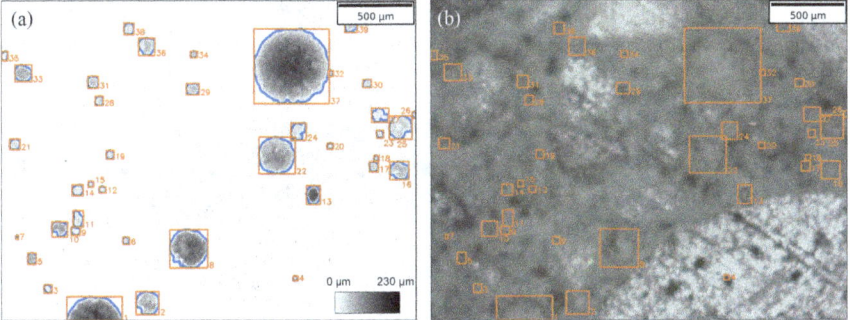

Fig. 2 Image pairs of the concrete specimen surface with edge lengths of 2.91 and 2.18 mm: **a** computed contours in a digital image generated by confocal laser scanning microscopy with their respective minimal rectangle. The segments represent the air voids and are labelled consecutively according to their location. **b** Result of the preprocessing algorithm: labelled air voids in the optical microscopic image by means of the transferred height information

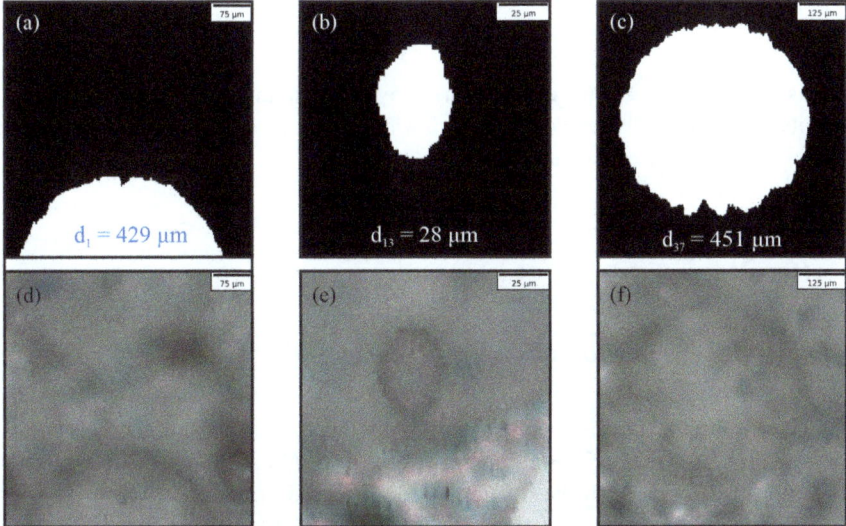

Fig. 3 Three exemplary image pairs of automatically labelled air voids with approximated diameters of $d_1 = 429\,\mu m$, $d_{13} = 28\,\mu m$ and $d_{37} = 451\,\mu m$: (**a**) to (**c**) binary masks to transfer the height information to the OM images (**d**) to (**f**). The presented images are part of the input dataset for the training of an object detector

the data collection process runs automatically for every specimen, we employed an autofocus for every image section. By slight adjustment of the autofocus, we integrated out of focus images to the dataset to improve the robustness referring to blurring in the test data.

The resulting dataset includes 32,750 examples of air voids in OM images, which were extracted from 1,044 single recordings out of three strip-shaped sections. Each

Table 1 Examples of the examined surface strips, the composed images and the automatically labelled air voids with their respective quantities. The final dataset consists of 32,750 labelled air voids with a range of diameters between 25 μm and 1.1×10^3 μm

Type	Image	Quantity
Strip from concrete specimen surface	$n_{1,1}$ ⋯ $n_{39,1}$ / $n_{1,9}$ ⋯ $n_{39,9}$	3
Composed image from strip		1,044
Single air void from composed image		32,750

strip is composed of 39 images $n_{1,1}$ to $n_{39,1}$ over its length and 9 images $n_{1,1}$ to $n_{1,9}$ over its width. Table 1 provides an overview about the specimens and the generated datasets, and illustrates the data collection process.

4 Results

4.1 Model Performance

The chosen architecture is a pre-trained object detector as a combination of *Reti-naNet* with both *ResNet-50-FPN* and *ResNet-101-FPN* backbone. We fed the one-stage object detector with the image dataset, described in Sect. 3, which contains 32,750 labelled air voids. The ratio of the data was 60% training data, 20% validation data and 20% test data. The results of two performed trainings are stated in Table 2, together with the training parameters. Computation was conducted using a *Nvidia RTX 2070 Max-Q* GPU with a *CUDA* compute capability of 7.5. For both trainings, the number of iterations was 5,000, the batch size was 10 and the number

Table 2 Object detector architectures with computing time, classification loss and regression loss related to the dataset of 32,750 air voids. The dataset is split into 60% training data, 20% validation data and 20% test data

Architecture	Computing time	Classification loss	Regression loss
RetinaNet ResNet50	360 min	0.3169	1.0298
RetinaNet ResNet101	420 min	0.3392	1.1574

of epochs was 10. Using these parameters, the architecture with *ResNet50* achieved a classification loss of 0.3169 and a regression loss of 1.0298. The architecture with *ResNet101* achieved a classification loss of 0.3392 and a regression loss 1.1574.

4.2 Discussion

A method based on 2D-image analysis qualifies the formulation of more precise air void characterization models compared to a 1D line as reduction of the evaluation space causes a dimensional loss of information. In general, this implies the enhancement of confidence within the subsequently applied probabilistic models by feeding them with information from section planes compared to traverse lines. However, the evaluated dataset is only useful to evaluate feasibility of the method due to its unidimensional character. Further augmentation is necessary for detailed analysis of the object detector performance and the computed accuracies are explicitly valid for a uniform set of images. Regardless of these limitations, the average precision for a test set is high despite blurry images within the dataset. Using a higher resolution lens is vital for the method to represent air voids in the microscale range. The effort to automatically label the air pores is little. Hence, the dataset is expendable with minimum effort.

The application of object detectors for the image dataset is based on information about the surface height collected from CLSM data. The suggested preprocessing procedure implicates the assumption that the air voids are uniquely identifiable by translating the height information to the optically recorded image. Further physical imperfections and noise from the measurement technology may affect the labelling method by producing incorrectly labelled data. Although, the spherical shape of the air void provides the option of integrating an additional geometrical boundary condition to improve the algorithm.

5 Conclusions

The developed preprocessing procedure of the digital image data enables the creation of an arbitrarily large dataset of air void examples in hardened concrete. Based on this dataset, we demonstrate the potential of evaluating the air void system by a one-stage object detector and subsequently computing its characteristic parameters. Feeding a *RetinaNet* architecture in combination with a *ResNet-50-FPN* backbone yielded the most accurate model performance regarding classification and regression loss for a low quality image dataset. Successively, the suggested procedure needs to be compared to standardized methods in terms of robustness towards data variation, accuracy in quantifying the relevant air void system parameters and, finally the cost and time efficiency. Further improvements are expected from explicit segmentation of the pixels of the air voids compared to the employed minimal rectangles, enlarging the dataset by application of modified optical recording conditions or digital correction of the pore edges, e.g. induced by mechanical specimen preparation. By optimizing the air void characterization quality, we contribute to a better understanding of the condition of structures and their properties, thus enhancing service life prediction and extension methods.

Acknowledgements This work has been supported by the German Research Foundation (DFG) in the framework of the Research Unit CoDA, FOR 2825. This support is gratefully acknowledged.

References

1. Mielenz, R.C., Wolkodoff, V.E., Backstrom, J.E., Flack, H.L.: Origin, evolution, and effects of the air void system in concrete. Part 1—entrained air in unhardened concrete. J. Am. Concr. Inst. **30**(1), 95–121 (1958)
2. Du, L., Folliard, K.J.: Mechanisms of air entrainment in concrete. Cement Concr. Res. **35**(8), 1463–1471 (2005)
3. Chung, C.W., Shon, C.S., Kim, Y.S.: Chloride ion diffusivity of fly ash and silica fume concretes exposed to freeze-thaw-cycles. Constr. Build Mater **24**, 1739–1745 (2010)
4. Powers, T.: The air requirement of frost-resistant concrete. In: Highway Research Board Proceedings 29th Annual Meeting. Washington, DC (1949)
5. Setzer, M.J., Auberg, R.: Freeze-thaw and deicing salt resistance of concrete testing by the CDF method. CDF resistance limit and evaluation of precision. Mater. Struct. **28**, 16–31 (1995)
6. Kessler, S., Thiel, C., Grosse, C.U., et al.: Effect of freeze-thaw damage on chloride ingress into concrete. Mater Struct. **50**(2), 121 (2017)
7. European Standard EN 480-11:2005-12. Admixtures for concrete, mortar and grout test methods—Part 11: determination of air void characteristics in hardened concrete (2005)
8. ASTM Standard C457/C457M. Standard Test Method for Microscopical Determination of Parameters of the Air-Void System in Hardened Concrete, ASTM International (2016)
9. Chatterji, S., Gudmundsson, H.: Characterization of entrained air bubble systems in concretes by means of an image analysing microscope. Cement Concr. Res. **7**, 423–428 (1977)
10. Jakobsen, U.H., Pade, C., Thaulow, N., Brown, D., Sahu, S., Magnusson, O., De Buck, S., De Schutter, G.: Automated air void analysis of hardened concrete—a Round Robin study. Cement Concr. Res. **36**, 14444–1452 (2006)
11. Peterson, K.W.: Automated air-void system characterization of hardened concrete: helping computers to count airvoids like people count air-voids-methods for flatbed scanner calibration. Ph.D. thesis, Michigan Technological University (2008)
12. Fonseca, P.C., Scherer, G.W.: An image analysis procedure to quantify the air void system of mortar and concrete. Mater. Struct. **45**, 3087–3098 (2015)
13. Zavrtanik, N., Prosen, J., Tusar, M., Turk, G.: The use of artificial neural networks for modeling air void content in aggregate mixture. Autom. Constr. **63**, 155–161 (2016)
14. Parichatprecha, R., Nimityongskul, P.: Analysis of durability of high performance concrete using artificial neural networks. Constr. Build. Mater. **23**(2), 910–917 (2009)
15. Austrian Standards Institute, ONR 23303:2010-09. Test methods for concrete—National application of testing standards for concrete and its source materials (2010)
16. Lin, T., Goyal, P., Girshick, R., He, K., Dollár, P.: Focal loss for dense object detection. In: IEEE Trans. Patt. Anal. Mach. Intel. **42**(2) (2017)
17. Girshick, R.: Fast R-CNN. In: The IEEE International Conference on Computer Vision, pp. 1440–1448 (2015)
18. He, K., Zhang, X., Ren, S., Sun, J.: Deep Residual Learning for Image Recognition. In: 2016 IEEE Conference on Computer Vision and Pattern Recognition (2016). https://doi.org/10.1109/CVPR.2016.90

Evaluation of Electromigration Desalination of Granite Contaminated with Salts–A Contribution to the Conservation of Architectural Surfaces

José N. Marinho, Eunice Salavessa, Said Jalali, Luís Sousa, Carlos Serôdio, and Maria J. P. Carvalho

Abstract Different cleaning technologies have been available for cleaning monumental façades, of stone or mortar. This is not only necessary for aesthetic reasons, but for the removal of salts that promotes conservation, as surface erosion may affect the long term preservation of buildings. The main advantage of eletromigration treatment is its efficacy in the salts removal or reduction in very porous materials, without damaging the original material. Many monuments of North of Portugal have granitic masonry and stonework structures that, sometime, as a result of continuous exposure to the rainwater infiltrations, or in the case of buildings near maritime cost, due to the marine-aerosols influence, salt efflorescence and disaggregation occurs.

J. N. Marinho
UTAD, Dto de Engenharias, Universidade de Trás-os-Montes e Alto Douro, Qta. de Prados, 5001-801 Vila Real, Portugal

E. Salavessa (✉)
CITAB-UTAD, Dto de Ciências Florestais e Arquitectura Paisagista, Universidade de Trás-os-Montes e Alto Douro, Qta. de Prados, 5001-801 Vila Real, Portugal
e-mail: eunicesalavessa@sapo.pt

S. Jalali
CTAC-UM, Dto de Engenharia Civil, Universidade do Minho, Centro do Território, Ambiente e Construção, Campus de Azurém, 4800-058 Guimarães, Portugal
e-mail: said@civil.uminho.pt

L. Sousa
CGEO-UC, Dto de Geologia, Universidade de Trás-os-Montes e Alto Douro, Qta. de Prados, 5001-801 Vila Real, Portugal
e-mail: lsousa@utad.pt

C. Serôdio
CITAB-UTAD, Dto de Engenharias, Universidade de Trás-os-Montes e Alto Douro, Qta. de Prados, 5001-801 Vila Real, Portugal
e-mail: cserodio@utad.pt

M. J. P. Carvalho
UTAD, Dto de Química, Universidade de Trás-os-Montes e Alto Douro, Qta. de Prados, 5001-801 Vila Real, Portugal

© The Author(s), under exclusive license to Springer Nature Switzerland AG 2022   217
J. Sena-Cruz et al. (eds.), *Proceedings of the 3rd RILEM Spring Convention and Conference (RSCC 2020)*, RILEM Bookseries 34,
https://doi.org/10.1007/978-3-030-76465-4_20

Hence, it is important to proceed in removal of chloride salts from the contaminated granite walls. This research work deals with electromigration desalination of contaminated "Mondim Yellow Granite" samples, presenting the physical properties, mineral composition, porosity and internal structure. This type of granite was selected because is frequently used in traditional buildings and monuments of the North of Portugal. The aim of the research was to evaluate the desalination effectiveness of electromigration. By its application chloride salts were completely removed and sulfate ions reduction was significant. Differences in effectiveness of the desalination were analyzed and are attributed to the shape of granite specimens. Results obtained demonstrate the viability of application of this electro kinetic procedure for removal of chloride and sulfate ions of granite ashlars of the monuments.

Keywords Salt efflorescences · Electromigration · Desalination · Granite · Efficacy

1 Introdution

In a previous study we developed on a monument situated near the Douro River delta, the Grilos' Church in Oporto, we could analyse the deterioration induced in stone and interior surfaces by soluble salts, which showed the necessity of the assessment of the appropriate techniques for their elimination in architectural heritage buildings. In the Grilos' Church (sixteenth-nineteenth centuries), the polychrome and decorative plasters of the walls and dome of Holy Sacrament Chapel (a neoclassic construction of this church), namely the decorative plaster panels, the pilasters, the cornice and the freeze, present visible signs of cracks and localized detachment of the coloured plasters. Rainwater penetration through the cracks of the masonries and dome, the lack of effective drainage system of the roof and rising damp through the ground (promoted by the absence of the rain-water drainage of the ground around the Church), are the principal factors of water retention in walls and decaying plasters. Infiltration waters, rich in salts, and repeated crystallization cycles have gradually promoted the decay of the plasterwork. The masonry retains a large amount of water which moves within the stone that evaporates through the plaster coating [1]. The white plasters of the H.S. Chapel present great concentrations of Cl^- (more than 1% m/m) which can result from the marine aerosols contamination. The conditions of exposure to the environmental humidity and the alkalis impact are responsible in the appearance of this phenomenon [2]. For that reason, the granite masonries contaminated with salts need an especial attention for removal of salts.

Several treatments have been tested to eliminate salts from stone and mortar monumental surfaces such as manual brushing, film poultice, including clay-based materials, which is held in close contact with the surface [3]. Many researchers have shown that the use of water to remove salt charge is not recommendable to architectural heritage, as the process may mobilize salts within the stone. Electromigration appears to be more efficient in extracting and reduce salts from very porous materials as soil, concrete and brick [4, 5]. Regarding granites, several published works refer to the influence of the type of granite in salt decontamination [6]. In our research work, the desalination by electromigration is applied to lightly deteriorated granite of NE of Portugal, artificially contaminated with a saline solution produced in laboratory, simulating the seawater. This study represents a preliminary contribution to the conservation of the monuments of the North of Portugal, which interior surfaces have valuable plasters, as in Grilos' Church.

2 Materials and Methodology

2.1 Granite Characteristics

As the origin of the granite of Grilos' Church was unknown, it was decided to choose a type of granite with similar physical and mechanical characteristics, especially the high porosity. We selected two-mica granite quarried in NE of Portugal, with the commercial designation of "Mondim Yellow", widely used in traditional buildings and monuments in the North of Portugal. This granite has a yellow colour and median grain and moderate level of meteorization. Its main physic-mechanic properties are indicated in the catalogue of ornamental stones, as follows: compressive strength: 1200 kg/cm^2; compressive strength after gelidity test: 1080 kg/cm^2; flexural strength: 86 kg/cm^2; bulk density: 2560 kg/m^3; water absorption P. At. N.: 1, 2%; accessible porosity: 3,4% [7].

2.2 Granite Specimens Preparation

For the electromigration test, two granite specimens, each measuring 80 mm x 80 mm x 30 mm, were used. Before salt contamination, one of them (identified as "Sample 4" in the text) was drilled at its centre, in thickness, and the resulting powder was used to constitute a soluble extract, previously prepared, in which, by ion chromatography, the chloride and sulphate contents of granite were measured to assess the effect of the artificial contamination, applied later.

The granite specimens were contaminated with synthetic seawater. For the preparation, based on Feijoo et al. [6], were used several salts in the follow quantities: $NaCl$ (140,30 g); $MgCl_2.6H_2O$ (35,60 g); $MgSO_4.6H_2O$ (28,60 g); $CaCl_2$ (6,66 g); K_2SO_4 (10,50 g) in 5 L of distilled water. The salt contamination of the specimens occurred in a 24 h cycle. At absorption period, the specimens were supported on the larger surface and partially immersed for one hour in the saline solution until the solution reached the top surface of the specimens, then they were fully immersed for two hours. After this period, while the specimens were drying, they were removed

Table 1 Content of Cl^- and $SO_4{}^{2-}$ in granite powder samples; pH values in granite powder in solution; (values before and after contamination and after treatment)

Granite powder of specimens	Cl^- (mg/l)	$SO_4{}^{2-}$ (mg/l)	pH
Before contamination (with synthetic seawater)	Not detected	1,5	7,5
Contaminated granite (after 18 contamination cycles)	4,3	1,6	7,5
After desalination treatment	Not detected	1,3	7,9

and left in room temperature for twenty one hours. The proceeding was repeated eighteenth times in order to maximize the saline contamination [6].

After the eighteenth salination cycles the chloride ion and sulphate ion content of contaminated stone were measured. The contaminated granite specimens were drilled and the collected powder was used to obtain the extract of its soluble components. The obtained powder was suspended in ultrapure water during two hours. The chloride ion and sulphate ion contents of the extract were determined by ion chromatography. The obtained values are presented in Table 1.

2.3 Electromigration Method. Materials and Experimental Proceeding

Electromigration is the main mechanism for transport of ions trough stone when an electric current is applied on it. The electric processes transform the current carried by electrons of metal electrodes in current carried by water ions which is in porous net. The electrolytic process depends on the material of which are constituted the electrodes, of the concentration and kind of ions which are in water which is in porous net and on the difference of the applied electric potential [8]. The principle of the electro-kinetic extraction of the ionic components of salts of a stone consists in movements dependent on the charge: positive ions (Na^+, K^+, Mg^{2+}, Ca^{2+}, Al^{3+}, Fe^{3+}) of porous solution of stone move towards the Cathode ($-$), and negative ions (Cl^-, $SO_4{}^{2-}$, F^-) move towards to Anode (+). In this way, the ions move from the stone nucleus and concentrate around the electrodes. For the extraction to be possible solubilisation of salts is needed. Further the migration of the solutions and transport of the ions in solution are able to begin the selective attraction. In theory, the process of salts removal will be incremental and so more crystallized salts which are in "porous space" of granite are dissolved and more ions are mobilized by electromigration.

In this research, the experimental procedures used to desalination the contaminated samples partially followed the conditions used by Feijoo et al. [6]. The electrolyte renewal was done every 24 h and the measurement of electric current 4 times a day with an interval of one hour, and in this way that renewal was done once a day. However, we must note that the desalination experimental procedures have the total duration of seven days as in [6] and from the obtained results (see Table 2) we can

Table 2 pH values of electrolytes poultice/sponges system applied in anode and cathode

pH values of electrolytes of system applied in sample 4								
Hours								
3		24	48	72	96	120	144	168
Anode	12,2	12,2	11,9	11,8	11,9	12,1	12,0	12,1
Cathode	7,7	9,8	11,1	10,8	9,7	11,2	11,2	11,3
pH values of electrolytes of system applied in sample 5								
Hours								
3		24	48	72	96	120	144	168
Anode	11,9	12,3	11,9	11,9	11,9	8,7	12,0	12,1
Cathode	8,1	10,5	11,2	9,1	10,1	12,0	9,6	10,4

conclude that the reduction in the number of renewal operation of the electrolytes was not relevant to the final result.

Two connected granite samples were used (see Fig. 1). The graphite electrodes (of 80 mm × 80 mm × 3mm) were placed on the granite surfaces (with the same dimensions). The contact material used was a composed system of sponge (of 3 mm thickness impregnated with electrolyte) and absorbent poultice (of 2 mm thick). The impregnated sponge had the objective of establishing the stable ionic contact with the graphite electrodes, hence, promoting the distribution of a homogeneous electric field and constitute a reservoir of the electrolyte. The electrolyte used is a solution of $0,2\ M$ sodium citrate and $0,2\ M$ citric acid buffered to a pH = 6. This solution avoid that the variations of pH in electrodes to alter significantly the pH of granite during the electromigration and, in this way, cause the alteration of the components more susceptible to oxygen-reduction processes such as the iron-magnesian minerals. On the other hand, reduces the electromigration of OH^- radicals formed throughout the cathodic reaction which promotes the transfer of the chloride ion during the electrochemical extraction [9].

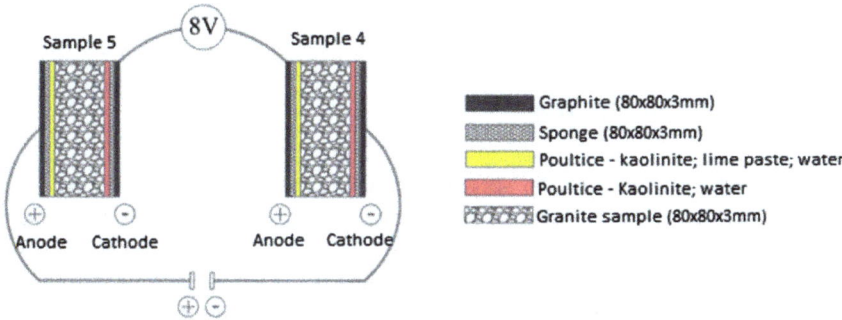

Fig. 1 Used systems in desalination test, connected to an electric circuit

To neutralize H^+ ions produced at the anode, a poultice obtained with a mixture of kaolinite, calcium carbonate putty and ultrapure water (in the volumetric ratio of 1:2:2) was used. To avoid drying, the poultice was in contact with the sponge impregnated with the above referred citric solution. This system was covered with an adhesive plastic film before its application. At the cathode it was used a kaolinite putty (1:2 of water, in volume), to guarantee the physical contact between the sample and the sponge, collecting there the mobilised ions. The function of those apposite is essentially to recollect the ions removed from the solution, besides guarantee a good contact with the granite specimens' surface.

To manage the fall in pH at the anode and keep the necessary mobilization of ions from the stone to the electrodes, at the end of 3, 24, 48, 72, 96, 120, 144 e 168 h, the poultices and the sponges with the electrolyte were replaced, being evaluated the pH of each sponge/poultice system used at the end of each cycle (see Table 2).

The graphite electrodes were connected to the two combinations which assemble the different elements in following sequence: graphite (+), sponge, poultice of lime putty, granite specimen, kaolin poultice, sponge, graphite (-) (see Fig. 1). In this experimental test of desalination an electric circuit of constant voltage of 8 V was applied and maintained by a Philips PE 1542 power supply. The test lasted eight days. Every day, each sponge/poultice system of the electrodes units was replaced by a new one, assembled to the granite specimen.

During the test, the following parameters were evaluated: the intensity of the electric current in the circuit and the voltage drop detected in each granite specimen; the amount of ions extracted and retained in poultices and sponges, in each electrolytes renewal; the pH values of the electrodes results are presented in Table 3. For analysing the concentration of Cl^- and SO_4^{2-} accumulated in the poultice/sponge systems at the end of each cycle their soluble extracts (by immersion in 100 ml of ultrapure water for one hour), were obtained. Then the extracted material was filtered and the reminiscent liquid was used to determine the chloride ion and sulphate ion by ion chromatography. After desalination test, the granite specimens were dried in the open air (during 21 h) and were perforated for collection of the powder samples and determination of relevant ions content. The same method was used for extraction and quantification.

3   Discussion of Results

The electrical resistance of the specimen/poultice/sponge varied during the test, probably in consequence of the concentration of dissolved ions within the granite specimens. Figures 2 and 3 and Table 3 indicate that, during the process of ions extraction and mobilization, the electrical resistance (Ω) (and resistivity) and electric current intensity (mA) measured for specimen/poultice/sponge systems increased. Ions migration from granite specimens in direction to the electrodes contributed to drop of electric current intensity and the reciprocal increasing of resistance to current flow.

Table 3 Desalination test results, at 0, 3, 24, 48, 72, 96, 120, 144, 168 and 192 h after electrolyte renewal

Granite sample	Time (hours)	Difference in voltage (V)	Electric current intensity (mA)	Electric Resistence (Ω)	Cl^- Concentr. (mg/l)	SO_4^{2-} Concentr. (mg/l)	pH
5	0(*)	1,94	6	323,33	4,3	1,6	7,5
4	0(*)	6,06	6	1010	4,3	1,6	7,5
5/Eletrolytes	3	3,93	2,56	1523,26			
Ânode	3				61,1	0,85	11,9
Cathode	3				44,1	3,3	8,1
4/Eletrólytes	3	4,07	2,56	1589,84			
Ânode	3				157,3	2,2	12,2
Cathode	3				21,3	0,6	7,7
5/Eletrolytes	24	4,86	2,95	1647,46			
Ânode	24				233,2	1,1	12,3
Cathode	24				10,3	0,66	10,5
4/Eletrólytes	24	3,14	2,95	1064,40			
Ânode	24				227,4	1,5	12,2
Cathode	24				9,0	0,90	9,8
5/Eletrolytes	48	2,92	5,22	559,39			
Ânode	48				278,2	1,8	11,9
Cathode	48				29,8	1,0	11,2
4/Eletrólytes	48	5,08	5,22	973,18			
Ânode	48				281,8	1,6	11,9
Cathode	48				18,9	0,87	11,1
5/Eletrolytes	72	3,81	3,04	1253,29			
Ânode	72				115,4	1,2	11,9
Cathode	72				21,5	1,0	9,1
4/Eletrólytes	72	4,19	3,04	1378,29			
Ânode	72				302, 2	2,2	11,8
Cathode	72				28,0	3,1	10,8
5/Eletrolytes	96	0,25	4,17	59,95			
Ânode	96				43,9	1,8	11,9
Cathode	96				33,7	1,2	10,1
4/Eletrólytes	96	7,75	4,17	1858,51			
Ânode	96				149,7	1,1	11,9
Cathode	96				14,5	1,8	9,7
5/Eletrolytes	120	3,89	3,85	1010,39			

(continued)

Table 3 (continued)

Granite sample	Time (hours)	Difference in voltage (V)	Electric current intensity (mA)	Electric Resistence (Ω)	Cl⁻ Concentr. (mg/l)	SO₄²⁻ Concentr. (mg/l)	pH
Ânode	120				57,8	4,1	8,7
Cathode	120				22,8	0,84	12,0
4/Eletrólytes	120	4,11	3,85	1067,53			
Ânode	120				86,6	3,8	12,1
Cathode	120				3,8	3,0	11,2
5/Eletrolytes	144	3,73	2,29	1628,82			
Ânode	144				69,8	0,79	12,0
Cathode	144				13,1	2,8	9,6
4/Eletrólytes	144	4,27	2,29	1864,63			
Ânode	144				31,7	2,1	12,0
Cathode	144				11,8	2,2	11,2
5/Eletrolytes	168	4,13	1,78	2320,22			
Ânode	168				123,7	1,8	12,1
Cathode	168				15,3	1,6	10,4
4/Eletrólytes	168	3,87	1,78	2174,16			
Ânode	168				26,5	0,91	12,1
Cathode	168				14,2	0,59	11,3
5	192				-	1,3	7,9
4	192				-	1,3	7,9

(*) Values in granite powder samples after contamination

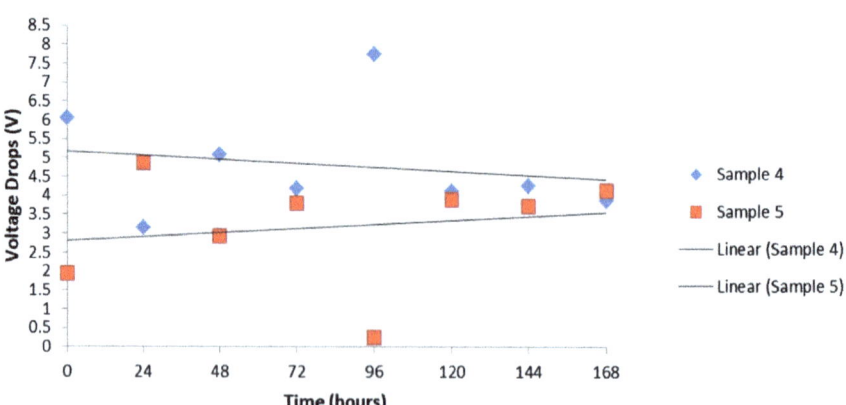

Fig. 2 Values of voltage drops (V)

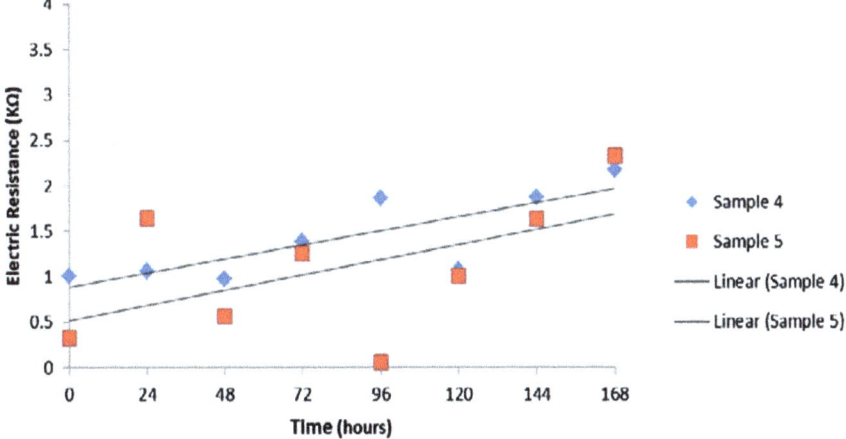

Fig. 3 Values of electric resistance (KΩ)

Figures 4 and 5 show the values measured for chloride ion content accumulated in poultice/sponge system at each cycle. The initial concentration of Cl^- was the greatest at the anode, in the first day, but in the following seven days a significant reduction was observed, indicating that the accumulation reduced. This indicates the decreasing of the amount of ions that were extracted from the granite specimens during the desalination process. On the contrary, results for SO_4^{2-} ion (Figs. 6 and 7) show a deficient extraction.

Having in consideration the initial contents in chloride and sulphate in the granite specimens, a drastic reduction in chloride ions was observed, that indicates the complete extraction from contaminated specimens saturated by synthetic sea water (see Table 3), and a slight reduction, although considered significant, of sulphate ions. The high concentration of sulphate ions in poultice/sponge systems affected, in some way, but didn't prevent extraction process of chloride ions (see Figs. 4, 5, 6

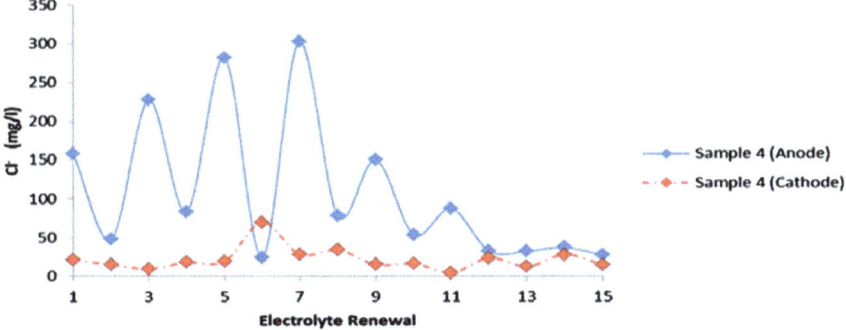

Fig. 4 Cl^- content in poultices and sponges measured at each electrolyte renewal (Sample 4)

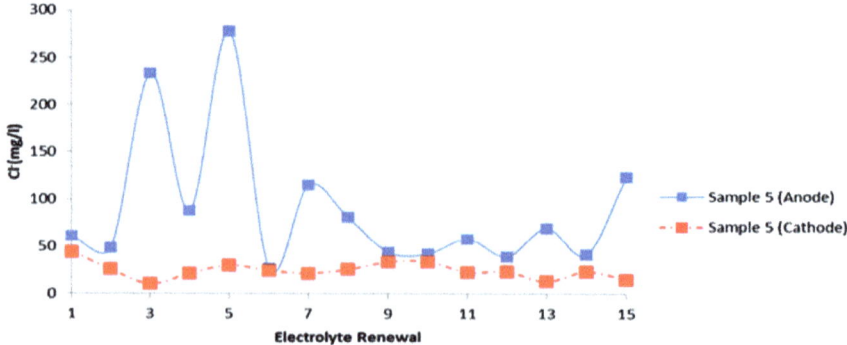

Fig. 5 Cl^- content in poultices and sponges measured at each electrolyte renewal (Sample 5)

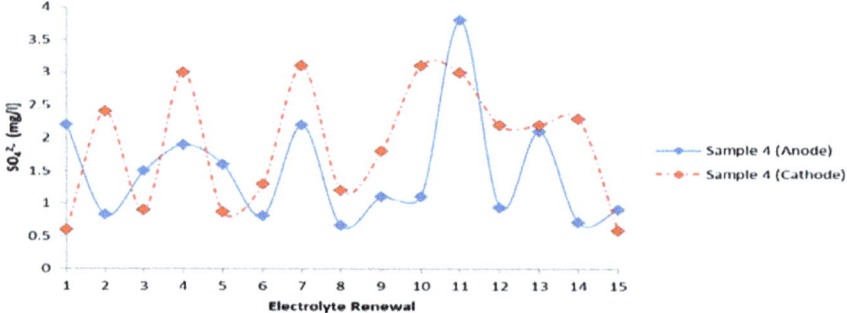

Fig. 6 SO_4^{2-} content in poultices and sponges measured at each electrolyte renewal (Sample 4)

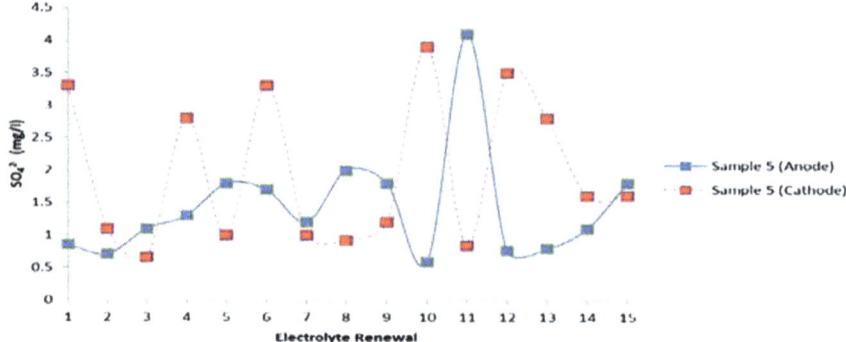

Fig. 7 SO_4^{2-} content in poultices and sponges measured at each electrolyte renewal (Sample 5)

and 7). In the beginning of the test, the ions are concentrated in a higher percentage at the anode, but this ionic concentration was reduced, after a certain number of renewal operations of the electrolytes, indicating that the chloride and sulphate ions have been extracted from rock in progressively smaller amounts.

The pH measurements at electrolytes throughout the test (see Table 2) reveals that, at the beginning, the anions were accumulated around the anode but ionic levels were decreasing after a certain number of cycles. The pH variations detected in renewal cycles were minimal. The oscillation in pH at the electrolytes indicates that they act as a buffer component, such as: the use of citrate of sodium and citric acid in the sponges prevented the fall in pH at the cathode, as the result of acidification at the anode; the calcium carbonate poultice, a mixture of kaolinite and $CaCO_3$, was efficient in neutralizing the acidity trough the anodic process and remained the granite pH stable (pH between 7 and 8) (see Table 1). After three hours: the electrolyte (poultice/sponge) inserted between the cathode and the granite specimen 5 had the pH value of 8,1; and the electrolyte (poultice/sponge) inserted between the cathode and the granite specimen 4 had the pH value of 7,7. After 24 h, the poultice/sponge systems became alkaline, the pH was 10,5 and 9,8, for the same granite specimens (Table 2).

The difference in average values of salts extraction between 4 and 5 specimens is clearly due to the differences in contact surfaces of the two specimens, the granite specimen 4 was perforated, presenting a hole of 5 mm diameter which crosses its thickness, at the centre of the greater surfaces. This hole in specimen 4 allows a greater movement of the fluid, carrying ions in migration towards an electric field.

4   Conclusions

The results of this research demonstrate the effectiveness of electrochemical methods in the extraction of soluble salts from granite stones and controlling the pH of stone surface. The specimens subjected to the salt electromigration test were not subjected to extreme acidification conditions, in order to ensure that the pH of the stone remained stable.

In the granite desalination test presented here the complete extraction of Cl- was achieved while the remove of $SO4^{2-}$ is not meaningful. The presence of $SO4^{2-}$ ions causes a decrease in the current carried by Cl^- ions, therefore a decrease in chloride extraction efficiency is observed. Differences in the desalination effectiveness are attributed to the different shape of the specimens.

The electrical characteristics of the circuit, current intensity (A), potential difference (V) and electrical resistance (Ω), have higher values for granite specimen 4 than for granite specimen 5. These differences may be explained by the aforementioned fact that the surfaces of the granite specimen 4 in contact with the electrodes were punctured in the centre.

Further research will be needed to determine the effectiveness of applying this in situ technique on masonry or masonry covered with plaster and stucco for the conservation of monuments.

Acknowledgements The authors would like to thank the support of UTAD Laboratories of Geology, Chemistry, Electrotechnics and Civil Engineering, as well from CITAB/UTAD, for development of this research work. This work is supported by National Funds by FCT–Portuguese Foundation for Science and Technology, under the project UIDB/04033/2020.

References

1. Marinho, J.N.L.B.: Avaliação de estuques e rebocos históricos: a Igreja de São Lourenço (Igreja dos Grilos), no Porto, e a Casa do Calvário, em Amarante (Historical plasters and stuccos assessment: the Church of St. Lawrence (Grilos' Church, at Oporto, and Calvary House, at Amarante). Master thesis in Civil Engineering of University of Tras-os-Montes e Alto Douro, Vila Real, Portugal, pp. 155–170 (2016)
2. Salavessa, E., Candeias, A., Mirão, J., Sousa, L.M.O., Duarte, N., Jalali, S., Salgueiro, J.: 19th coloured stuccos and plasters from Grilos' Church (Oporto, Portugal): materials and techniques employed. Color Res. Appl. COL22041 **41**(3), 246–251 (2016). https://doi.org/10.1002/col.22041. (Unique ID: 4863667–1552848)
3. Normandine, C., Slaton, D.: Cleaning techniques, In: Kemp, J., Henry, A. (eds.) Marble, stone conservation/principles and practice. Donhead Publishing Ltd, United Kingdom, pp. 127–157, 226–227 (2006)
4. Kamran, K.: Electrokinetic desalination of porous building materials. Eindhoven University of Technology, Department of Applied Physics (2012)
5. Rörig-Dalgaard, I.: Development of a poultice for electrochemical desalination of porous building materials: desalination effect and pH changes. Mater. Struct. 46, 959–970 (2013)
6. Feijoo, J., Ramón, X.N., Rivas, T., Mosquera, M.J., Taboada, J., Montojo, C., Carrera, F., Granite desalination using electromigration. Influence of type of granite and saline contaminant. J. Cultural Herit. 14, 365–376 (2013). https://doi.org/10.1016/j.culher.2012.09.004
7. LNEG, Catálogo de Rochas Ornamentais Portuguesas. (Catalogue of Portuguese Ornamental Stones). http://rop.lneg.pt/rop/?&lg=pt
8. Ottosen, L.M., Rörig-Dalgaard, I., Villumsen, A.: Electrochemical remove of salts from masonry-experiences from pilot scale. In: Proceeds from the "International Conference on Salts Weathering of Building and Stone Scultures", pp. 22–24, the National Museum, Copenhagen, Technical University of Denmark, Department of Civil Engineering, (Publishing), Denmark (2008)
9. Castellote, M., Andrade, C., Alonso, C.: Electrochemical removal of chlorides: modelling of the extraction, resulting profiles and determination of the efficient time of treatment. Cement Concr. Res. **30**, 615–621 (2000). https://doi.org/10.1016/S0008-8846(00)00220-9

Pathological Manifestations of Neoprene Support Devices in Infrastructure

Fernando R. Gonçalves, Mohammad K. Najjar, Ahmed W. A. Hammad, Assed N. Haddad, and Elaine G. Vazquez

Abstract The so-called works of special arts are constructions of high complexities that allow the advancement of widening gaps and overcoming obstacles previously unthinkable. With the increase in the magnitude of these structures, in addition to greater investments, the maintenance of these structures becomes an increasingly important factor for engineering. Among the elements of bridge structures, the support devices are components with important structural functions, being essential for their proper functioning and especially the durability of the entire structure. The culture of inspection and maintenance of road bridges, railroads, and viaducts in Brazil is recent. Only in 2016 was approved a specific standard for the inspection work on bridges, viaducts and concrete walkways, ABNT NBR 9452. This paper aims to evaluate the pathological manifestations in neoprene support devices so, according to inspections performed and the diagnosis of causes, define their best practices and treatments for the maintenance and mitigation of the pathologies found. In the practical study, the following steps were performed: survey and selection of the structures currently under maintenance of MetrôRio; selection of criteria for the evaluation of pathologies; carrying out inspections; comparative analysis between the viaducts to determine the priority order for negotiations; and definition of conduct. The results obtained were the result of the evaluation of the field analysis, diagnosis, and comparison with tests performed in support devices. Having as input the tests in the support devices, the best treatments and suggestions to avoid new pathologies were proposed. With the definition of the pillars that concentrated the largest number of critical devices, aiming at better use of the operation, the decision was made to replace 12 units, from the perspective of urgent replacement. The novelty of this work was the development of better informed and more systematic approaches to condition assessment, deterioration forecasting, and maintenance decision making over the life-cycle of the built asset.

F. R. Gonçalves · M. K. Najjar · A. N. Haddad · E. G. Vazquez (✉)
Escola Politécnica, Department of Civil Construction, (UFRJ), Rio de Janeiro, Brazil
e-mail: elaine@poli.ufrj.br

A. W. A. Hammad
Faculty of Built Environment of New South Wales University, Sydney, NSW, Australia

J. Sena-Cruz et al. (eds.), *Proceedings of the 3rd RILEM Spring Convention and Conference (RSCC 2020)*, RILEM Bookseries 34,
https://doi.org/10.1007/978-3-030-76465-4_21

229

Keywords Support device · Infrastructure · Pathological manifestation · Maintenance

1 Introduction

Throughout history, bridges have been built to exemplify the engineering prowess of a civilization, many enduring longer than the empires that built them [1]. Bridge infrastructure works are buildings of high complexities that allow the advancement of increasingly large spans and overcoming previously unthinkable obstacles.

With the increase in magnitude of these buildings, in addition to greater investments, the maintenance of these structures becomes an increasingly important factor for engineering. Designing, building and maintaining good quality and reliable physical infrastructure is fundamental to the smooth functioning and economic development of any society [2]. Since a high cost of deployment comes with extended life expectancy, durability and unchanged usage condition.

Bridges and infrastructures systems, due to their inherent vulnerability, are at risk from ageing, fatigue and deterioration process due to aggressive chemical attacks and other physical damage mechanisms [3]. Proper and continuous maintenance ensures longer service life of the structure, with satisfactory functional and structural performance. This preventive and corrective maintenance should be part of a comprehensive management process, including periodic surveys aimed at identifying any existing structural anomalies and failures, diagnosing them and then defining recovery and treatment actions, if necessary [4–6].

The culture of inspection and maintenance of road bridges, railroads and viaducts in Brazil is recent, being from the 80s the first studies of pathologies in the structures [7]. There is a specific standard for the inspection work on bridges, viaducts and concrete walkways, ABNT/NBR 9452/2016 [8].

A bridge bearing is the structural member to be installed in the connecting portion between superstructure and substructure. The purpose of the bearing is to transfer load from superstructure to substructure, to allow necessary rotation and expansion of superstructure by live load and temperature variations and to absorb relative motion of superstructure and substructure [9].

Support devices are components with structural functions essential for the proper functioning and durability of the structure, but not just a sample of the entire structure, the support device alone can already represent a maintenance point of the structure, so its monitoring Continuous inspections are considered very important in the bridge maintenance management process [10]. Bridge design is based on whole bridge analysis model, which the bearing functions are reflected. When the bearing functions deteriorate, it is possible for structural system of the bridge to be changed and have an effect on functions of superstructure and substructure [9].

It can be said that knowledge of the state of the bearing apparatus is a good sign and well represents the state of the bridges in their entirety in structural terms. Therefore, the analysis of their pathologies, causes and origins is of great importance in defining the treatment and maintenance of bridges, elevations and viaducts. Thus, the manifestation of pathologies in such an important element requires a comprehensive theoretical analysis so that the solutions taken are more effective both financially and technically [11].

Metal and concrete support devices expose some problems that discourage their use, either in terms of maintenance difficulty, poor property of materials or even the built-in cost. Therefore, over time, it was searched for elements that could cover all the needs of a support device, this way arose the support devices in elastomer, based on polychloroprene, whose widespread trade name is neoprene which as a product industrialized, it presents greater uniformity of physical characteristics, as well as exceptional resistance to light and ozone, thus providing durability significantly superior to that of other types of elastomers [12].

This paper aims to present approaches towards improving some specific infrastructure maintenance principles, strategies, models and practices, based on a recent study da manifestação patológica presente em aparelhos de apoio de neoprene, localizados nas estruturas das pontes do MetrôRio. A specific goal is to develop better informed and more systematic approaches to condition assessment, deterioration forecasting, and maintenance decision making over the life-cycle of the built asset.

2 Methodology

In the methodology the following steps were used: survey of the viaducts, elevations and bridges existing in the subway railway; selection of structures to be inspected; selection of criteria used in the evaluation of pathological manifestations; conducting visual inspections based on ABNT/NBR 9452/2016 [8]; and suggestions for future interventions.

2.1 Inspection of Bridges and Support Equipment

Inspections are paramount to characterize the bridge's constituent elements and, therefore, their classification according to criteria established in ABNT/NBR 9452/2016 [8]. Each element is evaluated according to specific visual aspects defined in the standard.

According to the same standard, the following types of inspections are considered: cadastral (knowledge of the entire structure of the work, among its essential elements, are: photographic record and classification, according to parameters established in the standard); routine (periodic, visual follow-up, with or without the use of special equipment and/or resources for analysis or access, performed within a period not exceeding one year); special (must be detailed and include graphical and quantitative mapping of anomalies of all apparent and/or accessible elements) and extraordinary (description of work and identification of anomalies including mapping, photographic documentation and recommended therapy) [8].

ABNT/NBR9452/2016 [8] provides in Annex A, a basic roadmap for tokens and cadastral inspections, the proposed initial documents are described such as project data, execution record and changes in the construction phase, previous inspections, among other elements that may provide more inputs for the definition of the causes and better dealings.

Inspection work on bridges and viaducts are fundamental activities in the verification of the conservation conditions in which these works are found and provide support to the planning of maintenance work. Also, the importance of the conditions of access to the places to be inspected is important, since in many of these structures there are no adequate accesses for the inspection works, especially in the regions of the expansion joints and the support devices, places that present a high incidence of anomalies [7].

Inspection of assistive devices may not be limited to the space in which they are positioned and to the element. It is necessary to identify the general functioning of the studied artwork and to verify the compatibility with the current behavior of the support devices [13].

Because of their location, support devices are structural elements that are difficult to inspect, but their behavior must be monitored by inspectors according to the following general procedures in Table 1 [14].

According to ABNT/NBR9452/2016 [8], Table 2 can be considered as a parameter for evaluating support equipment, given its condition and the scenario to which it is exposed.

Table 1 Items to be inspected on assistive devices [14]

Checks per DNIT recommendation [14]	
Visually inspect the accessible faces of the appliance; After a few years of service, small cracks 2 to 3mm deep and 2 to 3mm long are tolerable	Check that the support device has been correctly vulcanized and that there are visible and oxidized charter steel sheets
If there is displacement of the structure, measure the angles between the surfaces of the structures in contact with the support apparatus	Measure the heights of the support apparatus at the edges and center points
Measure distortions of the support apparatus	Check that the support device has been moved from its original position
Check for the presence of oils, greases or any other substance harmful to the elastomer	Check for defective expansion joints on the superstructure, very close to the support device or directly above the device

Table 2 Device Classification [8]

Condition	Description
Critical	Support devices and/or their surroundings present breakdowns at risk of structural collapse requiring repair intervention and/or immediate device replacement
Bad	Support devices and/or their surroundings present damage that compromises structural safety without risk of collapse, requiring repair intervention and/or short-term device replacement. All devices with breakage with charter exposure fall into this classification. Follow-up is recommended and interventions may be needed in the short term
Regular	Support devices and/or their surroundings present malfunctions that may generate some structural deficiency, but there are no signs of deterioration of the devices, nor compromise of the stability of the work. Follow-up is recommended and interventions may be necessary in the medium term
Good	Support devices and/or surroundings are not malfunctioning. Interventions may be necessary in the long run
Excelent	Support devices and/or their surroundings are not damaged and the devices were manufactured from 1987 following the recommendations of ABNT NBR 9783

2.2 Pathological Manifestation in Neoprene Support Devices

Although it has excellent performance compared to other types of support equipment, especially when not in need of maintenance, the neoprene device also requires some care. Table 3 shows recurrent pathological manifestations in neoprene supports.

The causes for the deterioration of structures can be as diverse as the natural "aging" of the structure to the irresponsibility of some professionals who choose to use materials that are out of specification [15].

The causes of pathologies in structures have their origins in two groups: intrinsic causes - referring to the processes of deterioration inherent in the structure itself, ie its origin is in execution, use, human failures, etc. and extrinsic causes - external to the material body, can be understood as factors that attack the structures from the

Table 3 Pathological manifestations in neoprene apparatus [12]

Most common pathological manifestations—neoprene apparatus
High neoprene distortion
Neoprene cracking or creep
Frame contact zone shutdown
High compression on neoprene
Loss of serviceability and distortion
Variations in rubber layer thickness
Unsticking of vulcanization of inner sheets
Degradation of sliding plates, guides or stops
Oxidation of steel elements

Table 4 Causes of pathological manifestations in support devices [17].

Most common causes of pathological manifestations	
Intrinsic damage not detected during installation	Irregular seating causing additional localized overload
Displacements, rotations and loads in service much higher than estimated	Unintended aggressiveness of the environment
Attack by chemicals	Badly nesting in the crib

Table 5 Neoprene support devices treatments–adapted [12].

Damage to support devices	Repair dealer
Corrosion, presence of dust and moisture	Cleaning and painting with protective paints. Water sealing and humidity control
Massive corrosion leading to section loss	Replacement
Offset or misalignment	Component replacement or total
Neoprene deterioration or wear	Replacement
Fissures	Crack sealing or replacement
Fragmentation of concrete in support	Removal and execution of new concrete

outside inwards, throughout the process of conception, execution or the useful life [16].

The most common causes of decreased service life in assistive devices are listed in the Table 4 [17].

The treatment of a pathological manifestation should be done according to the inspection report, condition and definition of the causes of the given manifestation.

As it is a synthetic structure, of specific manufacture, it is more common that if it presents anomalies, it will be replaced by a new device. Except in the case of incorrect positioning of support devices or displacement of a Teflon sliding plate, for example. Therefore, unless the cause of the condition is incorrect positioning in the project or an improper dislocation, without permanent damage to the support device, the approaches may be twofold: Element replacement or continuous monitoring until anomalies warrant intervention for replacement.

Table 5 presents some common types of repair methods according to each pathology in neoprene support devices [12].

3   Results−Case Study of Pathological Manifestations, Their Causes and Treatments in Neoprene

Given the importance of MetrôRio to society in Rio de Janeiro, it is essential that the entire system works continuously as there is no margin to support major service disruptions. The commercial operation of trains is the priority as the company's goal is to make life easier for its customers, providing quality transportation. However, it is a fact that maintenance is an essential element for the service to be provided with quality, punctuality and reliability. Thus, the maintenance of MetrôRio works today, on a continuous basis, 24 hours a day. For the case study in this paper, the road bridges are taken out of service condition to perform maintenance on the support devices.

All MetrôRio's assets are cataloged according to an asset tree [18], which aims to give an overview, keep all history of interventions, corrective or preventive, and maintenance plans in force combined with each group of systems and equipment. Thus, the support devices studied in this work are under the structures system.

Seeking greater efficiency, lower cost and higher system reliability, MetrôRio carries out preventive maintenance plans for each group of structures. Most maintenance plans concern visual inspections of assets and, identifying a possible pathological manifestation, a corrective note is opened to address the structure and mitigate its degradation.

The object of study of this work were the support devices and their surroundings. The elevations between São Cristóvão and the MetrôRio Maintenance Center and between the Triagem and Maria da Graça stations are the oldest in the system, and their construction dates back to the late 1970s, or about 35 years of operation.

The recommendations of DNIT inspections were followed [14]. Field information has been posted in a specific form for this service. A specific inspection campaign was carried out for the support devices of each structure. As an input for maintenance decision-making, 1228 neoprene support devices were performed along the entire railway structure. The inspections were performed by specialized technicians using the visual method, with the naked eye, and the access was made by stairs with a reach of 9.00 meters in height. The fieldwork was accompanied by a professional specialized in occupational safety, all safety procedures were complied with in accordance with current legislation. No device has been classified in the Excellent Class because it is over 30 years old and has not been manufactured [19].

The CNV–SCR Elevated is located between the Maintenance Center–MC and the São Cristóvão station (later, there was also the connection of the elevated with Cidade Nova station). The old elevation has a total length of 970 m and consists of 30 spans, 4 spans with 4 beam, 1 span with 3 beam and 25 spans with 2 beam. The trays are seated on 28 pillars and two staked joints at the longitudinal ends, constituting isostatic spans. In the beam × pillar and beam × encounter interface there are chartered neoprene support devices with regular dimensions of $700 \times 250 \times 40$ mm. In this elevation, 30 pillars and 138 support devices were evaluated, two of which were not inspected for being covered. It was found that most of the assistive devices (79 units) inspected at this stage were classified as being in a regular state of conservation (57%). Supporting devices fitted with poor condition total 57 units (41%), as shown in Fig. 1.

Elevated TRG -MGR is mostly located between the Triagem (TRG) and Maria da Graça (MGR) stations and starts after leaving the Bernold tunnel next to the Mangueira Olympic village. The Elevado has a total length of 2,925 meters and is formed by 84 spans, mostly with 4 precast beams each, with chartered neoprene

Fig. 1 Support device
conditions–CNV–SCR

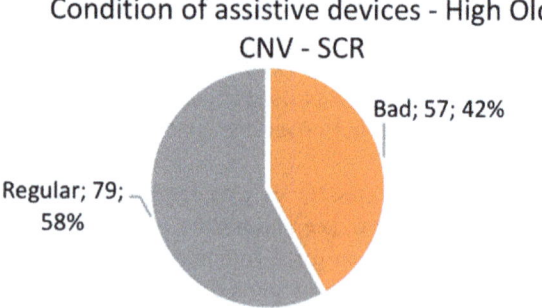

Condition of assistive devices - High Old
CNV - SCR

Bad; 57; 42%

Regular; 79;
58%

appliances with dimensions 750 × 200 × 40 mm. In this elevation, 361 support devices were evaluated, and 37 pillars were not inspected because they were in a risk area where access could endanger the inspection team. It was found that most of the support devices (269 units) inspected at this stage were classified as being in a regular state of conservation (72%). Poorly classified support devices total 78 units (21%), 9 units (3%) were considered in good condition and 5 critical support devices (1%), as shown in Fig. 2.

With the definition of the pillars that concentrated the largest number of critical devices, aiming at a better use of the operation, the decision was made to replace 12 units, from the perspective of urgent replacement. Tables 6, 7, 8, 9 and 10 present a summary of the main pathological manifestations, degree of risk, possible cause and indicated treatment.

Fig. 2 Support device
conditions–TRG–MGR

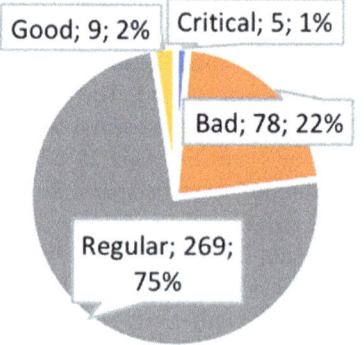

TRG - MGR High Support Appliance
Conditions

Good; 9; 2%   Critical; 5; 1%

Bad; 78; 22%

Regular; 269;
75%

Table 6 Pathological manifestations in the devices and support

Pathological manifestation—bojection	
Pathology	Volume expelled in the elastomer that takes the bulging shape on its sides.
Risk	Regular risk—although not a worrying pathological manifestation to the point of rapid degradation of the device or its surroundings, it is a first sign that the device is reaching its limit of use due to deformation
Cause	The most common cause of this pathological manifestation is the duration of use of the support device. Boosting on older appliances may crack due to rubber oxidation and dryness over time
Treating	Follow-up is recommended and interventions may be necessary in the medium term

Table 7 Pathological manifestations in the devices and support

Pathological manifestation—fissures	
Pathology	During the inspections, all the cracks found were measured to verify their advance in fronts of subsequent inspections.
Risk	Regular risk—although cracks are common and even tolerable, after a few years of service they can compromise the structure
Cause	The most common cause of this pathological manifestation is the duration of use of the support device. The dryness of the surface causes the bulge to evolve into a crack. Lack of cleaning or chemical attack due to pigeon droppings, for example, may potentiate surface rubber attack
Treating	Follow-up is recommended and interventions may be necessary in the medium term. DNIT itself [14] has a tolerance margin for cracks found: visually inspect the accessible faces of the device; After a few years of service, small cracks 2 to 3mm deep and 2 to 3 mm long are tolerable

Table 8 Pathological manifestations in the devices and support

Pathological manifestation—exposed chartering	
Pathology	The advancement of the cracks allows the environment aggression of the vulcanized metallic sheets next to the neoprene. Exposed charter plates are much more vulnerable to corrosion and will therefore lose their charter capacity along with neoprene plates, reducing the ability of the support apparatus to move and store the support conditions.
Risk	Bad risk—all handsets with breakthroughs with charter exposure fall into this classification
Cause	The most common cause is the presence of moisture that influences the corrosion process of the charters in the vicinity of the devices. This process is accentuated as the expansion joints are not sealed, and sometimes insufficient drainage system
Treating	Follow-up is recommended and interventions may be needed in the short term

Table 9 Pathological manifestations in the devices and support

Pathological manifestation—stringer beam degradation

Pathology	Concrete detachments with exposed reinforcement in corrosion process, located in the vicinity of the appliances that compromise the structural safety.
Risk	Critical risk—severe pathological manifestation with compromised structural safety, but can be treated if diagnosed before advanced reinforcement corrosion, without risk of collapse
Cause	One of the major reasons for the degradation of abutment heads and their elements is the deficiency in the drainage system. In this case, the cause of the pathological manifestation about the support device is no longer predominant for the treatment, since the most important and urgent treatment is the structure around the device, already condemning the conditions of use of the device in place
Treating	These pathological manifestations in the beam express the need for urgent treatment, as they can effectively compromise structurally a stringer beam, resulting in structural deficiency of all OAE. In neoprene support devices, it is not common to perform maintenance and treatment interventions on the element, but its immediate replacement

Table 10 Pathological manifestations in the devices and support

Pathological manifestation—cradle degradation (Plinth)	
Pathology	Anomaly around the support apparatus, but caused by malfunction of the element.
Risk	Critical risk–severe pathological manifestation with compromised structural safety
Cause	From a malfunction of the supporting bond between the superstructure and the mesostructure of the elevation
Treating	The treatment, in addition to the replacement of the support apparatus, should include reinforcement of the plinth with its recomposition at the bottom, called the cradle. As said, in neoprene support devices, it is not common to perform treatment interventions on the element, but rather its immediate replacement

4 Conclusions

The concern with the maintenance of structures such as those of special art works was motivating for the work and allowed to relate the pathological manifestations in support devices with the pathological manifestations of the structures, the intrinsic and extrinsic causes as a whole, allowing an analysis, albeit superficial in the field of subject matter, sufficient for decision

Supporting devices are structural connecting elements which allow forces to be transmitted between the superstructure of the artwork and its support and are therefore essential elements for the proper functioning of the structure into which they are inserted.

The inspection processes took place exactly in accordance with all the literature found, allowing great inputs for subsequent decision making. Thus, the pathological manifestations were quite explicit.

As for the causes of the manifestations, it is a complex study and although there is literature, it is not trivial to understand the reason why two support devices, theoretically manufactured under the same process, of the same age, suppose stored in the same form, are adjacent to each other, exposed to very close loads and exhibit behaviors so distinct in terms of behavior in service.

Even with breakage and chartering exposed in the corrosion process, the devices can still perform satisfactorily without causing movement restriction of the part. But in these cases, annual monitoring is essential to follow up on a case-by-case basis to make sure that the performance and operation of the chartered neoprene parts is still adequate.

As for the process of negotiations, in cases where no substitution was considered, monitoring will take place in accordance with the first inspection. Thus, the big point of the issue of support devices is to understand their operation not individually but in conjunction with adjacent structures, as it is evident that despite some anomalies found, support devices, except those that had signs of degradation around them, they were still able to remain in service, provided they were well monitored.

References

1. Wilson, K.: Building bridges: from simple stone spans to complex constructions of iron and concrete, Britain's abundance of bridges offers endless opportunities for photographic creativity. (GEO photo) Cengage Learning, Inc. Geographical, vol. 81(9), p. 82(3) (2009)
2. Yang, Y., Kumaraswamy, M.: Towards life-cycle focused infrastructures maintenance for concrete bridges. Facilities **29**(13–14), 577–590 (2011)
3. Biondini, F., Frangopol, D.M.: Design, assessment, monitoring and maintenance of bridges and infra structure networks. Struct. Infrastruct. Eng. **0311**(4), 413–414 (2015)
4. Vitório, J.: Vistorias, Conservação e Gestão de Pontes e Viadutos de Concreto. Anais do 48° Congresso Brasileiro do Concreto (2006)
5. Kainuma, S., Ahn, J-H., Jeong, Y-S., Imamura, T., Matsuda, T.: Applicability and structural response for bearing system replacement in suspension bridge rehabilitation. J. Constr. Steel Res. **95**(19), 172 (2014)

6. Vitório, J.: Pontes rodoviárias: fundamentos, conservação e gestão. Recife: CREA-PE (2002)
7. Araujoa, C.: Main aspects covered in ABNT NBR 9452: 2016, the importance of maintenance activities in bridges and viaducts and the difficulties of the conditions of access to the inspections. Revista IPT I Tecnologia e Inovação vol. 1, no. 5, ago. (2017)
8. Associação Brasileira de Normas Técnicas: ABNT NBR 9452: Vistorias de pontes, viadutos e passarelas de concreto, p. 48. ABNT, Rio de Janeiro (2016)
9. Sakano, M., Shivasaki, N.: Diadnosis of bridge based on field measurements. ISEC PRESS (2016). https://doi.org/10.14455/ISEC.res.2016.160. ISBN978-0-9960437-3-1
10. Freire, L., de Brito, J., Correia, J.: Inspection survey of support bearings i road bridges. J. Perform. Constr. Facil. **29**(4) (2015). Accessed 01 Aug 2015
11. Freire, L., de Brito, J., Correia, J.: Management system for road bridge structural bearings. Struct. Infrastruct. Eng. **10**(8), 1068–1086 (2014). Accessed 03 Aug 2014
12. Cordeiro, J.: Aparelhos de Apoio em Pontes Vida Útil e Procedimentos de Substituição. Dissertação de Mestrado. ISEL (Instituto Superior de Engenharia de Lisboa). Lisboa, PT, dezembro de (2014)
13. Da Silva, R.: Avaliação da resistência e funcionalidade dos aparelhos de apoio da ponte Rio-Niterói após 40 anos em serviço. Tese de Doutorado. Universidade Federal Fluminense Rio de Janeiro–RJ (2016)
14. DNIT, Departamento Nacional de Infra-Estrutura de Transportes. Manual de Inspeção de Pontes Rodoviárias, 2ª Edição, Rio de Janeiro–RJ (2004)
15. Souza, V., Ripper, T.: Patologia, Recuperação e Reforço de Estruturas de Concreto. 1 ed. São Paulo: Pini (1998)
16. Sartorti, A.: Identificação de patologias em pontes de vias urbanas e rurais no município de Campinas-SP. Campinas: Faculdade de Engenharia Civil–UNICAMP, p. 203 (2008)
17. Departamento de Estradas e Rodagem (DNER). Substituição de aparelhos de apoio e juntas de dilatação–Código 020. São Paulo (2006)
18. Associação Brasileira de Normas Técnicas. ABNT NBR ISO 55000:2014 Gestão de Ativos–Visão geral, princípios e terminologia. Rio de Janeiro (2014)
19. Associação Brasileira de Normas Técnicas: ABNT NBR 9783/87 - Inspeções em pontes e viadutos de concreto armado e protendido–Procedimento, p. 18. DNIT, Rio de Janeiro (2004)

Experimental Investigation on Organic Water Slaked Lime Mortar

Riya Susan George and Simon Jayasingh

Abstract Lime has been used as the most sustainable binder for the conservation and restoration of heritage buildings. The repair properties of lime based mortars are found superior to other mortars due to the compatibility issues of other mortars with ancient heritage structures. To increase the compatibility, the fresh lime mortar needs to be hardened by carbonation which is a slow process. Various researches showed that admixtures added advantage to lime mortar by increasing carbonation rate which increases the mechanical strength, weather and water resistance of the lime mortar. In India, plants, fruits and some part of vegetables were used as organic admixtures that increased the carbonation rate thereby promoting sustainable and eco-friendly repair mortar. This paper deals with how the slaking of lime putty with organic additives affect the carbonation and other properties of the repair mortar. To enhance the carbonation rate, organic was mixed with water and this organic water was added for slaking the lime putty. Raisins were added as the organic additive at an amount of 0, 2, 4 and 6 percentages. The process was monitored for a month and cubes were casted. Compressive strength test was performed to measure the mechanical resistance property of lime mortar. Analytical tests such as XRD analysis and FT-IR were performed to determine the chemical compatibility after curing it for 90 days. It was seen that the carbonation rate and the mechanical strength of lime was greatly influenced by the organic water slaked lime mortar.

Keywords Conservation · Lime Mortar · Carbonation · Slaking · Organic Additive

1 Introduction

Mortars are one of the oldest building materials we have been using which connect the different masonry units. Mortars act as a protective cover for the buildings; all the buildings and monuments in old centuries were discovered to build with lime mortar as the major ingredient [1]. Lime is a white alkaline material which consists of CaO

R. S. George (✉) · S. Jayasingh
Vellore Institute of Technology, Vellore, Tamil Nadu, India

© The Author(s), under exclusive license to Springer Nature Switzerland AG 2022
J. Sena-Cruz et al. (eds.), *Proceedings of the 3rd RILEM Spring Convention and Conference (RSCC 2020)*, RILEM Bookseries 34,
https://doi.org/10.1007/978-3-030-76465-4_22

and combines with water with the production of much heat. Natural hydraulic lime (NHL) is produced from limestone that contains clay and other impurities naturally. In 19th century, lime is mainly used for the restoration of heritage structures throughout the world [2]. Further Portland cement came in to role and it was used as a substitute for lime. But restoration works using other mortars didn't prove much effective due to compatibility issues such as low porosity, poor compatibility with ancient heritage structures and also formation of various soluble salts in the mortar [3]. Also production of cement requires huge amount of energy and the carbon dioxide produced during the cement production is 8%. However lime requires lower energy requirement and low CO_2 emission during its production and thus it is the most sustainable binder [4].

Carbonation process is an irreversible exothermic reaction where calcium carbonate is formed by the reaction between calcium hydroxide with carbon dioxide. This process alters the microstructure of the mortar thereby imparting more strength to the lime mortar [5]. The process can be explained through various stages- penetration of CO_2 through the pores in the mortar, a balance formation between the atmospheric CO_2 and dissolved CO_2 in the water present in the pores, reaction of CO_2 with water to form carbonic acid, dissolution of CO_2 to calcium and hydroxide ions which is affected by the microstructure of the mortar as well as the amount of water present in the pores, calcium carbonate precipitation as the water contains both calcium and carbonate ions [6].

Carbonation gives strength to the mortar which depends on the porosity in the structure of the mortar [7]. But more porous mortar has less compressive strength and also durability issues. So a balance is required between these factors which can be attained by adding organics as additives. Several studies are going on regarding the addition of organics to the lime mortar. It was seen that adding organic to lime increases its rate of carbonation. It also increases the mechanical strength, weatherresistance and water resistance compared to common lime mortar [8].

Quick lime (CaO) combines with water to form calcium hydroxide $Ca(OH)_2$ which can be used as binder in a mortar. This is the process of slaking. This calcium hydroxide further combines with carbon dioxide to form calcium carbonate. Slaking improves the surface area of the mix [9]. During ancient times, slaking was carried out in excavated soil pits–where the required amount of lime to be slaked was covered with a layer of sand. Water was sprinkled in all the directions .the lime will get dissociated inside the pit which was accompanied by smoke through the cracks inside the sand layer where it turned into fat lime which acts like a viscous and elastic mortar. Several others methods were done for slaking which includes adding additives [10].

During slaking the amount of water that was added to lime was kept constant and organic was added to increase the slaking rate of the lime putty. Various parameters like free flowing water, physically bound water and consistency of lime putty was found out. Free flowing water (w_s) is the drained water that can be measured after slaking the lime for 24 h [11]. The water that settled during slaking can be also found from the drained water during the days of slaking. Physically bound water (w_f) is the undrained or non settled water that can be removed only by heating. Consistency of lime putty was measured using a modified Vicat apparatus and the additional water required for each day of casting was found out [12].

In this research paper, raisins are added as additives to modify the properties of lime-mortar [13]. X-ray fluorescence test (XRF test) was conducted to categorise the quick lime for slaking purpose [14]. Slaking is done for a month and cubes are casted at a binder aggregate of 1:3. Compressive strength test is done after a curing period of 28 and 90 days. The mineralogical phase of the mortar is determined from XRD (X-ray diffraction) data. Infrared Spectroscopy test (FT-IR) is conducted on mortar samples to get the information on organic test. In FT-IR graphs, the peaks of carbohydrates and proteins gave the organic test results [15].

2 Materials and Sample Preparation

2.1 Raw Materials

Airlime was taken from Pollachi quarry. The composition of lime was determined as per IS 6932(Part I)-1973 [16]. River sand is used as aggregate for the mortar; sieve analysis is carried out as per IS: 2386 (Part I)-1963 and the sand passing through 4.75 mm sieve and a fixed percentage of each retaining on 2.36, 1.18 mm, 600, 300, 150, 80 μ sieves are taken to obtain well graded aggregates according to IS: 2116–1980 [17]. Pycnometer test is done to obtain the specific gravity of aggregate and is found out to be 2.65. Water having a pH value of 7.5 is used for mixing the lime mortar.

2.2 Organic Additive-Raisins

Raisins or dried grapes are soaked inside water for 24 h and blended for about 3 min to make it a pulp to get 2, 4 and 6% of organic water. It is then added while slaking to study the effect on slaking process.

2.3 Slaking Process

4 kg of air lime and an equal amount of water is mixed by adding 0, 2, 4 and 6% of organic water and kept for slaking separately in 4 different closed containers (Fig. 1). Lime putty is taken out at an interval of 5, 10 and 15 days and mixed with fine aggregates at a binder-aggregate ratio of 1:3 for casting the cube specimens [18].

Fig. 1 Slaked lime putty

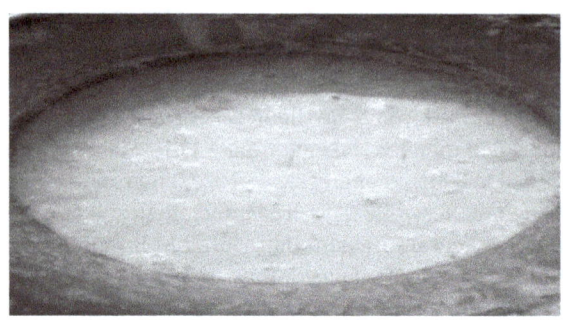

2.4 Casting the Cubes

Three samples each of size 50 mm × 50 mm × 50 mm are prepared for particular percentage of organic slaking and casted as per IS: 6932 (Part VII)-1973 [19]. Workability test is done to find out the water required for casting the cube specimen [20]. The mortar mix prepared is then filled in the moulds at three layers. After each layer, it is pressed gently using hands and finally after filling the third layer excess part is struck off and finished (Fig. 2).

Fig. 2 Casting the cubes

The mortar cubes are kept for 3 days to set and finally demoulded. The demoulded specimens are air cured for 28 and 90 days maintained at a temperature of 27 ± 2 °C and 90% RH [8].

3   Tests for Raw Materials

Sieve analysis and specific gravity test are done for the finding out the grain size distribution of fine aggregate. Workability and initial flow tests are performed to find out the appropriate water binder ratio for the lime paste and lime mortar respectively. Compression strength test is done for finding the mechanical strength of the sample.

3.1   Sieve Analysis

A 200 grams sample of fine aggregates is taken for performing the sieve analysis. The samples taken for sieve analysis corresponds to the well graded aggregates with reference to the percentages in IS: 2116–1980 [17]. "Figure 3 shows the particle size distribution graph for sample A".

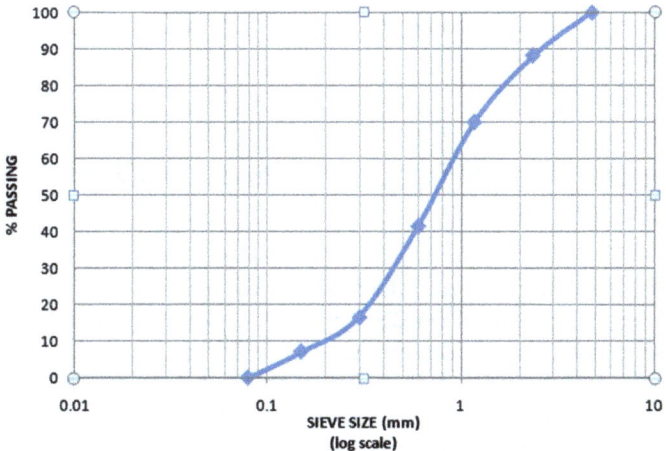

Fig. 3 Particle size distribution of fine aggregates

Fig. 4 XRD result of lime
sample

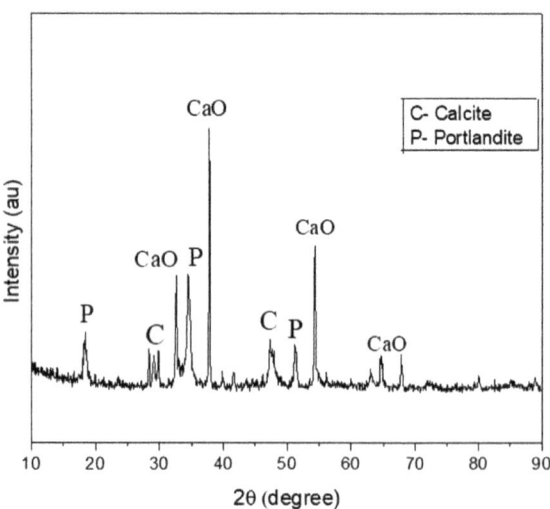

3.2   X-Ray Diffraction Test of Lime Sample

X-ray diffraction test is carried out for lime sample to find out the mineralogical composition of the raw material. "Figure 4 shows the mineralogical analysis of the lime sample. The presence of calcium oxide (denoted as CaO) can be seen as major compounds while traces of portlandite (denoted as P) as well as calcite (denoted as C) are found out".

3.3   Fourier Transform Infrared Spectroscopy (FT-IR) of Organic-Raisin

FT-IR is an analytical technique which is used for analyzing the organic matters depending on their chemical bonding properties. Generally in organic matters, chemical bonds which are weaker excite easier than stronger bonds.

"Figure 5 shows the graph of the FT-IR analysis of Raisins". The characteristic absorption bands of polysaccharides at 2931, 1249, 1099, and 1026 cm^{-1} and amide bands at 1631 cm^{-1} which indicates the presence of proteins are observed in the FT-IR analysis of Raisins. This shows that the organic is rich in polysaccharides and proteins.

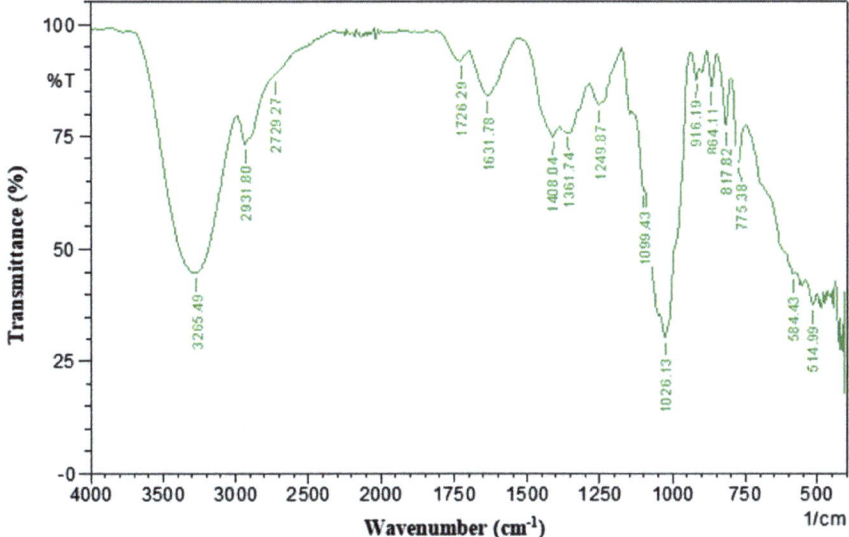

Fig. 5 FT-IR results of raisin

3.4 X-Ray Diffraction of Slaked Lime Putty

Figure 6 shows the XRD spectrum of lime sample with 4% organic added with 15 days of slaking. The spectrum shows predominant traces of portlandite (denoted as P) and small traces of calcium oxide (denoted as CaO) which represents better slaking of the lime putty.

Fig. 6 XRD spectrum of slaked lime sample

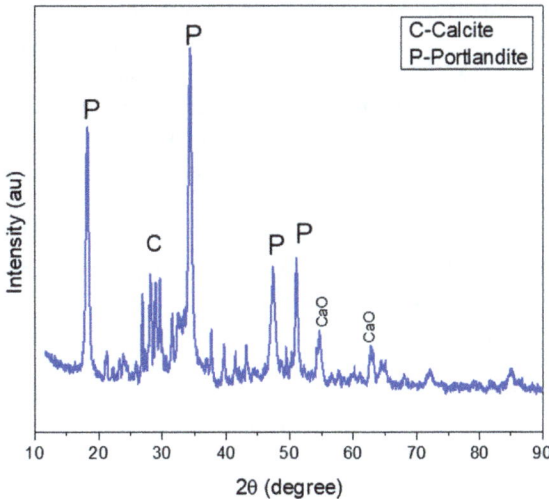

4   Preliminary Tests

4.1   Initial Flow Test and Workability

Water content is a main factor which affects the workability, which will also determine the durability of a mortar. It is the most reliable way to measure the correct amount of water that should be added to maintain the optimum initial flow of the mortar. The initial flow test was done according to the thesis done by Lawrence. Each of the lime binders was mixed with a suitable amount of water to attain the specific initial flow values of 165, 185 and 195 mm.

Workability test is done according to IS 6932 (Part8):1973.The workability is estimated by noting the number of bumps required attaining an average spread to 190 mm, the material is adjusted to normal consistency by the spread after one bump to 110 mm [20].

From the workability and initial flow test a water content of 0.65 is taken as optimum water content.

5   Secondary Tests

5.1   Compressive Strength

Compressive strength test is done to evaluate the mechanical properties of the cube specimen. The compressive strength is measured by crushing cube specimens or cylinder specimens using compression testing machine. The compressive strength is the ratio of failure load to the cross sectional area resisting the load. The specimens are tested on low capacity compression testing machine, INSTRON after 28 and 90 days of curing. The failure load of each cube specimens is noted down [19].

Figure 7a shows the 28 day compression results of the cube samples." There is a delay in strength gain at 28th day compression test due to the the presence of polysacchrides that are retarding agents which retains the dampness in the cube samples. It was seen that compressive strength increased with slaking days in the addition of 0 and 2% of organics. It shows a sudden decrease in the strength with slaking during the addition of 6% of organics while for 4%, it showed a decrease in strength from 5 days slaking to 10 days slaking and increase in strength from 10 days slaking to 15 days slaking.

"Figure 7b shows the 90 day compression results of the cube samples." There is a gain in strength during the 90th day compression results. It can be seen that compressive strength increased rapidly with slaking days in the addition of 4% of organics. A common trend in increase of compressive strength can be seen with slaking and addition of organics.

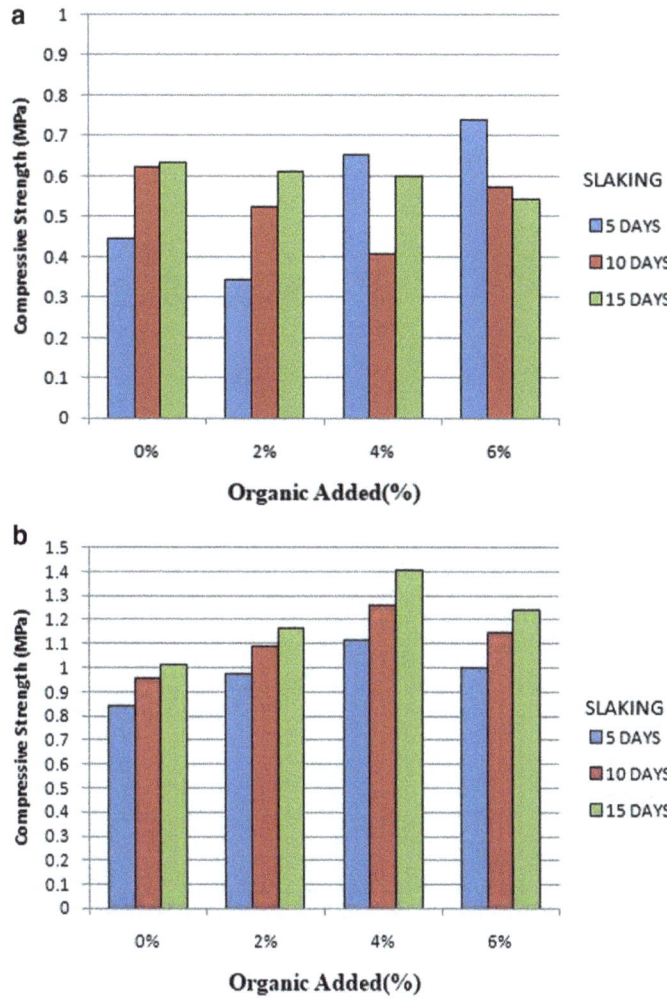

Fig. 7 a 28th day compressive strength result of organic water slaked lime mortar samples, **b** 90th day compressive strength results of organic water slaked lime mortar samples

5.2 X-Ray Diffraction Test

X-ray diffraction test shows the presence of minerals, organic and inorganic compounds present in the mortar samples. "Figure 8a, b and c show the XRD graphs of 90 days cured sample with 4% organic addition". The high intense peak of calcite (denoted as C) and lower peak value of portlandite (denoted as P) can be seen which shows better carbonation in the mortar samples.

Fig. 8 **a** XRD graph of 5
day slaked mortar specimen,
b XRD graph of 10 day
slaked mortar specimen,
c XRD graph of 15 day
slaked mortar specimen

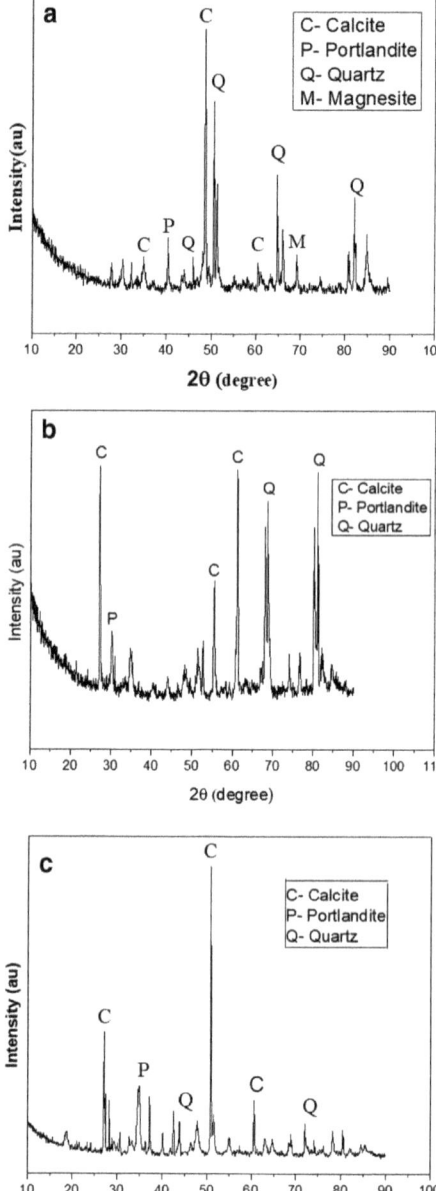

5.3 Fourier Transform Infrared Spectroscopy (FT-IR) of Mortar Samples

"Figure 9a, b and c show the FT-IR graphs of 90 days cured sample with 4% organic addition". The deeper intense bands of CO_3 can be seen at 871.82 cm^{-1} for the 5, 10 and 15 day slaked mortar specimens. The bands of 1795 cm^{-1} shows the presence of polysaccharide group(C–H group). In the samples the bands of 1408 cm^{-1} and 1404 cm^{-1} indicates aromatic C–C stretching band.

6 Conclusion

The project emphasized on improving the mechanical properties of aerial lime mortar which is rarely used in restoration works compared to natural hydraulic mortar due to less knowledge as well as lower performance of aerial lime based mortars. This paper explores the effect of organic content on slaking and carbonation properties of aerial lime mortar from which the following can be concluded:

The mortar having a dosage of 4% organics showed rapid improvement in strength compared to 2% and 6% of organics. The 5-day slaking using 4% and 6% organics showed more gain in strength compared to 10 and 15 day slaked reference mortar for the 28 day compressive test. This signifies that slaking rate increased with the addition of organics. The addition of organics initially reduced the mechanical strength due to the water retention characteristics of polysaccharides present in the organics.

The 90 day compressive strength showed greater mechanical strength for 4% organics. The greatest strength is achieved by 15 day slaking with 4% of organic water. This signifies the role of age on carbonation and mechanical strength. Polysaccharides also help in improving the carbonation of specimen by allowing using the carbon dioxide in the water retained inside the specimen. This can be confirmed by the presence of calcite in the XRD result of the mortar sample. The conversion of calcium oxide to portlandite has increased by organics. This can be observed in the XRD result of 15 day slaked lime putty added with 4% organic water. The FT-IR results of 5days, 10 days and 15 days organic slaked mortar samples showed the presence of polysaccharides and CO_3 bonding.

Thus it is observable that Raisins (organic) help to maintain the moisture in the mortar which made an early step to carbonation process by the carbon dioxide present in the retained moisture. This boosted the slaking process as well as the mechanical properties of lime mortar which is helpful in using the aerial lime mortar for the restoration of heritage structures.

Fig. 9 **a** FT-IR results of 5
day slaked mortar specimen,
b FT-IR results of 10 day
slaked mortar specimen,
c FT-IR results of 15day
slaked mortar specimen

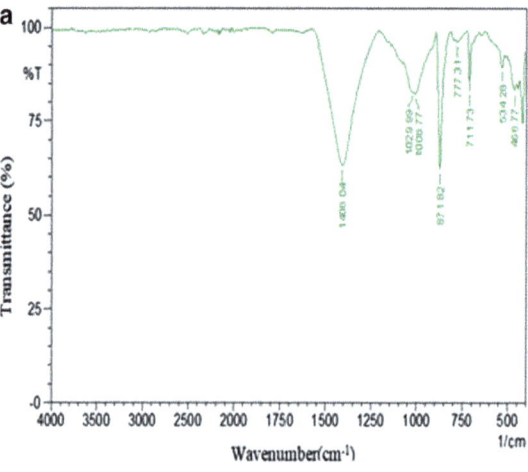

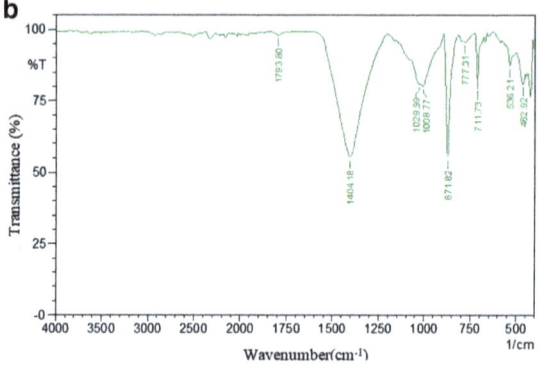

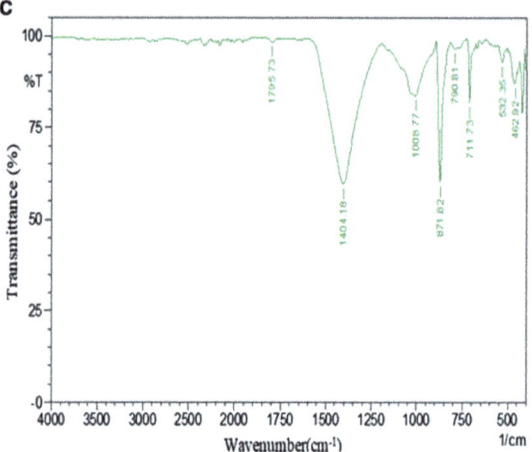

References

1. Lanas, J., Alvarez, J.I.: Masonry repair lime-based mortars: factors affecting the mechanical behaviour. Cement Concr. Res. **33**(11), 1867–1876 (2003)
2. Silva, B.A., Ferreira, A.P., Augusto, G.: Influence of natural hydraulic lime content on the properties of aerial lime-based mortars. Constr. Build. Mater. **72**, 208–218 (2014)
3. Kalagri, A., Karatasios, I., Kilikoglou, V.: The effect of aggregate size and type of binder on microstructure and mechanical properties of NHL mortars. Constr. Build. Mater. **53**, 467–474 (2014)
4. Ronak, C., Patel, J.M.: Critical review on lime mortar. J. Emer. Technol. Innovat. Res. 227–232 (2016)
5. Eleni, D., Thomas, S., Aurela, S., Frederik, V.: Literature study on the rate and mechanism of carbonation of lime in mortars. In: 9th International Masonry Conference 2014, pp. 1–12. International Masonry Society, Guimaraes Portugal, (2014)
6. Arandigoyen, M., Lanas, J.: Carbonation process in lime pastes with different water/binder ratio. Mater. Constr. **56**(281), 5–18 (2006)
7. Balen, K.V., Gemert, D.V.: Modelling lime mortar carbonation. Mater. Struct. **27**, 393–398 (1994)
8. Thirumalini, S., Ramdoss, R., Rajesh, M.: Experimental investigation on physical and mechanical properties of lime mortar: Effect of organic addition. J. Cultural Heritage **31**, 97–104 (2018)
9. Kilic, O.: Effect of slaking water properties to lime quality. Asian J. Chem. **19**, 3215–3220 (2007)
10. Thirumalini, S., Sekar, S.K.: Review on herbs used as admixture in lime mortar used in ancient structures. Indian J. Appl. Res. **3**(8), 295–298 (2013)
11. Zacharopoulou, G.: Correlation of initial and final available lime and free water in lime putties to carbonation rate and mechanical characteristics of lime mortars. In: 2nd Historic Mortar Conference 2010, pp. 1273–1281. Prague (2010)
12. Kerstin Elert, K., Navarro, C.R., Pardo, E.S., Hansen, E., Cazalla, O.: Lime mortars for the conservation of historic buildings. Stud. Conserv. **47**, 62–75 (2002)
13. Gour, K.H., Thirumalini, S., Ramdoss, R.: Revamping the traditional air lime mortar using the natural polymer–Areca nut for restoration application. Constr. Build. Mater. **164**, 255–264 (2018)
14. Sameh Abdelsalam Hassan, S.A., Zahrani, A.A.: Slaking lime for restoration and conservation of historical buildings and materials, criticism of an arabic historical manuscripts. Eng. Appl. Sci. **6**, 125–131 (2017)
15. Walker, P., Lawrence, R.M.H.: The impact of water /lime ratio on the structural characteristics of air lime mortars. In: Ayala, D., et al. (eds.) Structural analysis of historic construction, pp. 885–889 (2008)
16. IS 6932(Part I)-1973: Methods of tests for building limes, Determination of insoluble residue, loss on ignition, insoluble matter, silicon dioxide, ferric and aluminum oxide, calcium oxide and magnesium oxide
17. IS 2116–1980: Specification for sand for masonry mortars
18. Paiva, H., Velosa, A., Veiga, R., Ferreira, V.M.: Effect of maturation time on the fresh and hardened properties of air lime mortar. Cement Concrete Res. **40**, 447–451 (2010)
19. IS 6932 (Part VII)-1973: Methods of tests for building limes-determination of compressive and transverse strengths
20. IS 6932 (Part VIII)-1973: Methods of tests for building limes-determination of workability

Effect of HPFRCC on Cyclic Behavior of Diagonally Reinforced Coupling Beams

Sang Whan Han, Jongseok Jang, and Taeo Kim

Abstract Coupled shear wall system have been widely used in building construction due to their effectiveness in resisting lateral loads. In this system, individual shear walls are connected by concrete diagonally reinforced coupling beams (DRCBs), by which the coupled shear walls act as a single unit. However, difficulties are often arisen when placing reinforcement in DRCBs at construction sites because of congestion and interference of diagonal, transverse, and longitudinal reinforcement. To alleviate such difficulties, high performance fiber-reinforced cementitious composite (HPFRCC) can be used to construct DRCBs instead of concrete. In this study, two concrete and four HPFRCC DRCBs were made and tested under cyclic loads. Test variables were the amount of diagonal and transverse reinforcement as well as the application of HPFRCC. Test results showed that the HPFRCC DRCB specimen exhibited superior cyclic behaviour to the code-compliant standard concrete DRCB specimen, although HPFRCC specimens had only 50% of the transverse reinforcement placed in the standard specimen. However, the application of HPFRCC alone was not effective to reduce diagonal reinforcement.

Keywords Coupling beams · Cyclic behavior · Shear wall · Diagonal reinforcement · Transverse reinforcement · HPFRCC · Experiment · Reinforcement

1 Introduction

In the 1964 Alaska earthquake, concrete coupling beams reinforced with conventional beam reinforcement experienced a brittle sliding shear failure. To improve the seismic behavior of coupling beams, Paulay and Binney [1] developed diagonal reinforced coupling beams (DRCBs). Many experimental tests were conducted to demonstrate the superior seismic behavior of DRCBs [1–7].

S. W. Han (✉) · J. Jang · T. Kim
Hanyang University, Seoul, Korea

© The Author(s), under exclusive license to Springer Nature Switzerland AG 2022   259
J. Sena-Cruz et al. (eds.), *Proceedings of the 3rd RILEM Spring Convention and Conference (RSCC 2020)*, RILEM Bookseries 34,
https://doi.org/10.1007/978-3-030-76465-4_23

In recent years, slender DRCBs have been often used in building construction. However, slender DRCBs have less inner space to place reinforcement than deep DRCBs, resulting in significant reinforcement congestion and interference. Therefore, difficulties have been often arisen when constructing slender DRCBs in construction fields due to heavy reinforcement congestion and interference [10].

To reduce such difficulties, DRCBs constructed with high performance fiber reinforced cement composites (HPFRCC) and simple reinforcement configurations have been developed by previous experimental studies [5, 8–10]. They also reported that when using HPFRCC, a simple reinforcement arrangement with a reduced amount of reinforcement could be used without deteriorating the cyclic behavior of the DRCBs. In this study, the amount of diagonal and transverse reinforcement to be replaced by the application of HPFRCC was investigated. In order to explore the cyclic behavior of HPFRCC slender DRCBs with different amounts of diagonal and transverse reinforcement, this study conducted experimental tests with six slender DRCB specimens. The HPFRCC was made with polyvinyl alcohol (PVA) fibers. PVA fibers improve toughness and strain capacity in post-cracking response whereas steel fibers having high modulus and high tensile strength increase the strength of concrete. Thus, PVA was chosen in our study to improve the toughness and deformation capacities of coupling beams.

2 Details of Diagonally Reinforced Coupling

According to ACI 318–14 (https://www.concrete.org/store/productdetail.aspx?ItemID=318U19&Language=English) [11], coupling beams are classified into three groups according to their aspect ratio (l_n/h), where l_n and h are the length and height of a beam: (1) DRCBs with $l_n/h \leq 2$ should be reinforced with diagonal bars; (2) DRCBs with $l_n/h > 4$ should be reinforced with special moment frame (SMF) beam reinforcement; (3) DRCBs with $2 < l_n/h \leq 4$ could be reinforced with either diagonal reinforcement or SMF beam reinforcement.

ACI 318–14 [11] specifies two alternative confinement options for DRCBs. For the first option, each group of diagonal bars should be enclosed by rectilinear transverse reinforcement, whereas for the second option, transverse reinforcement should be placed around the entire beam perimeter. The second confinement option has less reinforcement congestion than the first confinement option. However, interference between diagonal and transvers bars still exists on the second confinement option.

To reduce the reinforcement congestion in DRCBs and to improve their cyclic behavior, DRCBs could be made of HPFRCC. HPFRCC does not generally contain coarse aggregates [12] and possesses a relatively high strain-capacity while generating multiple micro-cracks and a strain-hardening response [13, 14]. Previous studies reported that HPFRCC members had larger shear strengths, deformation capacities and damage tolerances than conventional RC members [15, 16]. HPFRCC members also require less transverse reinforcement due to the contribution of HPFRCC to concrete confinement. Therefore, HPFRCC could be applied to DRCB to reduce the amount of reinforcement without losing seismic performance compared to RC DRCBs [5, 8–10].

3 Experimental Test Program

This study focused on slender HPFRCC DRCBs. The contributions of HPFRCC, diagonal bars, and transverse bars to the cyclic behavior of DRCBs were investigated by conducting experimental tests. For this purpose, two concrete and four HPFRCC DRCB specimens were made (Fig. 1) and tested with quasi-static cyclic loading. The l_n/h value of all specimens was 3.5. The test variables were (1) the amount

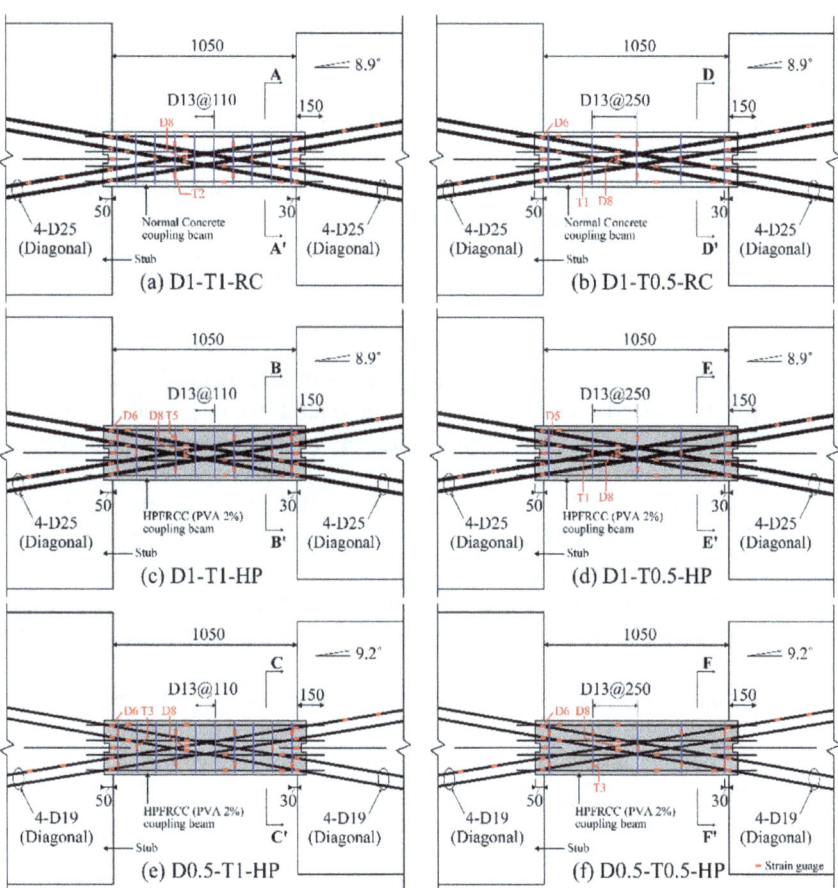

Fig. 1 Details of specimens

of transverse reinforcement, (2) the amount of diagonal reinforcement, and (3) the application of HPFRCC. Figure 1 shows the reinforcement details of the specimens.

The width (b), height (h) an length (l_n) of all specimens were 250, 300 and 1050 mm, respectively. The specified compressive strength of concrete and HPFRCC used for specimens was 40 MPa, and the specified yield strength of the reinforcement was 420 MPa. The shear strength (V_n) of a DRCB can be calculated using Eq. (1) (ACI 318–14 https://www.concrete.org/store/productdetail.aspx?ItemID=318U19& Language=English [11]).

$$V_n = V_{ACI} = 2A_{vd} f_{yd} \sin\alpha \leq 0.83\sqrt{f'_c} A_{cw} \qquad (1)$$

where A_{vd} is the total area of reinforcement in each group of diagonal bars (mm²) f_{yd} is the yield strength of the diagonal reinforcement (MPa), and α is the angle between the diagonal bars and the longitudinal axis of the coupling beam. In this study, the second confinement option was used to place the reinforcing bars in the DRCB specimens.

Specimen D1-T1-RC is the standard RC specimen (Fig. 1a) designed and detailed according to ACI 318–14 [11]. Specimen D1-T1-HP is an HPFRCC DRCB with the same reinforcement details as specimen D1-T1-RC (Fig. 1c). Specimen D0.5-T1-HP (Fig. 1e) is a specimen identical to specimen D1-T1-HP except for the amount of the diagonal reinforcement. The amount of diagonal reinforcement in this specimen was half that of specimen D1-T1-HP. To evaluate the effect of transverse reinforcement in DRCBs, specimens D1-T0.5-RC (Fig. 1b) and D1-T0.5-HP were made, which had only half the amount of transverse reinforcement as that in specimens D1-T1-RC and D1-T1-HP. In specimen D0.5-T0.5-HP (Fig. 1f), both diagonal and transverse reinforcement were reduced to half the amount of those reinforcement used in specimens D1-T1-RC and D1-T1-HP. Reinforcing details of each specimen are demonstrated in Fig. 1.

Polyvinyl alcohol (PVA) fibers are adopted into HPFRCC specimens to achieve high tensile strain capacity, toughness and structural integrity [17]. The same mix proportion of HPFRCC as that provided in [18] was also used in this study, which was the best mix proportion for HPFRCC found from coupon tests with various trial mixes. The volume fraction of PVA fibers was 2.0%, and calcium carbonate was used as a filler.

The average compressive strengths of the concrete and HPFRCC were 44 and 46 MPa, respectively. The average compressive strength of normal concrete used for loading concrete blocks was 69 MPa. Compressive strengths of all materials exceeded each design strength.

The average direct tensile strength of HPFRCC was 3.6 MPa. The HPFRCC exhibited ductile behavior with multiple events of microcracking on the surface before failure without crack localization. This could be attributed to fiber bridging effects that could efficiently transfer loads between cracks in the cement matrix through fibers, resulting in spreading cracks on the entire surface of the cement matrix.

4 Cyclic Curves of RC and HPFRCC DRCB Specimens

Figure 2 presents the cyclic and envelop curves of six DRCB specimens. For individual specimens, the yield drift ratio (θ_y), maximum strength (V_u) and drift ratios (θ_u) were determined according to the procedure proposed by Pan and Moehle [19] and denoted in Fig. 2.

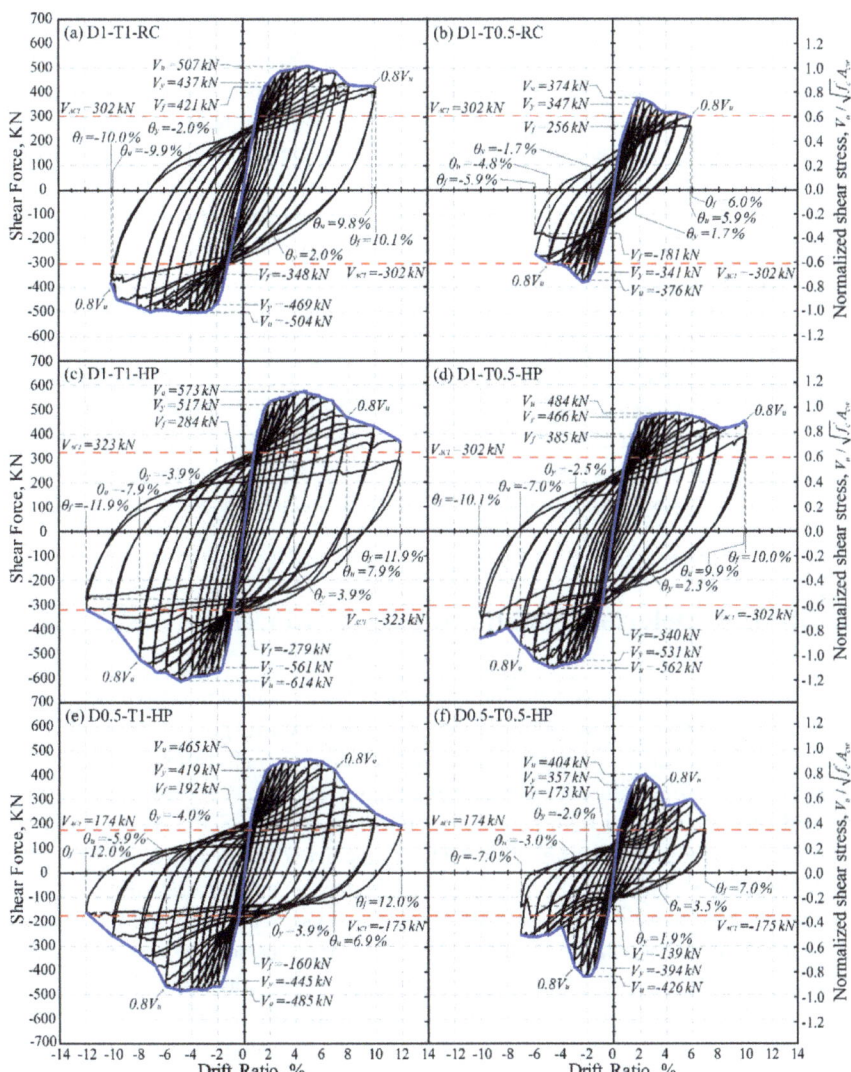

Fig. 2 Cyclic and envelope curves of specimens

The envelope of cyclic curves was idealized by an elastoplastic relation. The initial slope of idealized relation was a secant through the measured relation at a load equal to two-thirds of the measured strength (V_u). The plastic portion of the idealized relation passed through the maximum load (V_u). The intersection between these two lines defined yield drift ratio (θ_y). The yield strength (V_y) was determined as a strength corresponding to the incidence of yielding in diagonal reinforcement. The maximum drift ratio (θ_u) was defined as a drift when V_u is reduced by 20%. A point for shear strength and drift ratio at failure (V_f, θ_f) was defined as a point in the cyclic curve where a sudden drop in strength has occurred. The shear strength of specimens calculated using Eq. (1) is also denoted in Fig. 2 using a red dashed line.

The maximum strength (V_u) of all specimens was significantly larger than V_{ACI}. It is noted that in Eq. (1), the contribution of diagonal reinforcement in DRCBs is only considered to calculate their shear strength. Although concrete DRCB specimen D1-T0.5-RC had only 50% of transverse reinforcement required by ACI 318–14 [11], this specimen produced V_u larger than V_{ACI}. The observed shear strength (V_u) of D1-T0.5-HP and D0.5-T0.5-HP were significantly larger than V_{ACI}.

Standard specimen D1-T1-RC, detailed according to ACI 318–14 [11], had V_u of 507 kN and θ_u of 9.8%. This specimen exhibited stable cyclic behavior without a sudden strength drop (Fig. 2a), indicating that slender DRCBs designed according to the second confinement option had excellent cyclic behavior. All specimens had a maximum drift ratio (θ_u) larger than 5% except for D0.5-T0.5-HP ($\theta_u = 3.0\%$). A drift ratio of 5% is the acceptance criteria (θ_{ASCE}) for DRCBs for collapse prevention level specified in ASCE 41 [20]. Note that θ_{ASCE} for CP level is analogous to θ_u which was estimated when the shear strength of a DRCB was decreased by 20% [10].

5 Stiffness Deterioration and Energy Dissipation

At each drift ratio, stiffness was determined from the cyclic curves, which was the slope of a line connecting the points in the cyclic curves corresponding to the positive and negative peak drifts. Figure 3a shows the stiffness at each drift ratio normalized by the initial stiffness measured at a drift ratio of 0.25%. As shown in this figure, the stiffness deteriorated with an increase in the drift ratio. Stiffness deterioration can be used to measure the strength retention capacity for a specimen. As expected, the HPFRCC specimen D1-T1-HP exhibited the largest strength retention capacity among the specimens, whereas HPFRCC specimen D0.5-T0.5-HP had the smallest strength retention capacity. As pointed out in the previous section, the presence of HPFRCC alone could not completely compensate for 50% reduction in both diagonal and transverse reinforcement in the slender DRCB specimens.

Figure 3b shows the cumulative dissipated energy at each drift ratio. The energy dissipated in each loading cycle is the area enclosed by the cyclic curve. Specimen D1-T1-HP had the largest energy dissipation capacity among the specimens. However,

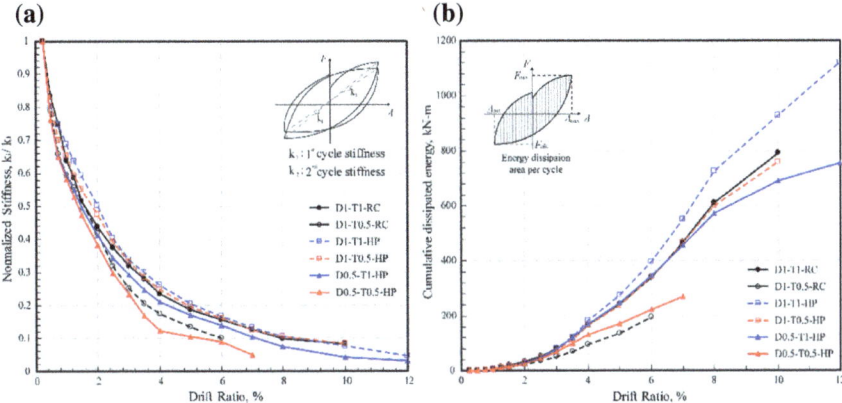

Fig. 3 **a** Stiffness degradation **b** Cumulative dissipated energy

unlike stiffness deterioration, RC specimen D1-T0 5-RC rather than HPFRCC specimen D0.5-T0.5-HP had the smallest energy dissipation capacity. HPFRCC specimen D1-T0.5-HP had an energy dissipation capacity slightly larger than standard RC specimen D1-T1-RC because of the HPFRCC.

6   Conclusion

Six specimens were made and tested to investigate the cyclic behavior of HPFRCC slender DRCBs. Test variables were the application of HPFRCC, the amount of diagonal reinforcement, and the amount of transverse reinforcement. Conclusions obtained from this study is as follows:

(1)   The shear strength and the deformation capacity of slender DRCBs were improved by using HPFRCC. HPFRCC slender DRCB specimens had energy dissipation capacity and strength retention capacity significantly larger than corresponding concrete DRCB specimens.

(2)   The cyclic performance of HPFRCC DRCB specimen D1-T0.5-HP with 50% of transverse reinforcement placed in standard concrete specimen was superior to the standard specimen. The amount of transverse reinforcement required for DRCBs in ACI 318–14 could be alleviated by using HPFRCC.

(3)   The application of HPFRCC to reduce the amount of diagonal reinforcement is not effective as that to reduce the amount of transverse reinforcement.

Acknowledgements The research was supported by a grant from the Korea Agency for Infrastructure Technology (20CTAP-C152179-02).

References

1. Paulay, T., Binney, J.R.: Diagonally reinforced coupling beams of shear walls. ACI Spec. Publ. **42**, 579–598 (1974)
2. Barney, G.B., Shiu, K.N., Rabbat, B.G., Fiorato, A.E., Russell, H.G., Corley, W.G.: Behavior of coupling beams under load reversals. Portland Cement Assoc. **10** (1980)
3. Tassios, T.P., Moretti, M., Bezas, A.: On the behavior and ductility of reinforced concrete coupling beams of shear walls. ACI Struct. J. **93**(6), 711–720 (1996)
4. Galano, L., Vignoli, A.: Seismic behavior of short coupling beams with different reinforcement layouts. ACI Struct. J. **97**(6), 876–885 (2000)
5. Canbolat, B.A., Parra-Montesinos, G.J., Wight, J.K.: Experimental study on seismic behavior of high-performance fiber-reinforced cement composite coupling beams. ACI Struct. J. **102**(1), 159–166 (2005)
6. Fortney, P.J.: The next generation of coupling beams Ph.D. Dissertation Cincinnati: University of Cincinnati (2005)
7. Naish, D., Wallace, J.W., Fry, J.A., Klemencic, R.: Reinforced concrete link beams: alternative details for improved construction. UCLA-SGEL Report 2009–06 Los Angeles: University of California (2009)
8. Lequesne, R.D., Setkit, M., Parra-Montesinos, G.J., Wight, J.K.: Seismic detailing and behavior of coupling beams with high-performance fiber-reinforced concrete. ACI Spec. Publ. **272**, 205–222 (2010)
9. Setkit, M.: Seismic behavior of slender coupling beams constructed with high-performance fiber-reinforced concrete. University of Michigan (2012)
10. Han, S.W., Lee, C.S., Kwon, H.W., Lee, K.H., Shin, M.S.: Behaviour of fibre-reinforced beams with diagonal reinforcement. Mag. Concr. Res. **67**(24), 1287–1300 (2015)
11. ACI 318–14. Building code requirements for structural concrete and commentary. Building code and commentary. American Concrete Institute (ACI) (2014)
12. Li, V.C.: Engineered cementitious composites (ECC)-material, structural and durability performance. In: Nawy, E., (ed.) Concrete construction engineering handbook. Boca Raton: CRC Press, chapter 24 (2007)
13. Naaman, A.E., Reinhardt, H.W.: Characterization of high performance fiber reinforced cement composites. International Symposium on High Performance Fiber Reinforced Cementitious Composites 1996, pp. 1–24 (2016)
14. Naaman, A.E.: High performance fiber reinforced cement composites: classification and applications. In: Karachi, P., (ed.) CBM-CI International Workshop 2007, pp. 389–401 (2007)
15. Fischer, G., Li, V.C.: Effect of matrix ductility on deformation behavior of steel-reinforced ECC flexural members under reversed cyclic loading conditions. ACI Struct. J. **99**(6), 781–790 (2002)
16. Parra-Montesinos, G.J.: High-performance fiber-reinforced cement composites: an alternative for seismic design of structures. ACI Struct. J. **102**(5), 668–675 (2005)
17. Akkari, A.: Evaluation of a polyvinyl alcohol fiber reinforced engineered cementitious composite for a thin-bonded pavement overlay. Report MN/RC 2011–11 St Paul (MN): Minnesota Department of Transportation (2011)
18. Han, S.W., Kang, J.W., Lee, C.S.: Seismic behavior of slender HPFRCC diagonally reinforced coupling beams with limited transverse bars. Earthq. Spectra **34**(1), 77–98 (2018)
19. Pan, A., Moehle, J.P.: Lateral displacement ductility of reinforced concrete flat plates. ACI Struct. J. **3**(86), 250–258 (1989)
20. American Society of Civil Engineers: ASCE/SEI 41-13 seismic evaluation and retrofit of existing buildings. American Society of Civil Engineers, Reston, Virginia (2014)

Production of Recycled Aggregate Concrete Using Construction and Demolition Waste

K. Oikonomopoulou, P. Savva, S. Ioannou, D. Nicolaides, and M. F. Petrou

Abstract The objective of this research was to evaluate the performance of concrete containing recycled aggregates of different sizes and replacement percentages. Mixtures were prepared using a combination of natural and recycled aggregates. The recycled aggregates were treated in a concrete mixer truck to partially remove the adhered cement paste in order to compare their performance in concrete as opposed to natural aggregates and non-treated recycled aggregates. One high-strength concrete series of mixtures was prepared to employ the concept of internal curing. The mixtures were tested for their mechanical properties and durability. The results showed that the addition of recycled aggregates in concrete up to a specific replacement percentage did not adversely affect the concrete properties. The specific research suggests that recycled aggregates can be a very good alternative source to reduce natural aggregates consumption and utilize waste material.

Keywords Adhered mortar · Treatment method · Mineral admixtures · Recycled concrete aggregate · Recycled aggregate concrete · Internal curing

1 Introduction

According to the World Commission on Environment and Development (WCED), "sustainable development is a development that meets the needs of the present without compromising the ability of future generations to meet their own needs" [1]. It is commonly accepted that following the common practice, inevitable exhaustion of natural aggregate resources will occur, while enormous amounts of construction and

K. Oikonomopoulou · S. Ioannou · M. F. Petrou
University of Cyprus, Nicosia, Cyprus

P. Savva (✉)
Latomia Pharmaka Plc, Nicosia, Cyprus
e-mail: psavva@pharmakas.com

D. Nicolaides
Frederick University, Nicosia, Cyprus

© The Author(s), under exclusive license to Springer Nature Switzerland AG 2022
J. Sena-Cruz et al. (eds.), *Proceedings of the 3rd RILEM Spring Convention and Conference (RSCC 2020)*, RILEM Bookseries 34,
https://doi.org/10.1007/978-3-030-76465-4_24

demolition waste yield [2]. Knowledge associated with the use of the recycled aggregate coupled with a projected overall increase in the durability of further construction would lead to a reduction of the environmental impact associated with repair or disposal of constructions when deemed unsafe for habitation. This would alleviate the issue of construction waste disposal, which is a prevalent issue all over the world, but especially in Cyprus due to the small areas effectively used for such purposes. Research on recycled concrete aggregates RCA allows for more recycling, further reducing waste as well as reducing the costs associated with disposal and storage of said waste. Successful implementation could also lead to marked improvements in the practical use of concrete. Effective usage of RCA would facilitate for optimization of processing which would preserve natural resources for a prolonged time. When considering the total amount of natural resources in a large scale, in Europe for example, the exploitation of resources might not be as significant as in a smaller scale, in Cyprus for instance, where the increased exploitation may reduce the resources to a crucial level, due to country's small size, compromising the needs of the following generations. It has to be noted though that great challenges have to be addressed before efficiently utilizing recycled aggregates in concrete. Specifically, recycled aggregates face notable disadvantages compared to natural aggregates such as lower density, higher water absorption, adhered weak mortar, poor mechanical properties and durability [2]. One of the foremost reasons causing these properties is the adhered mortar because of low inherent quality and double interfacial transition zone formation. Many researchers have attempted to reduce the adhered mortar with various methods such as resembled Los Angeles [3], carbonation and wrapping [2], microbial carnation precipitation [4], impregnation inside organic polymer solution and polyvinyl alcohol [5], low concentration hydrochloric acid [6, 7], etc. The research activity on recycled concrete aggregates along with the attention given on the RCA treatment both highlight the urge of developing methodologies able to adopt and re-utilize aggregate back in concrete. Kazemian et al. have reported that during the evaluation of flexural strength, the loose mortar of the recycled aggregate was the most influential factor in reducing it [7]. The introduction of 25 and 50% of RCA has resulted in a reduction of 10% and 14% in flexural strength respectively. On the contrary, identical replacement using treated recycled aggregate has resulted in a 3% reduction. Ismail and Ramli evaluated the combination of aggregate treatment using hydrochloric acid to remove the loose adherent and calcium metasilicate to fill the remaining voids [6]. They concluded that the treated recycled aggregates performed better than the non-treated recycled aggregates in terms of compressive and flexural strength and Young's modulus. The old mortar includes micro-cracks and has high porosity [8], thus it becomes the weakest link in RAC and its strength is the upper limit of the strength of concrete. As the mortar-aggregate bond strength increases, the concrete strength also increases [9, 10].

The objective of this research is to determine whether the incorporation of treated and non-treated RCA aggregates and non-treated RCA sand in high-strength concrete would have a beneficial effect on mechanical properties and durability. The mixture design yielded after the calculation of optimal treatment of aggregates.

2   Experimental Program

2.1   Materials and Testing Methods

Two fractions of recycled coarse aggregates were used in this work, namely 0–4 and 4–10 mm (sieve analysis was applied to divide the material into the different gradings). The natural aggregates (coarse and fine) were obtained from Latomia Pharmakas Plc. The same particle size ranges for both natural and recycled aggregates were used in the project. The coarse aggregates were tested on their grading (EN 933–1, 2), shape (EN 933–4), flakiness index (EN 933–3), abrasion resistance (EN 1097–2), density (EN 1097–6) and water absorption (EN 1097–6).

2.2   Aggregates Treatment

The adhered mortar in addition to the unknown age, source, and origin of the recycled aggregates are the main reasons why RCA cannot be easily characterized and cause high variation in the results. In an effort to improve the RCAs performance, the adhered mortar was partially removed. A certain amount of RCAs was exposed to heat treatment according to an established technique, to remove entirely the adhered mortar and determine its mass loss percentage [3, 11, 12]. Such a methodology cannot be applied for large quantities of RCA. The aggregates mass loss was also compared to the aggregate circularity extracted from an image analysis investigation.

RCAs intended to be utilized in concrete mixtures were added in a concrete truck mixer for a specific time interval, which was determined based on the image analysis results and the mass loss. This method, which resembles a prolonged Los Angeles test, has evidently decreased significantly the adhered mortar, and discarded the weaker or fractured aggregates, keeping only the stronger and sounder ones [3]. The treatment method is also expected to affect the shape of the aggregates by making them rounder. The specific was tested using image analysis to determine the circularity of the aggregates for each time interval for 1–5 h. The optimum treatment time was decided based on the results and adopted for the corresponding mixtures. Concrete mixtures were produced using field recycled aggregates RCA-F and treated recycled aggregates RCA-T. The performance of the specific mixtures was compared with each other and with one normal weight aggregate NA mixture.

Mechanical treatment The aggregates were initially water treated using a 150 L inclined concrete mixer for various intervals and correlated to their circularity based on image analysis. Five random samples of RCA (samples of 4–10 mm) were dried to constant mass at a temperature of 110 °C. Around 3 kg of each sample was selected and sieved using the 2.36 mm sieve. The samples were then weighted (m1). Each sample was placed in the mixer drum at a speed of 0.5 rps for 1, 2, 3, 4 and 5 h respectively with water. The water was added to improve the effectiveness of the method by weakening the adhered mortar [3]. When the treatment was completed the

RCAs were sieved and washed using the 2:36 mm. Then, they were dried to constant mass at a temperature of 110 °C to gain their final mass (m2). The mass reduction of the RCA was calculated by using the following equation:

$$Residual\,mortar(\%) = \frac{m1 - m2}{m1} \tag{1}$$

m1   dried sample mass before treatment (g).
m2   dried sample mass after treatment (g).

Circularity. The external morphology of aggregates can affect the properties of fresh and hardened concrete (e.g. workability, compressive strength, tensile strength, and durability). Therefore, by studying the circularity of the RCA particles, concrete properties may be correlated and result in mixtures with improved properties. Digital image processing was applied to characterize the circularity of the aggregates. Using pictures on a digital form, mathematical procedures calculate the necessary information [13]. In this paper, the software GIMP and ImageJ were used. GIMP is an open-source image editor. ImageJ is an open-source program that can display, edit, analyze, process, save and print images. Although ImageJ has editing abilities, image editing was done using GIMP to achieve more accurate results. ImageJ was used to analyze the aggregates and calculate their circularity. The software first calculates the area and the perimeter of each aggregate and then uses the following equation to calculate the circularity of each aggregate:

$$Circularity = \frac{4x\pi x A}{P^2} \tag{2}$$

A   Area of the aggregate (mm^2).
P   Perimeter of the aggregates (mm^2).

Circularity values range between 0.0 to 1.0, 1.0 being a perfect circle. As the value approaches 0.0 it indicates an increasingly elongated shape. The circularity value, however, is not as accurate for smaller particles as it is for the bigger particles [14].

Large scale treatment. A low-cost treatment method was utilized by Latomia Pharmakas Plc to remove part of the adhered mortar. The specific methodology was presented in another publication of the research team with slight modifications [3]. Dimitriou et al. have treated the aggregates for 5 h whereas the treatment time in this research was decided after the small-scale treatment and image analysis investigation. RCA-F aggregates were placed into an 8 m^3 concrete mixer. The mixer was rotated at a speed of 10 rpm, and during this process, water was added to remove the smaller particles, dust and the weaker adhering mortar. At the end of the treatment period, the aggregates were sieved through a modified sieve to discard aggregates with sizes lower than 4 mm.

Table 1 Experiments, age of testing and standards

		Test	Standard	Age of testing (days)	Specimens
Hardened concrete	Mechanical	Compressive strength	EN 12390-3	1, 3, 7, 28	3 cubes
		Tensile splitting strength	EN 12390-6	7, 28	2 cylinders
		Capillary water absorption	Reference [15]	1, 3, 7, 28	3 cubes
	Durability	Porosity	Reference [15]	1, 3, 7, 28	3 cubes
		Chloride resistance	ASTM C-1202	28	3 cylinders

2.3 Experiments

The current investigation included the evaluation of the mixtures' mechanical properties and durability. Test type, age of testing and relevant standards are summarized in Table 1. For the determination of mechanical properties, compressive strength (EN 12,390–3) and splitting tensile strength (EN 12,390–6) were conducted. Durability investigation focused on the transport properties of produced concrete and material's performance when subjected to major degradation processes. Specifically, capillary water absorption (RILEM TC-116), porosity (vacuum saturation method [15]) and chloride resistance (ASTM C-1202) were performed.

2.4 Concrete Mixtures

A total of 7 mixtures were cast. The mixture design used in the experimental work are presented in Table 2. The same mixture design was adopted for all mixtures. The effective w/c ratio was kept constant at 0.25. The objective was to add RCA in high strength concrete and evaluate whether their inferior quality could be offset due to the employment of internal curing, the high cement content, and low water/ cement ratio.

Table 2 Mixture proportions (kg/m^3)

Cement I 52.5 N	864
Coarse aggregates 4/10 mm	680
Fine aggregates 0/4 mm	650
Water	216
Superplasticizer	1.35
w/c ratio	0.25

Table 3 Mixture design with replacement ratios

	Code	Replacement (%)	RCA (mm)	NA (mm)
Mixture 1	NA	0	–	0–4 + 4–10
Mixture 2	RFF25	25	0–4	4–10
Mixture 3	RFF50	50	0–4	4–10
Mixture 4	RFC25	25	4–10	0–4
Mixture 5	RFC50	50	4–10	0–4
Mixture 6	RTC25	25	4–10	0–4
Mixture 7	RTC50	50	4–10	0–4

In the designed mixtures there was a replacement of NA with different types of RCA for six mixtures. The replacement ratios of the mixtures are presented in Table 3. The code names of each mixture consist of "R" for RCA, "F" for fine aggregates replacement, "C" for coarse aggregates replacement, "F" for field, "T" for treated, "50" or "25" represent the replacement percentage of NA. Cubes of 100 mm were used to determine the compressive strength, the sorptivity, and the porosity. Cylinders $\Phi 100 \times 200$ mm were used to determine the rapid chloride permeability, the splitting tensile strength and the modulus of elasticity.

2.5 Casting and Curing

Concrete mixtures were prepared using a 200L planetary mixer. Aggregates were in air-dried condition and the required water corrections were made, according to aggregates' moisture and absorption values. Slump tests were performed with a target slump class S3, aiming to create a relatively fluid and workable concrete. The desired slump was achieved by adjusting the amount of superplasticizer added in the concrete mixture. Compaction was achieved using a vibrating table. The vibration time was kept constant for all mixtures. The samples were removed from the molds after 24 h and were placed in water tanks at a constant temperature of 22 ± 2 °C.

3 Experimental Results and Discussion

3.1 Properties of Aggregates

The mechanical and physical properties of the three types of aggregates used for this research are presented in Table 4.

NAs in Cyprus present higher absorption values, reaching up to 4.5% [16]. From the tests performed on aggregates, it is concluded that:

Table 4 Mechanical and physical properties of aggregates

Properties	NA	RCA-F	RCA-T
Los angeles coefficient (LA) (%)	29	32	15
Apparent particle density (Mg/m^3)	2.69	2.72	2.74
Particle density (Mg/m^3)	2.52	2.28	2.49
Particle density in SSD (Mg/m^3)	2.58	2.44	2.58
Water absorption (WA) (%)	3.8	6.5	4.5
Flakiness	16	6	4
Shape index	9	7	5

- Treated aggregates had significantly lower LA value than the RCA-F and NA [3]
- The aggregate treatment leaded to a more cubical/ spherical shape (shape index)
- The RCA had significantly higher water absorption
- The RCA-T had lower water absorption than the RCA-F

Table 5 presents the aggregates' sieve analysis.

Table 5 Aggregates' sieve analysis

Sieve aperture size (mm)	NA 4–10	RCA-F 4–10	NA 0–4	RCA 0–4	RCA-T 4–10
20	100.0	100.0	–	–	100.0
14	98.8	87.3	–	–	72.5
12.5	98.2	76.7	–	–	57.0
10	87.2	48.5	–	–	26.8
8	58.2	20.3	100.0	100.0	9.4
6.3	35.3	5.4	100.0	100.0	2.6
5.60	–	–	99.9	94.9	–
4.0	5.5	3.6	98.8	89.7	0.9
3.35	–	–	96.7	86.0	–
2.0	2.5	3.5	78.8	78.1	0.6
1.0	2.3	3.4	52.1	54.9	0.3
0.50	–	–	33.8	41.5	–
0.25	–	–	21.0	32.5	–
0.125	–	–	12.2	24.6	–
0.063	1.8	3.3	8.2	19.3	0.0

3.2   Aggregates Treatment and Image Analysis

According to the specific analysis, the rate of mass loss due to the aggregate treatment was the highest up to the third hour of treatment. The rate of mass loss was much lower after the third hour of treatment. The experimental treatment procedure was compared to thermal treatment and it was concluded that the energy consumption of the chamber furnace was approximately 15 times more than the mechanical treatment method. It was therefore decided that the large-scale treatment was going to be conducted for a time interval of 3 h.

3.3   Mechanical Properties

Table 6 presents the results of the mechanical properties of all concrete mixtures tested.

Compressive Strength. The compressive strength test results indicate that the use of RCA affected negatively the strength of the concrete (Table 6). Regarding the early strength, all RCA mixtures showed an average reduction of 28.7% on day 1 and 24.2% on day 3 compared to the NA mixture. Concrete mixtures with field aggregates exhibited slightly higher compressive strength than the mixtures with treated aggregates at an early age. It is worth noticing that the coefficient of variation of RCA mixtures was significantly high (8.0%) compared to the NA mixture (2.8%). Comparing the replacement percentage, no difference was observed in the early age results. Despite the low initial compressive strength, the NA mixture had the lowest rate of strength increase up to the age of 28 days. Specifically, the NA mixture had an increase of 9.2% from day 1 to day 28 where the RCA mixtures observed an average 27% increase. Interestingly, mixture RFC25 had the lowest initial compressive strength and the highest 28-day compressive strength, slightly lower than the NA mixture (1.8%). Finally, regarding the compressive strength results, the NA mixture observed the highest performance (99.2 MPa), with the 25% replacement mixtures

Table 6 Mixtures mechanical properties

Mixture	$f_{em.}$ (MPa)				f_{ct} (MPa)	
	1 day	3 days	7 days	28 days	7 days	28 days
NA	90.0	93.9	94.6	99.2	4.75	5.24
RFF25	71.7	75.3	83.0	86.6	4.92	3.70
RFF50	62.9	82.1	76.2	89.8	4.56	4.23
RFC25	53.5	84.8	86.5	97.4	5.23	4.56
RFC50	73.4	65.7	84.2	81.7	4.15	4.97
RTC25	62.8	57.4	91.3	96.7	4.83	3.15
RTC50	60.7	61.7	79.8	82.1	3.83	5.10

Table 7 Mixtures durability indicators

Mixture	Porosity (%)				ρ (kg/m^3)	Sorptivity (mm/$\sqrt{}$min)	RCP (Coulombs)
	1 day (%)	3 days (%)	7 days (%)	28 days (%)			28 days
NA	8.8	6.1	6.1	3.7	2359	0.0044	1959
RFF25	9.2	3.6	8.9	8.6	2277	0.0110	1872
RFF50	9.0	6.7	8.3	8.7	2271	0.0088	1739
RFC25	9.7	7.6	8.9	8.0	2302	0.0095	2169
RFC50	7.3	7.1	7.2	10.2	2273	0.0096	2041
RTC25	7.4	7.5	5.3	11.2	2267	0.0084	2201
RTC50	8.3	8.5	9.3	6.9	2243	0.0098	3480

exhibiting the second higher performance (97.1 MPa) and the fine RCA with the 50% replacement mixtures the lower performance (83.5 MPa).

Splitting tensile strength Although the NA, RFC50 and RTC50 mixtures did not exhibit the highest early tensile strength (7 days), it was the mixtures that exhibited increased strength on the 28th day. Concrete is a brittle material with very low capabilities of carrying tensile loads without reinforcement. The key element of increasing concrete tensile strength is the interlocking between the constituent materials. In the compressive strength test, the aggregate surface roughness is not such a determining factor as in the tensile test. Matias et al. concluded that the rough and angular contact area allows better adherence [17].

3.4 Durability Properties

The durability test results are presented in Table 7. The mixtures with RCA had lower density due to the lower density of the adhered mortar. Porosity experiments showed that the early age values of the NA and RCA mixtures were similar, and by the age of 28 days, the NA mixtures exhibited a significantly lower porosity (3.7% compared to average 8.9% of the RCA mixtures). On the contrary, the RCP resistance of the RCA mixtures were not significantly different, although in several cases RCA mixtures were characterized by lower values than the NA mixture. The only exception was the RTC50 mixture which exhibited a 174% higher than the average of the rest of the mixtures. The specific was not reflected in the sorptivity experiments. Despite the benefits obtained from the mechanical treatment, it seems that some negative effects also occur, however, since treated aggregates evidently improve concrete properties [3], we will not rush into drawing premature conclusions. Following this, the specific properties will be evaluated in a long-term period (up to 540 days) since it is known that various mechanisms such as internal curing may contribute to the improvement of such properties [3, 18, 19]. The increase of the sorptivity coefficient of RAC can also be attributed to the higher absorption capacity of the recycled aggregates compared to natural aggregates.

4 Conclusions

This work investigated the addition of RCA in concrete, in an effort to reduce the demand in natural aggregate and utilize an otherwise waste, hard to dispose material. No significant difference has been observed with the use of field and treated aggregates, however, it is anticipated that the treated aggregates effect will be more prominent in later ages, due to the effect of internal curing. Similar behavior has been observed in the splitting tensile strength experiments. The chloride resistance of the RCA mixtures was similar to the NA mixtures, where the sorptivity measurements exhibited significant differences. The investigation has concluded that the addition of RCA in concrete is possible without significantly affecting the material's properties, however, one has to be fully aware of the effects of the RCA type (treated or non-treated) and RCA percentage replacement. Finally, the effect of the aggregate treatment was not that prominent up to the tested age, however, the investigation will continue up to the age of 560 days.

Acknowledgements The authors would like to express their sincere gratitude to the Republic of Cyprus, the Cyprus Research Promotion Foundation (RPF) and the European Regional Development Fund, for funding the research project entitled "Recycled Aggregates for the Production of Concrete" (ENTERPRISES/0618/0034).

References

1. Commision, Brundtland, Report of the World Commission on Environment and Development, United Nations (1987)
2. Wang, J., Zhang, J., Cao, D., Dang, H., Ding, B.: Comparison of recycled aggregate treatment methods on the performance for recycled concrete. Constr. Build. Mater. **234**, 117366 (2020). https://doi.org/10.1016/j.conbuildmat.2019.117366
3. Dimitriou, G., Savva, P., Petrou, M.F.: Enhancing mechanical and durability properties of recycled aggregate concrete. Constr. Build. Mater. **158**, 228–235 (2018). https://doi.org/10.1016/j.conbuildmat.2017.09.137
4. Qiu, J., Sheng, Q., Yang, E.-H.: Surface treatment of recycled concrete aggregates through microbial carbonate precipitation. Constr. Build. Mater. **57**, 144–150 (2014). https://doi.org/10.1016/j.conbuildmat.2014.01.085
5. Spaeth, V., Djerbi Tegguer, A.: Improvement of recycled concrete aggregate properties by polymer treatments. Int. J. Sustain. Built Environ. **2**, 143–152 (2013). https://doi.org/10.1016/j.ijsbe.2014.03.003.
6. Ismail, S., Ramli, M.: Mechanical strength and drying shrinkage properties of concrete containing treated coarse recycled concrete aggregates. Constr. Build. Mater. **68**, 726–739 (2014). https://doi.org/10.1016/j.conbuildmat.2014.06.058

7. Kazemian, F., Rooholamini, H., Hassani, A.: Mechanical and fracture properties of concrete containing treated and untreated recycled concrete aggregates. Constr. Build. Mater. **209**, 690–700 (2019). https://doi.org/10.1016/j.conbuildmat.2019.03.179

8. Xiao, J., Li, W., Fan, Y., Huang, X.: An overview of study on recycled aggregate concrete in China (1996–2011). Constr. Build. Mater. **31**, 364–383 (2012). https://doi.org/10.1016/j.conbuildmat.2011.12.074

9. Mindess, S., Young, J.F., Darwin, D.: Concrete, Second. Pearson Education Inc., New Jersey (2003)

10. Poon, C.S., Shui, Z.H., Lam, L.: Effect of microstructure of ITZ on compressive strength of concrete prepared with recycled aggregates. Constr. Build. Mater. **18**, 461–468 (2004). https://doi.org/10.1016/j.conbuildmat.2004.03.005

11. Sánchez de Juan, M., Gutiérrez, P.A.: Study on the influence of attached mortar content on the properties of recycled concrete aggregate. Constr. Build. Mater. **23**, 872–877 (2009). https://doi.org/10.1016/j.conbuildmat.2008.04.012

12. Demetriou, G., Petrou, M.F.: Enhancing mechanical and durability properties of recycled aggregate concrete. In: 27th Biennial National Conferences Concrete Institute Australia Conjuction with 69th RILEM Week, Melbourne, pp. 1472–1481 (2015)

13. Uthus, L., Hoff, I., Horvli, I.: Evaluation of grain shape characterization methods for urban aggregates (2005)

14. Ferreira, T., Rasband, W.: ImageJ user guide, IJ 1.46r. (2012) 137. http://rsbweb.nih.gov/ij/docs/guide/146-30.html. Accessed 10 Feb 2015

15. Kanellopoulos, A., Petrou, M.F., Ioannou, I.: Durability performance of self-compacting concrete. Constr. Build. Mater. **37**, 320–325 (2012). https://doi.org/10.1016/j.conbuildmat.2012.07.049

16. Kanellopoulos, A., Nicolaides, D., Petrou, M.F.: Mechanical and durability properties of concretes containing recycled lime powder and recycled aggregates. Constr. Build. Mater. **53**, 253–259 (2014). https://doi.org/10.1016/j.conbuildmat.2013.11.102

17. Matias, D., De Brito, J., Rosa, A., Pedro, D.: Mechanical properties of concrete produced with recycled coarse aggregates-Influence of the use of superplasticizers. Constr. Build. Mater. **44**, 101–109 (2013). https://doi.org/10.1016/j.conbuildmat.2013.03.011

18. Savva, P., Nicolaides, D., Petrou, M.F.: Internal curing for mitigating high temperature concreting effects. Constr. Build. Mater. **179**, 598–604 (2018). https://doi.org/10.1016/j.conbuildmat.2018.04.032

19. Savva, P., Petrou, M.F.: Highly absorptive normal weight aggregates for internal curing of concrete. Constr. Build. Mater. **179**, 80–88 (2018). https://doi.org/10.1016/j.conbuildmat.2018.05.205

Contribution of Thermodynamic Modeling to the Understanding of Interactions Between Hydrated Cement Pastes and Organic Acids

Cédric Roosz, Marie Giroudon, Laurie Lacarrière, Matthieu Peyre-Lavigne, and Alexandra Bertron

Abstract In numerous contexts, concrete can be exposed to chemically aggressive conditions that can damage their microstructure and reduce their lifespan. The concrete facilities from the agricultural and agro-food industries dedicated to the storage or the treatment of effluents are more particularly exposed to organic acids coming from the microbial activity naturally occurring in such media. This biodeterioration leads mainly to mineralogical transformations, such as hydrated and anhydrous phase dissolution, and to ion exchanges between acidic effluents and cement-based materials. The poorly crystalline mineralogy of hydrated cement pastes and their reactivity makes the geochemical behavior of such materials difficult to investigate and thus to predict over large periods of time and wide variety of chemical conditions. The degradation of cementitious materials in these aggressive conditions mainly leads to the leaching of calcium and the precipitation of amorphous secondary phases. The purpose of this work is (i) to assess the stability of the cement phases involved in such chemical conditions as well as to identify the alterations products, and (ii) to understand the evolution of concentration and the behavior of elements in solution such as aluminum or silicon. A thermodynamic model of cement pastes subjected to acid attacks has been developed, in order to reproduce, experimental data also presented here. Our model reproduces a major part of the behaviors shown by the experiments, i.e. a progressive decalcification of solid matrix (successive dissolution of portlandite, aluminates hydrates and C-S-H) during acid degradation and the identification of alteration zones in agreement with the experimental observations.

Keywords Biodeterioration · Organic acid · Cement paste · Thermodynamic modeling

C. Roosz · M. Giroudon · L. Lacarrière (✉) · A. Bertron
LMDC, 135 Avenue de Rangueil, 31077 Toulouse Cedex 4, France
e-mail: laurie.lacarriere@insa-toulouse.fr

M. Giroudon · M. Peyre-Lavigne
TBI, UMR INSA/INRA 792, 135 av. de Rangueil, 31077 Toulouse Cedex 4, France

© The Author(s), under exclusive license to Springer Nature Switzerland AG 2022
J. Sena-Cruz et al. (eds.), *Proceedings of the 3rd RILEM Spring Convention and Conference (RSCC 2020)*, RILEM Bookseries 34,
https://doi.org/10.1007/978-3-030-76465-4_25

1 Introduction

Interaction of organic acids with hydrated cement phases can affect their microstructure and durability especially in the agro-food and nuclear waste industries. The concrete facilities exposed to organic acids are subject to severe chemical and biological attack [3, 4], leading to ion exchanges between acidic effluents and cement-based materials, decalcification of cement pastes, and dissolution of hydrates and anhydrous phases [3, 4, 6, 19]. Different types of structures can be degraded (storage silos, collecting systems, biogas silos, etc.) sometimes with high kinetics [2], which could have an influence on their safety and performances.

Considering these effects, several studies have been carried out to model these interactions, testing different organic acids and different cement compositions [6, 7, 11]. Although these studies succeeded in modeling most of the mechanisms observed experimentally, some missing data did not allow them to go further, particularly in terms of C-S-H stability (low pH and degrading chemical conditions with the presence of organic acids), aluminous phases stability (comprising Al^{3+} uptake by C-S-H), and kinetic modeling.

The aim of the work presented here is to carry out an experimental study, allowing to eliminate the kinetic factors as much as possible, in order to establish the missing data in the actual thermodynamic databases. In addition, a thermodynamic model applied to the interactions between cement pastes and acetic acid has been developed, in order to reproduce, experimental data and validate a new version of the database previously proposed by De Windt et al. [6] in light of new thermodynamic data published recently [14, 17].

2 Materials and Methods

2.1 Batch Experiments

2.1.1 Specimen Fabrication and Degradation Protocol

Cement pastes of ordinary Portland cement, CEM I 52.5 R (denominated CEM I) were made with a water/binder mass ratio of 0.3. The composition of CEM I cement is given in Table 1. The mixing of pastes was carried out according to a procedure adapted from the standard EN 196-1 [9] and cast in cylindrical moulds 70 mm high and 35 mm in diameter. Each specimen was sealed and the pastes were removed from their molds after an endothermic cure in a tempered room at 20 °C for 28 days. Finally, two specimens were crushed to a particle size of less than 80 μm to duplicate the batch experiment detailed below.

The aim of the experiment was to understand the chemical reactions occurring during the equilibration of CEM I paste in a solution of acetic acid, with a solid liquid ratio of 20. In this view, the powdered CEM I was added in 50 steps of 2 g in

Table 1 Chemical composition of CEM I 52.5 R (LOI = Loss of ignition)

wt.%	CaO	SiO$_2$	Al$_2$O$_3$	Fe$_2$O$_3$	MgO	TiO$_2$	Na$_2$O	SO$_3$	LOI
CEM I	66.2	20.0	4.85	2.64	1.06	0.27	0.14	3.01	1.64

a 2 L of acetic acid solution (0.28 mol/L) buffered with NaOH (50.12 mmol/L) with an initial pH of 3.95. The solution was constantly stirred and the pH was monitored continuously to check the stabilisation of the pH, and thus the achieving of chemical balance in solution, before the next addition. The reactor was flushed with nitrogen to prevent carbonation. The liquid/solid ratio after the 50th addition was 20. This experiment was carried out twice, and the aqueous solution was sampled at different steps for chemical analyses.

2.1.2 Chemical Characterization

The total concentrations of Ca, Al, Si, Mg, Fe, SO$_4$ ions were determined by using ion chromatography (Dionex Ion Chromatography System, ICS-3000). Before analyses, all solutions were diluted by a factor of 100 using ultrapure water (18 MΩ.cm).

2.2 Modeling Approach

2.2.1 Thermodynamic Equilibrium Calculations

All calculations were carried out using PhreeqC code version 3.0 [15]. In this code, mineral-solution equilibria are calculated considering the extended Debye-Hückel activity coefficient model, and using the Debye-Hückel B-dot relation ([13]). The thermodynamic properties used in this work came from the database used by De Windt et al. [6] and relying on *MINTEQ* 3.00 database [12] for organic acids and aqueous complexes, and *Thermoddem* database ([5]) for cement phases. An enrichment with new phases from *Thermoddem* database was carried out for the present work. The database obtained was tested by drawing activity diagrams, to assess its reliability.

2.2.2 Thermodynamic Data and Modeling Procedures

The equilibrium calculations were made using the initial solid composition (Table 1) and the solution composition (described above). As for batch experiments, calculations were based on the addition of 2 g of CEM I in 50 steps (0.04 g per step). Each step was calculated at the chemical equilibrium.

Table 2 Cement phases selected in the different calculations. [A] are data used in De Windt et al. [6], and [B], [C] correspond to data used in the present work and considering or not C-A-S-H phases. All phases are extracted from *Thermoddem* database

Name	[A]	[B]	[C]
C-S-H 0.8	X		
C-S-H 1.2	X		
C-S-H 1.6	X		
Ettringite	X	X	X
Gibbsite	X	X	X
Portlandite	X	X	X
$SiO_{2(am)}$	X	X	X
Hydrotalcite		X	X
C_3AH_6		X	X
C-S-H 0.7 to 1.6		X	X
C-A-S-H 0.7 to 1.5			X

This model was used to compare three databases. These databases were based on the work of De Windt et al. [6] and included organic acid and aqueous complexes data from *MINTEQ* database as presented in De Windt et al. [6]. The differences between these databases only concern cement phases. They are highlighted in Table 2. Finally, iron phases were neglected in all calculations.

3 Results and Discussion

3.1 Experiments Results

Figure 1 shows the evolution of the solution composition function of the step of powder addition in the acetic acid solution.

These results show that calcium concentration (Fig. 1a) increases linearly with the amount of crushed CEM I paste added, from the first to the 14th step. This suggest that almost all the calcium of the powder is dissolved for these steps. From the 15th step, calcium begins to precipitate in newly formed phases, and the calcium concentration seems to peak and stabilize around 120–130 mmol/L, showing that the chemical equilibrium has been reached.

Silicon, aluminium and iron (Fig. 1b) are showing the same behaviour in solution. The concentrations first increase proportionally with the addition of CEM I crushed paste, (steps 1–7) reflecting a total dissolution of these elements at each step. The maximum concentrations are observed around steps 8–9. These maxima are around 10 mmol/L for silicon, 2.5 mmol/L for aluminium and 1 mmol/L for iron. A significant decrease in concentrations is observed for these three elements from step 10

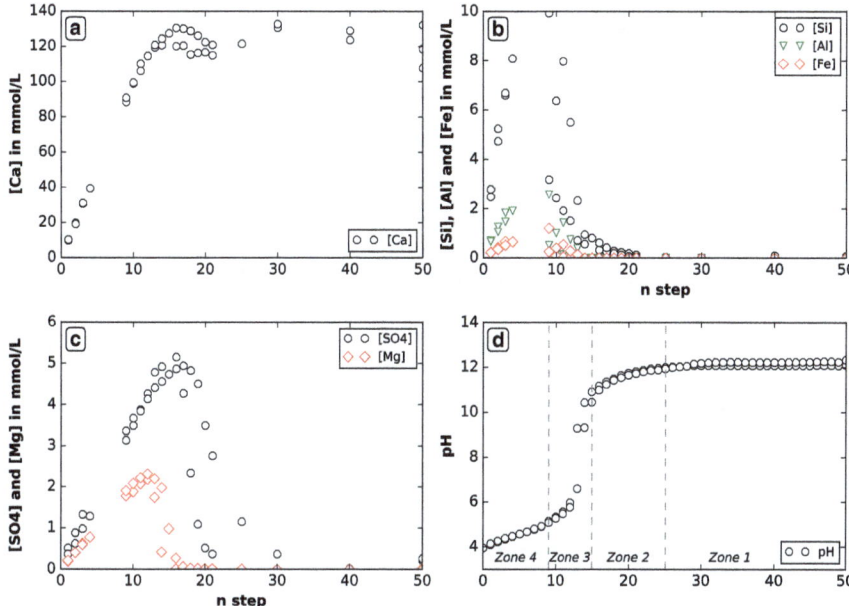

Fig. 1 Experimental data of concentration in calcium (**a**), silicon, aluminium and iron (**b**) and sulfate and magnesium (**c**), and pH values (**d**) in equilibrium solutions at each step of the batch experiment

showing the precipitation of one or more phases including these elements. Detection limit is reached rapidly for aluminium and iron (steps 13–14) while a slope break is observed for silica between steps 13 and 21 (64 to 1 mmol/L).

Finally, the concentrations of sulfate and magnesium (Fig. 1c) are showing a slightly different comportment, with for sulfates a linear increase from the first to steps 16–17 (5 mmol/L) and a significant decrease from step 19 to steps 30–35 (near detection limits), and for magnesium a linear increase until step 12 (2.3 mmol/L) and a significant decrease from steps 13 to 17 (less than 0.1 mmol/L).

The pH values (Fig. 1d) are reflecting the previous observations, with an increase from 4.1 for the first step to 5.5 for step 11. The most significant increase takes place between steps 12 and 17 with values evolving from 5.8 to 11.3. The neutralisation of the acid is thus achieved between steps 12 and 13.

In the first linear part of their trend, concentrations of Ca, Si, Al, and Mg (from first step to steps 8–12) represent more than 80% of the quantities of elements entered in the system by dissolution, showing that equilibrium is almost reached for these steps. Iron is showing a somewhat different evolution. The concentration of iron in the first step represent around 70% of the proportion of iron in the CEM I paste, and this part of dissolved iron compared to the total of iron added to the system rapidly drops to 50% at the third step. This can be explained by the C4AF low reactivity [10, 16, 18].

These observations are in agreement with previous works on the degradation of monolithic cement pastes by organic acids, where a mineralogical and chemical zonation of the altered cement paste was highlighted [3, 4, 6, 19]. In these studies, from the sound core of the specimen to the surface in contact with the acid, the first altered zone shows a slight decalcification caused by the dissolution of the portlandite. A second, more advanced alteration zone, is characterized by the progressive dissolution of the C-S-H, other hydrates phases and anhydrous phases. Finally, the third and most altered zone is described as being totally decalcified and mainly composed by an Si-Al gel. Reported to our study, these three zones could be respectively assimilated (i) to steps 50 to 25 ("Zone 1" in Fig. 1d), then (ii) to steps 20 to 15 ("Zone 2"), and finally (iii) to steps 15 to 9 ("Zone 3"). The steps 9 to 1 of our experiment corresponding to a complete dissolution of the compounds entered in the system ("Zone 4"). However, a solid characterization of our samples are necessary in order to go further.

3.2 Thermodynamic Modeling

3.2.1 Thermodynamic Properties and Stability Diagrams

Figure 2 presents simplified stability diagrams of (i) calcium salts and conjugated bases (Fig. 2a), and (ii) synthetic C-S-H from [17], with respect to pH. These phase diagrams are used to determine the aqueous species and solid phases that are stable when the pH or the acid concentration evolves. The activity approximately corresponds to the concentration for activity $\leq 5.10^{-2}$. Above, activity corrections would be required. In addition, only the predominant species (with maximum activity) are represented on these diagrams.

The main difference between the database used here and the one developed by De Windt et al. 2015 [6] being C-S-H composition, it was important to verify that new C-S-H compositions from [17] have the same comportment regarding to organic acids salts. C-S-H composition used here, from Roosz et al. 2018 [17], are discretized every 0.1 C/S from C/S = 0.7 to 1.5 compared to 3 compositions (C/S = 0.8, 1.2 and 1.6) used in De Windt et al. [6]. Figure 2a shows that the use of these new C-S-H compositions does not modify the observations made by De Windt et al. [6], with regard to the stability of the Ca(Acetate)$_2$(aq). Figure 2b shows the evolution of the stability domains of the different C-S-H with regard to their C/S ratio.

3.2.2 Equilibrium Calculations

The modeled composition of solution are presented in Fig. 3, for the different databases used, and superimposed to the experimental datas. These compositions are relative to calculated solid proportions shown in Fig. 4. The pH values are fairly well reproduced by the three database versions (Fig. 3a). In terms of calcium concen-

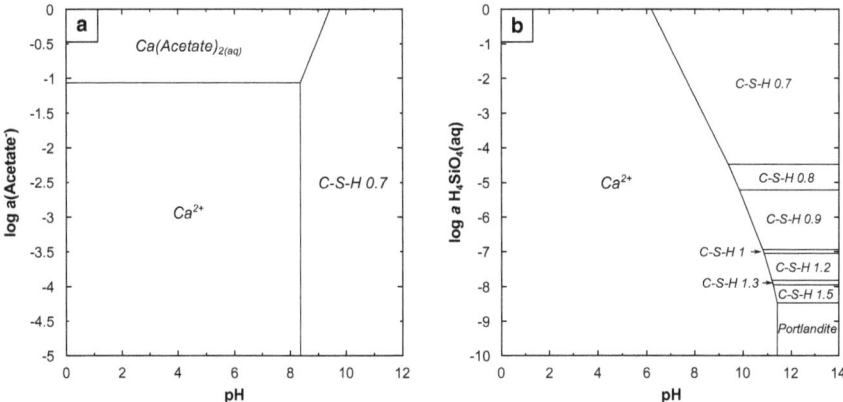

Fig. 2 Simplified speciation predominance diagrams of calcium whith respect to pH and activity of acetate (**a**) or H4SiO4 (**b**)

trations (Fig. 3b), the first observed slope between steps 1 and 14 is due to calcium solubility and is thus not affected by the changes made in our databases. The only modifications can be seen between steps 15 and 50 and are mainly due to the modification of C-S-H compositions, as showed by the differences of C/S ratio between databases on Fig. 4d. The addition of C-A-S-H to the database (Fig. 4c) has no impact on the calculated calcium concentrations (Fig. 3b), but the absence of C_3AH_6 in De Windt et al. [6] database leads to predict larger quantity of portlandite.

The aluminium concentrations (Fig. 3c) are more complicated to reproduce due to the competition between multiple phases. The maximum of modeled concentrations is reached some steps (step 5) before the experimental data (step 9). This is mainly due to the high stability of $Al(OH)_3$ in the database that is too quick to precipitate. For steps 15 to 50, two types of behaviours are modeled. The database from De Windt et al. [6] leads to an increase of the aluminium concentration from steps 20 to 30 until the Al concentration reaches a plateau between steps 30 and 50 at 2.25 mmol/L, due to the absence of C_3AH_6 and hydrotalcite (Fig. 4a). In this case, gibbsite and ettringite alone cannot fix all the aluminum in the system. The database presented in this work shows a better reproduction of experimental data, with only one maximum at step 21, corresponding to the passage between the $Al(OH)_3$ stability domain to those of ettringite and C_3AH_6 (Fig. 4b). The incorporation of C-A-S-H in this database does not change the results much (Fig. 4c).

Silicon concentrations (Fig. 3d) are driven by multiples phases. Below pH 9, amorphous silica is the only phase that can precipitate for the three databases (Fig. 4). The maximum observed experimentally at step 9 is not reproduced due to the saturation in $SiO_{2(am)}$ in the model. These experimental data can be explained by kinetic factors induced by the solubility of silicon [1, 8] neglected in our models. After step 10, the small differences observed between [6] database and ours is due to the difference of

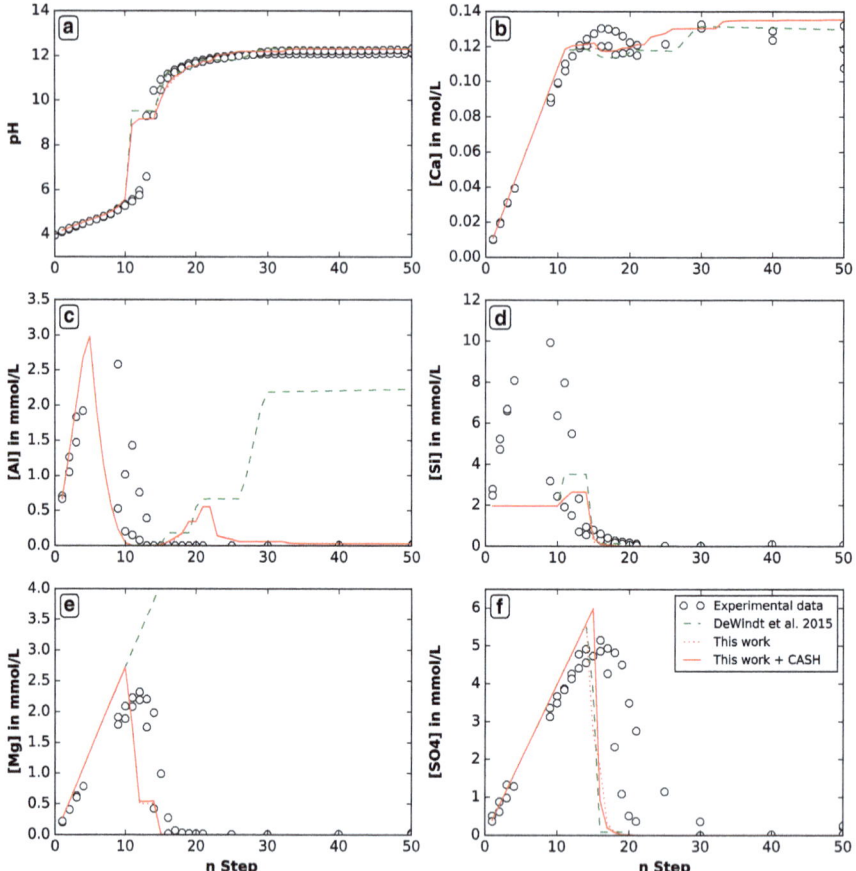

Fig. 3 Calculated solutions composition with regard to experimental data. **a** pH values, and concentration in calcium **b**, aluminium **c**, silicon **d**, magnesium **e** and sulfate **f** at each step of the batch experiment

stability of low C/S C-S-H (Fig. 4d), and the presence of C-A-S-H in the database has no impact on the calculated silicon concentrations.

In term of magnesium concentrations (Fig. 3e), experimental data are well reproduced with our databases. This can be explained by the fact that magnesium is only driven by hydrotalcite here, which was not considered in De Windt et al. [6] database (Fig. 4a).

Finally, sulfate concentrations, presented in Fig. 3f, show that calculations are close to experimental data. Ettringite is the only phase in the three databases to catch sulfate, and thus the only differences between this work and the database from De Windt et al. [6] is due to the change of other aluminous phases (C-A-S-H, C_3AH_6, hydrotalcite), that are in competition for aluminium (Fig. 4).

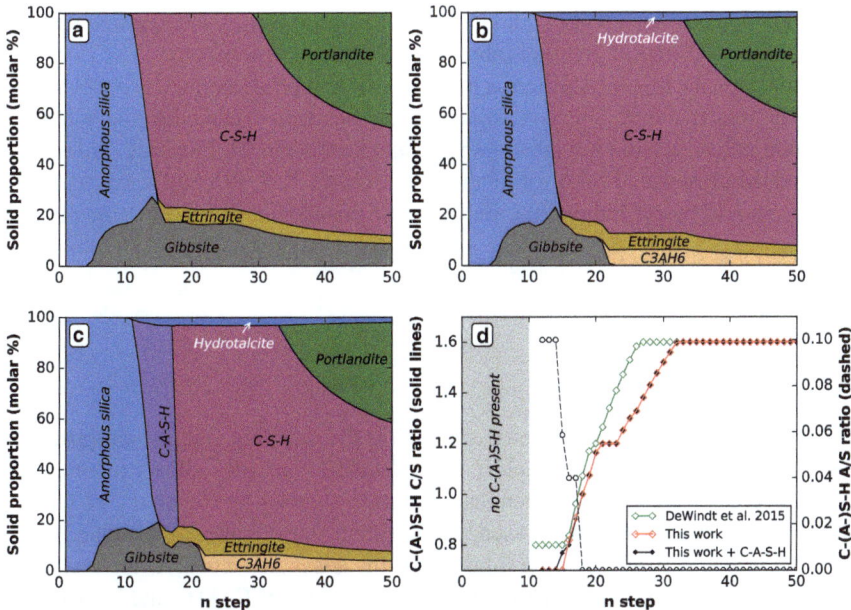

Fig. 4 Calculated solid proportions in molar % regarding to De Windt et al. [6] database (**a**), "This work" database (**b**), "This work + C-A-S-H" database (**c**), and associated C/S (solid lines) and A/S (dashed line) of C-(A-)S-H phases (**d**)

3.2.3 Properties of Aluminous Phases

The database presented here allows a better understanding of the mechanisms controlling aluminum leaching in cement pastes subjected to attack with acetic acid. Most differences noticed between [6] database and those developed in this paper are in the incorporation of aluminous phases.

The addition of the new C-S-H compositions allows a better representation of the evolution of the C/S ratio even if solid analyzes are required to go further in the interpretation. The addition of C-A-S-H hardly changes anything, except at pH between 8 and 10 where the added phases are the most stable, which is in agreement whit observations made in a previous study [17] reporting a higher stability of aluminium in C-A-S-H of low C/S ratios. Moreover, the fact that no radical modification of the C/S ratio have been seen between our database with or without C-A-S-H between steps 8 and 20 confirm the consistency of the database.

3.2.4 Limits—Next Steps

Although these results reproduce a major part of the behaviors shown by the experiments, there are still some points that can be improved. A lack of thermodynamic data for C-A-S-H phases, especially for those with higher aluminum concentrations is still to be filled. Such phases could allow better reproduction of experimental data between steps 15 and 25 for aluminium concentrations in solutions.

In addition, the major issue here is the difficulty to reproduce silicon concentration between steps 0 and 9. This can be due to kinetics factors linked to the solubility of silicon at low pH and not taken into account in this model. Previous studies also reported the presence of Si-Al gel in this pH range [3, 4, 6, 19], and solid characterizations should be carried out in order to better understand the phases precipitating here.

4 Conclusion

The work carried out here enabled evaluating the database initially proposed by De Windt et al. [6] and enriching it. This database currently allows representing in an acceptable manner the chemical evolution showed by the attack of CEM I pastes by acetic acid, with the presence of different steps of alterations linked to the progressive decalcification of the solid during the attack. Moreover, compared to the database proposed by De Windt et al. [6], the addition of C_3AH_6 and hydrotalcite allows a better prediction of the composition of CEM I pastes from the first stages of deterioration (steps 25–50) which could for example affect the prediction in terms of mechanical properties.

The next step is to develop a chemical and mineralogical characterization protocol (i) to evaluate the evolution of the C/S and A/S ratio of C-(A-)S-H at the different steps of our batch experiment, and also (ii) to characterize the degraded phases, likely a silico-aluminous gel, absent from the databases so far.

Acknowledgements The authors gratefully thank the French National Research Agency (ANR) for funding the project BIBENdOM—ANR—16—CE22—001 DS0602.

References

1. Alexander, G.B., Heston, W.M., Iler, R.K.: The solubility of amorphous silica in water. J. Phys. Chem. **58**(6), 453–455 (1954). https://doi.org/10.1021/j150516a002
2. Alexander, M., Bertron, A., Belie, N.D.: Performance of cement-based materials in aggressive aqueous environments, RILEM state-of-the-art reports, vol. 10. Springer Netherlands, Dordrecht (2013). https://doi.org/10.1007/978-94-007-5413-3
3. Bertron, A., Duchesne, J., Escadeillas, G.: Attack of cement pastes exposed to organic acids in manure. Cement Concr. Compos. **27**(9–10), 898–909 (2005). https://doi.org/10.1016/j.cemconcomp.2005.06.003
4. Bertron, A., Duchesne, J., Escadeillas, G.: Degradation of cement pastes by organic acids. Mater. Struct. Materiaux et Constr. **40**(3), 341–354 (2007). https://doi.org/10.1617/s11527-006-9110-3

5. Blanc, P., Lassin, A., Piantone, P., Azaroual, M., Jacquemet, N., Fabbri, A., Gaucher, E.: Thermoddem: a geochemical database focused on low temperature water/rock interactions and waste materials. Appl. Geochem. **27**(10), 2107–2116 (2012). https://doi.org/10.1016/j.apgeochem.2012.06.002

6. De Windt, L., Bertron, A., Larreur-Cayol, S., Escadeillas, G.: Interactions between hydrated cement paste and organic acids: thermodynamic data and speciation modeling. Cement Concr. Res. **69**, 25–36 (2015). https://doi.org/10.1016/j.cemconres.2014.12.001

7. De Windt, L., Devillers, P.: Modeling the degradation of Portland cement pastes by biogenic organic acids. Cement Concr. Res. **40**(8), 1165–1174 (2010). https://doi.org/10.1016/j.cemconres.2010.03.005

8. Dove, P.M., Han, N., Wallace, A.F., De Yoreo, J.J.: Kinetics of amorphous silica dissolution and the paradox of the silica polymorphs. Proc. Natl. Acad. Sci. United States Am. **105**(29), 9903–9908 (2008). https://doi.org/10.1073/pnas.0803798105

9. EN, N.: 196-1, méthodes d'essais des ciments-partie 1: détermination des résistances mécaniques. French Standard (2006)

10. Fukuhara, M., Goto, S., Asaga, K., Daimon, M., Kondo, R.: Mechanisms and kinetics of C4AF hydration with gypsum. Cement Concr. Res. **11**(3), 407–414 (1981). https://doi.org/10.1016/0008-8846(81)90112-5

11. Grandclerc, A., Dangla, P., Gueguen-Minerbe, M., Chaussadent, T.: Modelling of the sulfuric acid attack on different types of cementitious materials. Cement Concr. Res. **105**(2017), 126–133 (2018). https://doi.org/10.1016/j.cemconres.2018.01.014

12. Gustafsson, J.P.: Visual MINTEQ version 3.0. KTH, Dept. of land and water resources engineering. Stockholm, Sweden (2010)

13. Helgeson, H.C.: Thermodynamics of hydrothermal systems at elevated temperatures and pressures. Am. J. Sci. **267**(7), 729–804 (1969)

14. Lothenbach, B., Kulik, D.A., Matschei, T., Balonis, M., Baquerizo, L., Dilnesa, B., Miron, G.D., Myers, R.J.: Cemdata18: a chemical thermodynamic database for hydrated Portland cements and alkali-activated materials. Cement Concrete Res. **115**, 472–506 (2019). https://doi.org/10.1016/j.cemconres.2018.04.018

15. Parkhurst, D., Appelo, C.: Phreeqc (version 3.0. 4)–a computer program for speciation, batch speciation, one-dimensional transport, and inverse geochemical calculations. US Geol. Surv. Techn. Methods Book **6**, 497 (2013)

16. Plowman, C., Cabrera, J.: Mechanism and kinetics of hydration of C3A and C4AF. Extracted from cement. Cement Concr. Res. **14**(2), 238–248 (1984). https://doi.org/10.1016/0008-8846(84)90110-8

17. Roosz, C., Vieillard, P., Blanc, P., Gaboreau, S., Gailhanou, H., Braithwaite, D., Montouillout, V., Denoyel, R., Henocq, P., Madé, B.: Thermodynamic properties of C-S-H, C-A-S-H and M-S-H phases: results from direct measurements and predictive modelling. Appl. Geochem. **92**, 140–156 (2018). https://doi.org/10.1016/j.apgeochem.2018.03.004

18. Rose, J., Bénard, A., El Mrabet, S., Masion, A., Moulin, I., Briois, V., Olivi, L., Bottero, J.Y.: Evolution of iron speciation during hydration of C4AF. Waste Manag. **26**(7), 720–724 (2006). https://doi.org/10.1016/j.wasman.2006.01.021

19. Voegel, C., Giroudon, M., Bertron, A., Patapy, C., Matthieu, P.L., Verdier, T., Erable, B.: Cementitious materials in biogas systems: Biodeterioration mechanisms and kinetics in CEM I and CAC based materials. Cement Concr. Res. **124**(2018), 105815 (2019). https://doi.org/10.1016/j.cemconres.2019.105815

An Analytical Approach for Pull-Out Behavior of TRM-Strengthened Rammed Earth Elements

A. Romanazzi, D. V. Oliveira, and R. A. Silva

Abstract Rammed earth constructions, beyond being largely spread in the built heritage, are known for their high seismic vulnerability, which results from high self-weight, lack of box behavior and low mechanical properties of the material. Hence, to mitigate this seismic vulnerability, a compatible textile reinforced mortar (TRM) is here proposed as a strengthening solution, because of its reduced mass and high ductility. The few research about the structural behavior of TRM-strengthened rammed earth elements addresses the global behavior, overlooking the local behavior of the system. An analytical approach to infer the bond stress-slip relationship following the direct boundary problem is proposed. Based on a previous series of pull-out tests, an adhesion-friction constitutive law is portrayed considering also a damage model that considers the degradation of the reinforcing fibers due to friction.

Keywords Rammed earth · Textile reinforced mortar · Bond · Analytical approach

1 Introduction

Raw earth is one of the most ancient building materials and its related building techniques are spread worldwide, counting about 10% of the built UNESCO World Heritage and between 20 and 30% of the global population living in earthen dwellings [1, 2]. Among the different building techniques based on the use of soil, rammed earth consists in compacting a mixture of moistened earth within a formwork, which is directly supported on the wall and moved horizontally once a block is completed [1]. This technique is used since ancient times to build both monuments [1, 3] and affordable dwellings [4]. Nonetheless, rammed earth buildings are also well known

A. Romanazzi (✉) · D. V. Oliveira · R. A. Silva
ISISE & IB-S, University of Minho, Guimarães, Portugal

D. V. Oliveira
e-mail: danvco@civil.uminho.pt

R. A. Silva
e-mail: ruisilva@civil.uminho.pt

for their high seismic vulnerability, which is due to low mechanical properties of the material, high self-weight and poor connection between structural elements. Thus, moderate to intense earthquakes are expected to produce in-plane cracking of the walls, formation of out-of-plane mechanisms and collapse of the roof and floors [5]. For this reason, textile reinforced mortar (TRM) has been proposed as a solution to mitigate the high seismic vulnerability of rammed earth dwellings, due to its low self-weight, tensile strength and ductility, as demonstrated for masonry buildings [6–8].

Since the TRM is a composite system [6, 9], understanding the mechanical response of the matrix-fiber interface is a key point to predict the overall performance of a strengthened structure. In this context, different test setups have been implemented to deduce the interaction between the two components [10–12]. Among these tests, the pull-out test is the most accepted, and consists in pulling out a single fiber or a mesh embedded in a specimen representing the matrix, while the corresponding load-displacement relationship $P(u)$ is recorded. However, the $P(u)$ curve is a response of a system with a specific geometry, not representing a material property of the tested composite. For obtaining material parameters to define the shear force transmission independently from the geometric properties, known as the bond stress-slip relationship (BSR), numerical or analytical models can be applied to the experimental $P(u)$ curve [13, 14]. In the case of cement-based matrix composites, an interface zone is assumed with properties different from the matrix and fiber [15]. As a consequence, if the stiffness of the interface is much smaller than that of the constituents, the deformation in this zone might be higher than that of the fiber u_f or matrix u_m. Therefore, the difference between the deformation of the components represents the interface deformation and it is defined as slip $[s = u_f - u_m]$. While, the bond stress-slip law is the transferred shear stress $\tau(s)$ as function of the slip (s) at the interface matrix-fiber at any coordinate of the fiber (x) [15–18].

In order to deduce the analytical law of the mortar-mesh interaction for imperfect interface models, two approaches have been implemented so far, namely a direct boundary problem (DBP) and an inverse boundary problem (IBP). In the case of the former approach, the load versus displacement relation $P(u)$ of a pull-out test is calculated on the basis of an assumed constitutive law $\tau(s)$ [15, 19–23]; therefore, its parameters must be supposed and the pull-out curve simulated to be compared with the experimental data. By means of fitting or an optimization process, the bond-slip law is verified once the best approximation is achieved.

The abovementioned approach is followed here to propose a method to infer a BSR of the experimental program conducted in Romanazzi et al. [24]. At first, the materials and the pull-out results of the experimental program are reported and discussed to hypothesize a BSR. Subsequently, the problem statement of the analytical model with the implementation of a novel damage model is derived.

2 Experimental Program

The material properties and the results of a series of pull-out tests used to derive the BSR are reported in Romanazzi et al. [24]. The mechanical properties of the selected earth-based mortar were characterized according to EN 1015-11 [25]. The average flexural strength f_b is 0.5 MPa (CoV= 14%), while the compressive strength f_c is 1.2 MPa (CoV = 12%). The Young's modulus E_m was evaluated by means of axial compression tests on three cylindrical specimens of casted mortar with 90 mm of diameter and 175 mm high. The Young's modulus E_m was computed by linear fitting of the stress-strain curves in the range 0–30% of f_c, which provided an average value of about 4915 MPa (CoV = 20%). The tensile behaviour of the selected low-cost fiber glass mesh was evaluated according to the procedure prescribed in ASTM D6637 [26] and RILEM TC-250 CSM [27]. Five specimens were prepared with width of about 50 mm and with free length of 300 mm, considering the direction along which they show higher tensile capacity, as found in Oliveira et al. [28]. The resulting average maximum linear force $P_{w,p}$ is 18.4 kN/m (CoV 11%), while the average tensile strength f_t of a single yarn and the peak axial strain ε_{peak} are 626 MPa (CoV 11%) and 0.021 (CoV 10%), respectively. In addition, the average Young's modulus E_y is of about 32181 MPa (CoV 6%), as computed by linear fitting of the tensile stress-strain curve in the range 0–30% f_t. Table 1 summarizes the material properties.

The pull-out specimens consisted of a glass fiber mesh band embedded in earth-based mortar cylinders with diameter of ±150 mm and height corresponding to the bonded length L_b. In the present study, only the specimens with bonded length 90 mm and 150 mm are considered. The specimens were casted ensuring the correct filling of the mold and perfect alignment of a single mesh band of 50 mm wide, while the unbonded part of the mesh was kept vertically to avoid any damage due to bending. The drying period of the specimens was of 28 days under constant hygrothermal conditions (T = 20 ± 2 °C and RH = 60 ±5 %), after which they were subjected to displacement controlled pull-out tests. The displacements of the mesh were recorded by means of one LVDT set at the free end and two LVDTs set at the loaded end close to the mortar surface (see Fig. 1). Further details are presented in Romanazzi et al. [24].

Figure 2a–d present the response curves in terms of force per width, displacement at the loaded end and displacement at the free end for the two different bonded lengths. Based on the literature [17, 20–23, 29, 30], the experimental pull-out curves can be divided into two zones corresponding to different shear stress distributions along the interface. In particular, an initial linear response is observed, in which the load is transmitted by adhesion. When the shear strength is achieved, micro-cracks

Table 1 Properties of the selected glass fibre mesh and earth-based mortar

Material	$P_{w,p}$ (kN/m)	ε_{peak} (–)	f_t (MPa)	E_y (MPa)	A_y (mm^2)	f_c (MPa)	f_b (MPa)	E_m (MPa)	A_m (mm^2)
Glass fiber mesh	18.4	0.021	626	32181	0.294	–	–	–	–
Earth mortar	–	–	–	–	–	1.2	0.50	4915	2.355

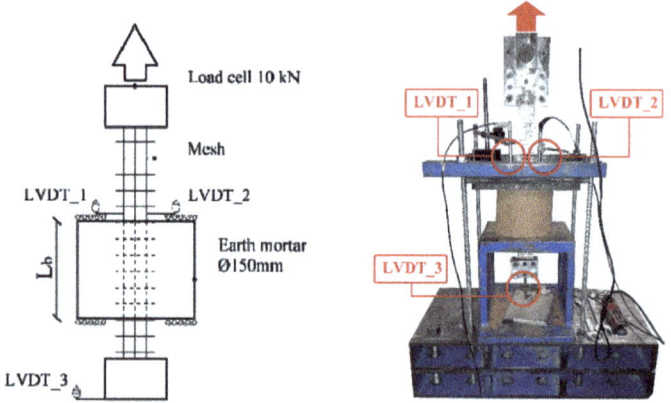

Fig. 1 Setup of the pull-out tests

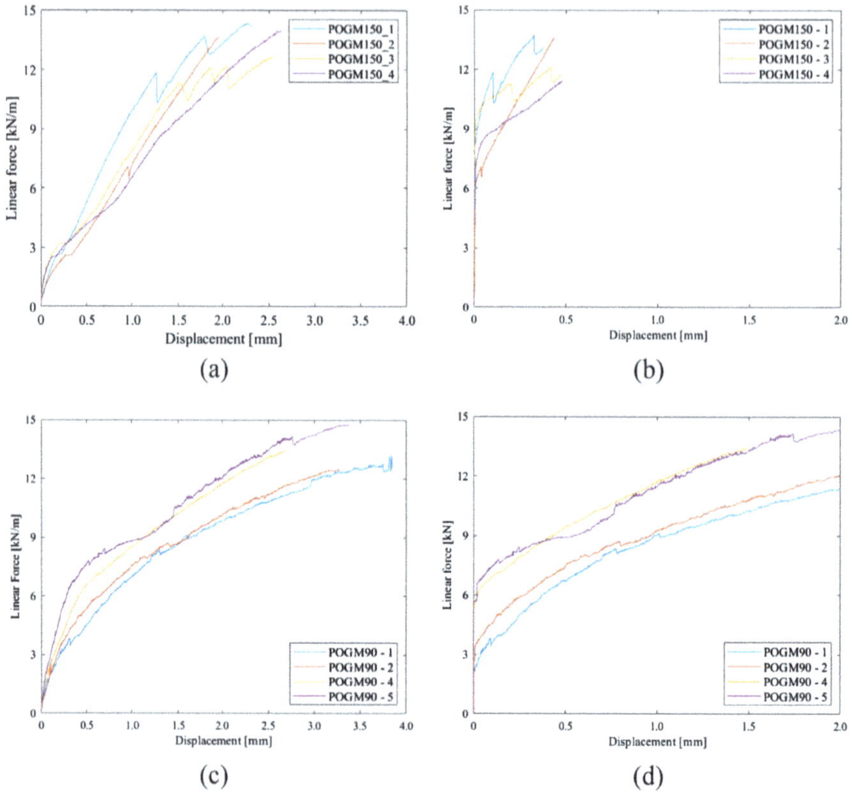

Fig. 2 Pull-out experimental curves: **a** loaded end L_b 150 mm; **b** free end L_b 150 mm; **c** loaded end L_b 90 mm; **d** free end L_b 90 mm

are developed and the response becomes non-linear. In this stage, adhesion is still at the interface of the bonded fibers and friction between the two components is found in the detached part. As the shear strength is attained at the free end, friction becomes the only resistant mechanism.

3 Analytical Model

The direct approach was implemented for processing the experimental pull-out data and derive an analytical bond stress-slip law. Therefore, an adhesion-friction constitutive law was assumed with a linear response up to the maximum shear strength τ_{Max} and elastic slip s_{El}; subsequently, the strength drops to the shear friction resistance τ_{Fri} until failure (see Fig. 3).

Given the static equilibrium along the embedded length (Fig. 4), the tensile force in the yarn F is transferred to the matrix M through the interface. Considering the infinitesimal interface dx, the equilibrium can be expressed as

$$\frac{dF}{dx} = -\frac{dM}{dx} = p\tau(x) \tag{1}$$

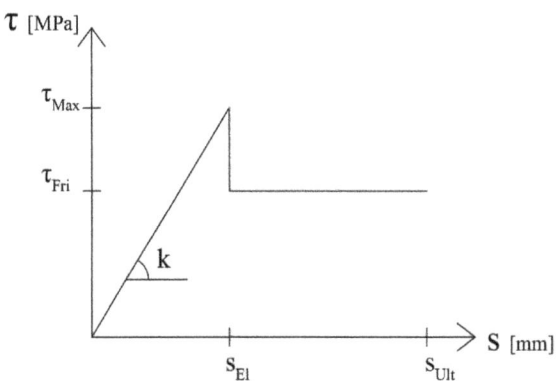

Fig. 3 Assumed adhesion-friction bond stress-slip relationship

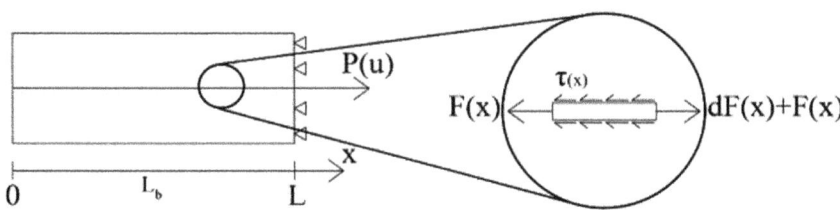

Fig. 4 Static scheme of the interface during the pull-out test

where p is the perimeter of the yarn and τ is the shear stress at the yarn-matrix interface.

As the axial elongation in the fiber and in the matrix can be defined as $\varepsilon_y = \frac{F}{A_y E_y}$ and $\varepsilon_m = \frac{F}{A_m E_m}$, the slip in the section x is the difference between the elongation of the components $s(x) = \frac{du}{dx} = \varepsilon_y - \varepsilon_m$, hence:

$$\frac{du}{dx} = \frac{F(x)}{A_y E_y} + \frac{F(x)}{A_m E_m} \tag{2}$$

where A_m, A_y, E_m and E_y are the cross section areas and Young's moduli of the matrix and fiber respectively. Substituting (2) into (1), one obtains:

$$\frac{dF}{dx} = \frac{d^2 u}{dx^2} Q = p\tau \tag{3}$$

where $Q = \frac{1}{A_y E_y} + \frac{1}{A_m E_m}$ is the relative axial stiffness between the two components (3) represents the analytical problem statement of the pull-out test to be solved according to the stage in which the section is, as described in the next section.

3.1 Linear Stage

During the adhesion phase, the assumed interface stress-slip relationship is linear with $\tau = ku$ (Fig. 3), which substituted in (3) leads to:

$$u'' - \lambda^2 u = 0 \tag{4}$$

with $\lambda = \sqrt{pkQ}$. The general solution of the second differential Eq. (4) is:

$$u(x) = C_1 e^{\lambda x} + C_2 e^{-\lambda x} \tag{5}$$

which substituted in (2) leads to:

$$F(x) = \frac{du}{dx} \frac{1}{Q} = \frac{1}{Q}\left(C_1 \lambda e^{\lambda x} - C_2 \lambda e^{-\lambda x}\right) \tag{6}$$

Considering as the boundary conditions the force in the fiber at the free-end, which is null $F(0) = 0$, and the force in the fiber at the loaded-end, which is equal to the pull-out force $F(L_b)=P$, the coefficients C_1 and C_2 result as $C_1 = C_2 = \frac{PQ}{\lambda(e^{\lambda L} - e^{-\lambda L})}$; which replaced in (6) gives the force distribution along the fiber $F(x)$ as:

Fig. 5 Shear stress
distribution along the
interface at elastic load

$$F(x) = P\frac{(e^{\lambda x} - e^{-\lambda x})}{(e^{\lambda L_b} - e^{-\lambda L_b})} = P\frac{sinh(\lambda x)}{sinh(\lambda L_b)} \tag{7}$$

while the shear $\tau(x)$ and slip $s(x)$ distribution along the interface are respectively:

$$\tau(x) = \frac{dF}{dx}\frac{1}{p} = P\frac{\lambda}{p}\frac{cosh(\lambda x)}{sinh(\lambda L_b)} \tag{8}$$

and

$$u(x) = \int_0^x F(x)Qdx = PQ\frac{1}{sinh(\lambda L_b)}\int_0^x sinh(\lambda x)dx \tag{9}$$

For pull-out loads lower than the elastic limit load $F(L_b)=P < P_{El}$, the shear stress at the interface is less than the shear strength τ_{Max} and the yarn and the matrix are full bonded. Once the pull-out force achieves the elastic load $F(L_b) = P = P_{El}$, the shear strength τ_{Max} is attained at the loaded end $x = L_b$ and the debonding onsets. Hence, in such configuration the shear stress distribution is illustrated in Fig. 5 and (8) becomes (10).

$$\tau(L_b) = \tau_{Max} = P_{El}\frac{\lambda}{p}\frac{cosh(\lambda L_b)}{sinh(\lambda L_b)} \tag{10}$$

While the slip at the loaded-end results:

$$s(L_b) = u_{El} = \int_0^{L_b} F(x)Qdx = P_{El}Q\frac{1}{sinh(\lambda L_b)}\frac{1}{\lambda}[cosh(\lambda L_b) - 1] \tag{11}$$

Therefore, given the experimental elastic pull-out load and displacement (P_{El} and u_{El}), the shear strength $[\tau_{Max}]$ and the shear stiffness of the interface $[k]$ are obtained by solving the system of Eqs. (10) and (11) at the coordinate of the loaded end ($x = L_b$).

3.2 Nonlinear Stage

For loads beyond the elastic limit $F(L_b) = P > P_{El}$, the micro-cracks propagate along the interface toward the free-end. Consequently, the fiber and mortar are detached in the length L_d, while they are still bonded in the remaining length L_b-L_d. The resulted shear stress distribution is composed by constant frictional stress τ_{Fri} in the debonding length $L_d < x < L_b$ and adhesion $\tau = ku$ along the bonded length $0 < x < L_b-L_d$, while the maximum shear strength τ_{Fri} is achieved at the coordinate $x=L_b-L_d$ (Fig. 6).

Therefore, the pull-out force is the sum of the forces resulting from adhesion and friction as:

$$F(L_b) = P = F_{I_{Max}} + F_{II} = \frac{\tau_{Max}}{\lambda} tanh[\lambda(L_b - L_d)] + \tau_{Fri} p L_d \qquad (12)$$

In this case, the boundary conditions are:

- $F(0) = 0$
- $F(L_b - L_d) = \frac{\tau_{Max} p}{\lambda} tanh[\lambda(L_b - L_d)]$

- $F(L_b) = P$

Which, placed in (3), lead to the force distribution $F(x)$ in the elastic length ($0 < x < L_b-L_d$) as:

$$F_I(x) = \frac{\tau_{Max} p}{\lambda} tanh[\lambda(L_b - L_d)] \frac{sinh(\lambda x)}{sinh[\lambda(L_b - L_d)]} \qquad (13)$$

And in the debonded length ($L_b-L_d < x < L_b$):

$$F(x) = F_{I_{Max}} + F_{II} = \frac{\tau_{Max} p}{\lambda} tanh[\lambda(L_b - L_d)] + \tau_{Fri} p(x - L_b + L_d) \qquad (14)$$

The slip along the yarn $s(x)$ can be evaluated as the sum of the slips according to the different stage in which the two parts of the yarns are. Therefore, the total slip at the loaded end results

$$s(L_b) = s_{El} + s_{Fri} = \frac{F_{I_{Max}} \lambda Q}{sinh[\lambda(L_b - L_d)]} [cosh[\lambda(L_b - L_d)] - 1]$$

Fig. 6 Shear stress distribution along the interface during nonlinear response

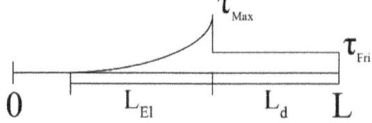

$$+ L_d Q \left(F_{l_{Max}} + \frac{\tau_{Fri} p L_d}{2} \right) \qquad (15)$$

In this configuration, the debonded length L_d and the shear friction τ_{Fri} can be obtained by solving the system of Eqs. (12) and (15), in which the inputs are the experimental pull-out force P and slip u in the nonlinear branch of the curve.

3.3 Damage Model

Observing the experimental pull-out curve, a damage in the yarn due to the friction between mortar and fiber is deemed, as discussed in previous investigation [24]. In view of that, a damage model that considers the reduction of the cross-section of the yarn is introduced as function of the sliding. The damage is defined as $\xi(u) = \frac{A_y - A_{yRed}}{A_y} x 100$, where A_y and A_{yRed} represent the initial cross-section area (undamaged state) and the reduced cross section area of the yarn due to friction action, respectively. Considering the experimental ultimate load, the reduced section is evaluated as $A_{yRed} = \frac{P_{Ult}}{n^\circ f_t}$, where f_t is the tensile strength of the dry mesh and n is the number of yarns. Afterwards, a correlation between the average value of damage and the bonded length (Fig. 7a) and the sliding during the nonlinear stage (Fig. 7b) was found and expressed in (16).

$$\xi = 2.3772e^{1.0497(u - u_{El})} X \frac{1}{100} \qquad (16)$$

Therefore, the value of $[1 - \xi(u)]$, with $\xi(u) = 2.3772e^{1.0497(u - u_{El})} X \frac{1}{100}$, is introduced as factor to reduce the cross section of the yarn to evaluate the relative stiffness

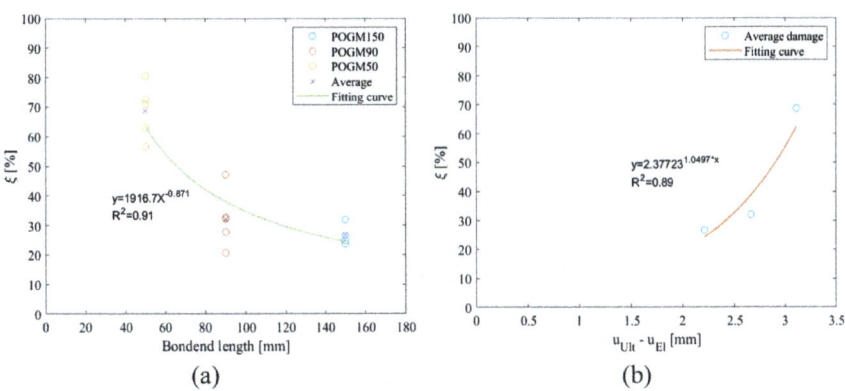

(a)   (b)

Fig. 7 Experimental correlations: **a** damage-bonded length; **b** damage-sliding in the non-linear stage

$Q(u)$ in (17), which is then considered in the system of equations composed of (12) and (15) for the nonlinear stage.

$$Q(u) = \frac{1}{A_y(1 - \xi(u))E_y} + \frac{1}{A_m E_m} \tag{17}$$

4   Conclusions

A method to infer a bond stress-slip relationship (BSR) of a TRM-based solution for strengthening rammed earth is here proposed following the direct boundary problem. At first, the properties of the materials used in the composite and the pull-out tests are presented. Based on the test evidences, an adhesion-friction BSR is hypothesized. Consequently, the equations to describe the stress transmission along the interface are inferred for the linear and non-linear stage. Therefore, the BSR parameters will be obtained considering the experimental results. Observing a correlation between the level of sliding and the loss of resistance with respect to the dry mesh, a novel damage model is assumed, which reduces the axial stiffness of the interface.

The proposed method is being implemented in an algorithm for simulating the pull-out test, in which a sensitivity analysis on the BSR parameters will be conducted. Though, the presented method is expected to be an approach for different combination of materials, rather than a thorough model for any composite with the use of earth-based matrixes.

Acknowledgment This work was partly financed by FEDER funds through the Operational Programme Competitiveness Factors (COMPETE 2020) and by national funds through the Foundation for Science and Technology (FCT) within the scope of project SafEarth - PTDC/ECM-EST/2777/2014 (POCI-01-0145-FEDER-016737). The support from grant SFRH/BD/131006/2017 is also acknowledged.

References

1. Houben, H., Guillaud, H.: Earth construction: a comprehensive guide. Intermediate Technology Publication, London (2008)
2. Silva, R.A., Oliveira, D.V., Miranda, T., Cristelo, N., Escobar, M.C., Soares, E.: Rammed earth construction with granitic residual soils: the case study of northern Portugal. Constr. Build. Mater. **47**, 181–191 (2013)
3. Jaquin, P.A., Augarde, C.E., Gerrard, C.M.: A chronological description of the spatial development of rammed earth techniques. Inter. J. Architect. Herit. **2**, 377–400 (2008)
4. Silva, R.A., Mendes, N., Oliveira, D.V., Romanazzi, A., Domínguez-Martínez, O., Miranda, T.: Evaluating the seismic behaviour of rammed earth buildings from Portugal: from simple tools to advanced approaches. Eng. Struct. **157**, 144–156 (2018)

5. Costa, A.A., Varum, H., Rodrigues, H., Vasconcelos, G.: Seismic behaviour analysis and retrofitting of a row building. In: Correia, M., Lourenço, P.B., Varum, H. (eds.) Seismic retrofitting: learning from vernacular architecture. Taylor and Francis Group, London (2015)
6. De Felice, G., De Santis, S., Garmendia, L., Ghiassi, B., Larringa, P., Lourenço, P.B., Oliveira, D.V., Paolacci, F., Papanicolaou, C.G.: Mortar-based systems for externally bonded strengthening of masonry. Mater. Struct. **47**, 2021–2037 (2014)
7. Ghiassi, B., Marcari, G., Oliveira, D.V., Lourenço, P.B.: Numerical analysis of bond behavior between masonry bricks and composite materials. Eng. Struct. **43**, 210–220 (2012)
8. Valluzzi, M.R., Modena, C., De Felice, G.: Current practice and open issues in strengthening historical buildings with composites. Mater. Struct. **47**, 1971–1985 (2014)
9. Righetti, L., Edmondson, V., Corradi, M., Borri, A.: Fiberglass grids as sustainable reinforcement of historic masonry. Materials **9**, 1–17 (2016)
10. Dalalbashi, A., Ghiassi, B., Oliverira, D.V., Freitas, A.: Effect of test setup on the fiber-to-mortar pull-out response in TRM composites: experimental and analytical modeling. Compos. Part B **143**, 250–268 (2018)
11. D'Ambrisi, A., Feo, L., Focacci, F.: Experimental analysis on bond between PBO-FRCM strengthening materials and concrete. Compos. Part B **44**, 524–532 (2013)
12. Donnini, J., Lancioni, G., Corinaldesi, V.: Failure modes in FRCM systems with dry and pre-impregnated carbon yarns: experiments and modeling. Compos. Part B, **140**, 57–67 (2018)
13. Soranakom, C., Mobasher, B.: Geometrical and mechanical aspects of fabric bonding and pullout in cement composites. Mater. Struct. **42**, 765–777 (2009)
14. Li, Y., Bielak, J., Hegger, J., Chudoba, R.: An incremental inverse analysis procedure for identification of bond-slip laws in composites applied to textile reinforced concrete. Compos. Part B, **137**, 111–122 (2018)
15. Banholzer, B., Brameshuber, W., Jung, W.: Analytical simulation of pull-out tests: the direct problem. Cement Concr. Compos. **27**, 93–101 (2005)
16. Banholzer, B.: Bond of a strand in a cementitious matrix. Mater. Struct. **39**, 1015–1028 (2006)
17. Sueki, S., Soranalom, C., Mobasher, B., Member, A.S.C.E., Peled, A.: Pullout-slip response of fabrics embedded in a cement paste matrix. J. Mater. Civil Eng. **19**, 718–727 (2007)
18. Banholzer, B., Brameshuber, W., Jung, W.: Analytical simulation of pull-out tests: the inverse problem. Cement Concr. Compos. **28**, 564–571 (2006)
19. Naaman, A.E., Member, A.S.C.E., Namur, G.G., Alwan, J.M., Najm, H.S.: Fiber pullout and bond slip. I: analytical study. J. Struct. Eng. **117**, 2769–2790 (1991)
20. Carozzi, F.G., Colombi, P., Fava, G., Poggi, C.: A cohesive interface crack model for the matrix–textile debonding in FRCM composites. Compos. Struct. **143**, 230–241 (2016)
21. Ferreira, S.R., Martinelli, E., Pepe, M., Silva, F.D., Filo, R.D.T.: Inverse identification of the bond behavior for jute fibers in cementitious matrix. Compistes Part B **95**, 440–452 (2016)
22. D'Antino, T., Colombi, P., Carloni, C., Sneed, L.H.: Estimation of a matrix-fiber interface cohesive material law in FRCM-concrete joints. Compos. Struct. **193**, 103–112 (2018)
23. Focacci, F., D'Antino, T., Carloni, C., Sneed, L.H., Pellegrino, C.: An indirect method to calibrate the interfacial cohesive material law for FRCM-concrete joints. Mater. Design **128**, 206–217 (2017)
24. Romanazzi, A., Oliveira, D.V., Silva, R.A.: Experimental investigation on the bond behavior of a compatible TRM-based solution for rammed earth heritage. Int. J. Architect. Herit. (2019). https://doi.org/10.1080/15583058.2019.1619881
25. BS EN 1015: Methods of test for mortar for masonry. Part:11 determination of flexural and compressive strength of hardened mortar (1999)
26. ASTM D6637: Standard test method for determining tensile properties of geogrids by the single or multi-rib tensile method (2015)
27. De Felice, G., Aiello, M.A., Caggegi, C., Ceroni, F., De Santis, S., Garbin, E., Gattesco, N., Hojdys, L., Krajewski, P., Kwiecień, A., Leone, M., Lignola, G.P., Mazzotti, C., Oliveira, D., Papanicolaou, C., Poggi, C., Triantafillou, T., Valluzzi, M.R., Viskovic, A.: Recommendation of RILEM technical committee 250-CSM: test method for textile reinforced mortar to substrate bond characterization. Mater. Struct. **51**, 95 (2018)

28. Oliveira, D.V., Silva, R.A., Barroso, C., Lourenço, P.B.: Characterization of a compatible low cost strengthening solution based on the TRM technique for rammed earth key. Eng. Mater. **747**, 150–157 (2018)
29. Zhang, X.B., Aljewifi, H., Li, J.: Failure behaviour investigation of continuous yarn reinforced cementitious composites. Constr. Build. Mater. **47**, 456–464 (2013)
30. D'Antino, T., Carloni, C., Sneed, L.H., Pellegrino, C.: Matrix-fiber bond behavior in PBO FRCM composites: a fracture mechanics approach. Eng. Fract. Mech. **117**, 94–111 (2014)

Self-Healing Concrete Research in the European Projects SARCOS and SMARTINCS

Nele De Belie, Kim Van Tittelboom, Mercedes Sánchez Moreno, Liberato Ferrara, and Elke Gruyaert

Abstract Self-healing concrete and preventive repair of structures will slow down the development of cracks and/or arrest the ingress of aggressive agents. When the cracks are closed or a decrease in crack width is achieved, this will be associated with improved durability of the structure. This paper describes the literature review and inter-laboratory comparison carried out within the COST Action CA15202 (SARCOS), as well as the research planned within the recently started International Training Network SMARTINCS.

Keywords Self-healing concrete · Repair · Durability · Healing efficiency

1 COST Action CA 15202 SARCOS

1.1 Introduction

The search for smart self-healing materials and preventive repair methods is justified by the increasing sustainability and safety requirements of structures. The appearance of small cracks in concrete is unavoidable, not necessarily causing a risk of collapse for the structure, but certainly impairing their serviceability, accelerating its degradation and diminishing the service life and sustainability of constructions. That loss of performance and functionality promotes an increasing investment on maintenance

N. De Belie (✉) · K. Van Tittelboom
Magnel-Vandepitte Laboratory for Structural Engineering and Building Materials, Ghent University, Ghent, Belgium
e-mail: nele.debelie@ugent.be

M. Sánchez Moreno
Inorganic Chemistry Department, University of Córdoba, Córdoba, Spain

L. Ferrara
Department of Civil and Environmental Engineering, Politecnico di Milano, Milan, Italy

E. Gruyaert
Department of Civil Engineering, KU Leuven, Leuven, Belgium

© The Author(s), under exclusive license to Springer Nature Switzerland AG 2022
J. Sena-Cruz et al. (eds.), *Proceedings of the 3rd RILEM Spring Convention and Conference (RSCC 2020)*, RILEM Bookseries 34,
https://doi.org/10.1007/978-3-030-76465-4_27

and/or intensive repair/strengthening works. Therefore, the COST Action CA 15,202 (SARCOS) [1] firstly focuses on preventive repair solutions including self-healing approaches and innovative external repair methods for existing concrete elements. Despite the promising potential of the developed healing technologies, they will only find their way to the market when sound characterization techniques for performance verification are developed, being the SARCOS's second focus. The third focus deals with modelling the healing mechanisms taking place for the different designs and with predicting the service life increase achieved by these methods.

The added value of networking in SARCOS mainly lays in the expertise exchange between the most recognized international research groups working in the topic. The discussion at international level between worldwide leading groups covers a wide range of approaches, methodologies and applications, ensuring the objective of creating guidelines and recommendations for real applications.

From the sustainability point of view, self-healing concrete and preventive repair of structures may slow down the development of cracks, even achieving a decrease in crack width with associated benefits for the structures. Filling of the cracks with healing agent will furthermore reduce their negative effects regarding penetration of aggressive substances in the concrete. These approaches are also associated to the optimal application of different repair technologies, involving environmentally friendly technologies, and to the minimization of the use of raw materials and of the worker's risk associated to traditional repair operations, which will be a benefit in the long-term. The new functionalities included in these advanced cement based materials and the higher performance, associated to the developed methodologies, are expected to delay the ageing of concrete structures built with these materials, thus leading not only to higher service life but also to better performance in more aggressive environments and under more demanding conditions.

1.2   Literature and Research Needs

A recent review paper by SARCOS members [2] reveals the key challenge that the self-healing additions up-to-date are produced at lab scale and self-healing efficiency is only shown at paste/mortar level. The few existing demonstrators at the concrete level often show insufficient or not yet proven self-healing efficiency. Further, factors such as effect of variability in the design parameters and longevity of the embedded system may be questioned. One of the reasons for reduced efficiency after upscaling to concrete is the significant dilution of the additives when maintaining the dosage relative to cement weight; however, keeping the same dosage in proportion to the total volume, results mostly in an unacceptable strength decrease and high healing agent cost. Another challenge is that the durability of self-healing concrete elements has only been scarcely investigated. No long-term durability results of self-healing concrete are available, and even results for accelerated durability testing in lab conditions are scarce. Mostly, durability is assessed indirectly through parameters such as gas and water permeability, surface resistivity and capillary absorption. It is suggested that further research should focus on durability of the healed structures e.g. resistance to chloride diffusion and carbonation, corrosion, freeze/thaw, salt crystallization, etc. [1].

A second SARCOS review paper [3] relates to the experimental methods and techniques, which have been employed to characterize and quantify the self-sealing and/or self-healing capacity of cement-based materials, together with the methods for the analysis of the chemical composition and intrinsic nature of the self-healing products. This article also addresses the correlation between crack closure and recovery of mechanical properties. Especially the experimental characterization of the self-healing capacity under sustained loads has been highlighted as an important research need to provide a basis for incorporation of self-healing concepts and to allow incorporation of self-healing functionalities in predictive models and durability based design approaches.

The third review paper [4], which is related to SARCOS' third objective, discusses research progress on numerical models for self-healing cementitious materials. This article provides a summary of self-healing techniques and discusses mechanical models for self-healing, transport processes in materials with embedded healing systems, fully coupled models and other modelling techniques used to simulate self-healing behaviour. The models discussed include those based on continuum-damage-healing mechanics, micro-mechanics, as well as models that use discrete elements and particle methods. The article also covers transport models and the simulation of carbonation in concrete since this mechanism governs self-healing based on calcite precipitation in cracks. The article highlights the lack of fully-coupled models, although approaches that couple some aspects of transport and mechanical healing behaviour are discussed. One bottleneck pointed out is that many models are presented with only very limited experimental validation. It seems that there has been insufficient interaction between numerical and experimental research teams. Other limitations include that often only one cycle of healing can be simulated, healing takes place under zero-strain conditions, damage and healing are not coinciding and healing takes place instantaneously. The statistical variations have only been considered in a few exploratory investigations.

1.3   Interlaboratory Tests to Define Methods for Self-Healing Efficiency

Currently, the SARCOS partners are conducting a Round Robin Test campaign to verify the different characterization techniques and evaluation criteria for the preventive repair approaches. The outcome of this test campaign will be used as input for recommendations and guidelines. Important in this regard is to standardize a procedure including pre-cracking and multiple healing measurements. The self-healing evaluation includes microscopic crack closure quantification, regain of water tightness (through water absorption and water flow tests), regain of resistance against chloride ingress, and in some cases recovery of mechanical properties. The tested types of self-healing specimens include (1) concrete with mineral additions, (2) concrete with MgO, (3) concrete with crystalline admixtures, (4) high performance fibre reinforced concrete with crystalline admixtures, (5) mortar and concrete with macrocapsules containing a polymeric healing agent and (6) concrete with encapsulated bacteria. For some test series, first results are already available. For instance, the inter-laboratory test on self-healing mortar with macrocapsules showed that the target crack width of 300 μm could be obtained with great accuracy as a result of the applied active crack width control technique. This resulted in similar results for the water permeability test, yet additional specimens will be investigated to eliminate all critical points. It was much more difficult to obtain the target crack width in the concrete specimens used for capillary water absorption testing, since no active crack width control was foreseen in this case and due to the presence of large aggregates. This resulted in a high variability regarding the measured capillary water absorption.

2 Marie Curie ITN SMARTINCS

To fill in the need for further research in the area of self-healing concrete and repair materials, a team of SARCOS participants has taken the initiative to establish an International Training Network (ITN) with the name SMARTINCS (Smart, Multifunctional, Advanced Repair Technologies In Cementitious Systems) [5]. The project was launched on 1 December 2019.

SMARTINCS will implement new life-cycle thinking and durability-based approaches to the concept and design of concrete structures, with self-healing concrete, repair mortars and grouts as key enabling technologies. This will create a breakthrough in the current practice of the construction industry, which is characterized by huge economic costs related to inspection, maintenance, repair and eventually demolition activities and additional indirect costs caused by traffic congestions during maintenance and environmental effects.

SMARTINCS will train a new generation of creative and entrepreneurial early-stage researchers in prevention of deterioration of (i) new concrete infrastructure by innovative, multifunctional self-healing strategies and (ii) existing concrete infrastructure by advanced repair technologies. The project brings together the complementary expertise of research institutes pioneering in smart cementitious materials, strengthened by leading companies along the SMARTINCS value chain, as well as certification and pre-standardization agencies. They will intensively train 15 early stage researchers to respond to the clear demand to implement new life-cycle thinking and durability-based approaches to the concept and design of concrete structures, minimizing both the use of resources and production of waste in line with Europe's Circular Economy strategy. The new generation of researchers will be immediately employable to support the introduction of the novel technologies allowing the expected spectacular growth of the self-healing materials market to take place. By combined experimental research and enhanced coupled multiscale numerical models for prediction of the self-healing behaviour, SMARTINCS strives to move beyond the state-of-the-art.

The scientific objectives are attained by joint Ph.D. research and envisage:

(i)　To develop and model innovative self-healing strategies for bulk and local application, including optimization of mix designs and development of multi-functional self-healing agents with attention to cost, applicability and environmental impact.

(ii)　To scientifically substantiate and model the durability of self-healed concrete and repaired systems for an accurate service life prediction and to integrate self-healing into innovative service-life based structural design approaches to foster the market penetration through an innovative life-cycle thinking.

(iii)　To quantify and prove the eco-efficiency of newly developed smart concrete / mortars by life cycle assessment modelling.

Acknowledgements The author(s) would like to acknowledge networking support by the COST Action CA15202 "SARCOS".

 This project has received funding from the European Union's Horizon 2020 research and innovation programme under the Marie Skłodowska-Curie grant agreement No 860006

References

1. SARCOS homepage. https://www.sarcos.eng.cam.ac.uk/. Accessed 23 Jan 2020
2. De Belie, N., Gruyaert, E., Al-Tabbaa, A., Antonaci, P., Baera, C., Bajare, A., Darquennes, A., Davies, R., Ferrara, L., Jefferson, T., Litina, C., Miljevic, B., Otlewska, A., Ranogajec, J., Roig, M., Paine, K., Lukowski, P., Serna, P., Tulliani, J.-M., Vucetic, S., Wang, J., Jonkers, H.A.: Review on self-healing concrete for damage management of structures. Special Issue "Self-healing materials". Adv. Mater. Interf. **1800074**, 28 (2018)
3. Ferrara, L., Van Mullem, T., Alonso, M.C., Antonaci, P., Borg, R.P., Cuenca, E., Snoeck, D., Jefferson, A., Peled, A., Ng, P.I., Roig Flores, M., Serna Ros, P., Sanchez, M., Schroefl, C., Tulliani, J.M., De Belie, N.: Experimental characterization of the self-healing capacity of cement based materials and its effects on the material performance: A state of the art report by COST Action SARCOS WG2. Constr. Build. Mater. **167**, 115–142 (2018)
4. Jefferson, A., Javierre, E., Freeman, B., Zaoui, A., Koenders, E., Ferrara, L.: Research progress on numerical models for self-healing cementitious materials. Adv. Mater. Interf. **170378**, 19 (2018)
5. SMARTINCS homepage. http://www.smartincs.eu. Accessed 21 Jan 2020

Strengthening of Hollow Core Slabs to Reduce Excessive Vibrations: A Case Study

Thanongsak Imjai, Surin Suthiprabha, Fabio P. Figueiredo, and Reyes Garcia

Abstract The LB Building 6 is the main logistic building of the Total Agility Logistic Co. Ltd located in Bang Phi district, Samut-Prakan province, Thailand. The building consists of a 2 storey steel structure having steel beams with a typical span of 6.0 m supported by reinforced columns. After a few months of service, cracking and spalls was found on the concrete overlay on the 2nd floor. Preliminary investigations were first performed by the building engineer and it was reported that deflections occurred at the 1-ton forklift moving path zone. Therefore, the capacity of structural elements in the building had to be re-assessed and field measurements of vibration levels were also taken during the peak operation period. To reduce the vibration level as well as deformation of the floor under normal operation, a structural strengthening intervention was made by adding additional steel beams under the hollow core slab as a span shortening technique. Field measurements have shown that the vibrations of the strengthened concrete slab reduced by up to 35% compared to the pre-strengthening measurements, thus confirming the effectiveness of the strengthening intervention. The work reported in this paper also sets the scene for identifying the major decisions for a design engineer starting from on-site inspection to the post-construction supervision.

Keywords Vibration · Human perception · Hollow core slab · Strengthening · Logistic building

T. Imjai (✉)
Walailak University, Nakhon Sri Thammarat 80161, Thailand
e-mail: thanongsak.im@wu.ac.th

S. Suthiprabha
Rajamangala University of Technology Tawan-Ok, Bangkok 10330, Thailand

F. P. Figueiredo
University of Minho, 4800-058 Guimarães, Portugal

R. Garcia
The University of Warwick, Coventry CV4 7AL, England

© The Author(s), under exclusive license to Springer Nature Switzerland AG 2022
J. Sena-Cruz et al. (eds.), *Proceedings of the 3rd RILEM Spring Convention and Conference (RSCC 2020)*, RILEM Bookseries 34,
https://doi.org/10.1007/978-3-030-76465-4_28

1 Introduction

Precast elements are widely used in the construction of medium and large structures around the world. Precast elements are convenient as they help reduce construction time, whilst at the same time guarantee that the concrete and reinforcement detailing (many times done off site) comply with minimum code provisions. In particular, floor systems of industrial facilities in South East Asia are often built using precast hollow slabs so as to maximize the use of floor space and to minimize structural mass and therefore foundation size/depth [1–3]. These slabs are commonly supported on concrete or steel beams, which in turn support on columns. In an attempt to produce very light floor systems, the cross sections and thicknesses of such slabs are minimized to an extent that the slabs can present structural issues (e.g. excessive deflections and cracking) even under service loads. Previous research has indicated that the level of damage in slabs can be assessed by examining the level of vibration and dynamic characteristics of the slab [4–6]. In all cases, a thorough assessment is necessary to determine the potential causes of structural distress, as well as possible remedial solutions to solve the issues and prevent further damage.

This paper discusses the assessment and strengthening of a real existing heavy logistic industrial building. The 2-storey steel building is an integral part of the Total Agility Logistic Co. Ltd, which is the main logistics company in Thailand. The elements of the building developed cracks and spalling of concrete at several locations just after six months of construction. As a result, design documents of the building were thoroughly reviewed to determine the potential causes of distress. An on-site structural inspection also identified the location and nature of the structural defects. A strengthening intervention using span shortening of the hollow core slabs was selected as the preferred strengthening solution to reduce both excessive deflection and vibration level. The work reported in this paper also sets the scene for identifying the major decisions for a design engineer starting from on-site inspection to the post-construction supervision.

2 Project Background

The LB Building 6 is the main logistic building of the Total Agility Logistic Co. Ltd located in Bang Phi district, Samut Prakan province, Thailand. The building consists of a 2 storey steel structure having steel beams with a typical span of 6.0 m supported by reinforced concrete columns (Fig. 1). The construction of the LB building was completed in 2015. The design live load is induced by heavy forklift vehicles moving on the 2nd floor during the normal operating period (24 h). After a few months of service, cracking and spalling were found on the concrete overlay on the 2nd floor. Preliminary investigation was first performed by the building engineer and it was reported that deflections occurred at the 1-ton forklift moving path zone. Moreover, during the peak operation time (12:00–15:00 hrs), staff and clients reported that

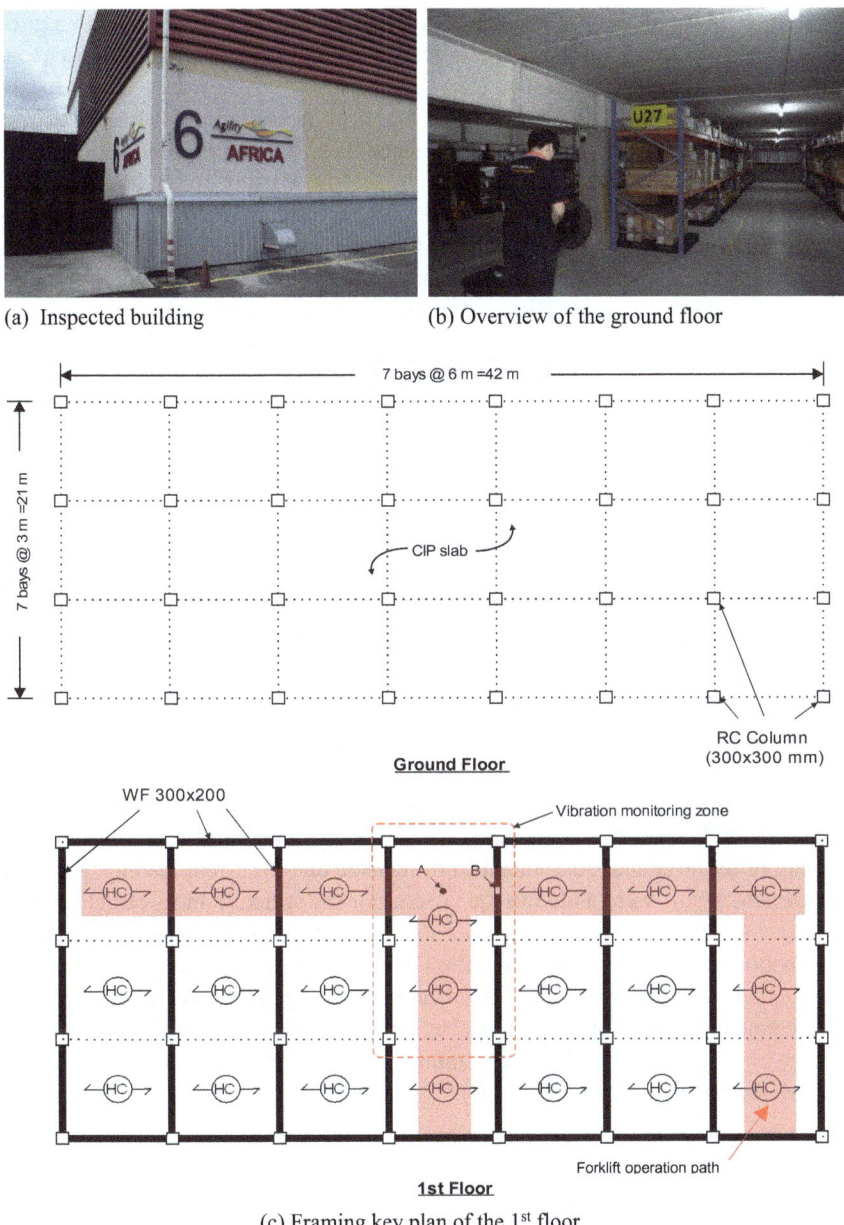

(a) Inspected building   (b) Overview of the ground floor

(c) Framing key plan of the 1st floor

Fig. 1 Overview of LB6 Building framing key plan and vibration monitoring zone

they felt the floor shaking when the forklifts were in operation. The manager of the building maintenance department request a further investigation and subsequent engineering solution for this issue.

At the time of inspection, the building had been in operation for five years and annual inspections have been carried out since then. It was proposed to repair the cracked and spalls of concrete overlay in the 2nd floor using epoxy injections and epoxy floor coating prior to carrying out any strengthening work. Several strengthening schemes were proposed to increase flexural capacity of the hollow core slabs and steel beams, including externally bonded steel plate (EBS), externally bonded fibre reinforced polymers (EBR) and addition of steel beams under the existing hollow core slab to reduce the span length. These solutions aimed to reduce the vibration level as well as deformation of the floor under normal operation. Currently, the building LB6 is still in operation and periodic inspections are carried out every year.

3 Structural Engineering Assessment

3.1 Structural Damage Assessment of the Building

The condition survey examined and assessed the current level of damage and deterioration of the structure, and determined the preliminary serviceability condition such as excessive deflection and/or vibration of the structural components of the building. A rating point system was used to evaluate serviceability, and to compare the current and newly-built structural condition only, without determining the load capacity of the structural components. For damage due to current loading, repair or replacement, terms were added to the criteria for the structure in its current condition, i.e. a structural element that shows small defects, good maintenance and good construction practice was given a condition rating of 2 or "Fair condition". If the same element had some deffects/damage from current loading to the level the replacement of the element will be of more beneficial to the whole structure than its rehabilitation, then the term "Replacement" was recommended for "Fair condition" (adopted from RILEM Technical Committee 104 [7]).

3.2 Design Verification

During the construction of the building, documents such as as-built drawings, calculation report and a material testing report were produced and these were still available at the time of inspection. Detailed reviews on the documentation was carried out prior taking any further actions. A yield strength of 392 and 245 MPa were used for the steel reinforcement and hot-rolled steel beam, respectively. The original design floor

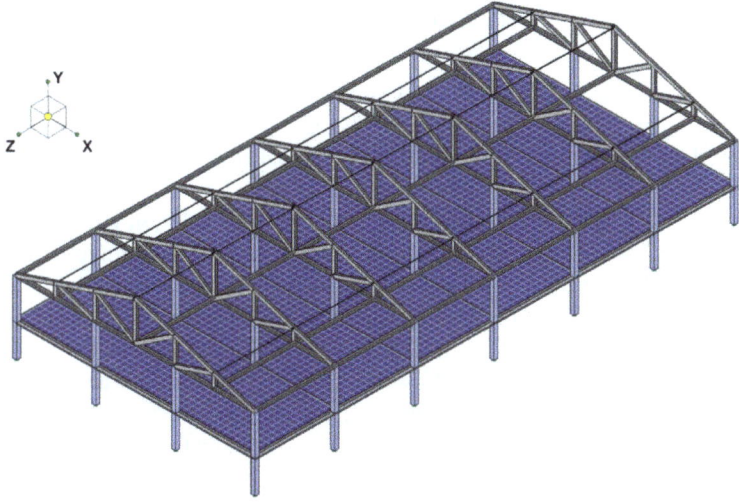

Fig. 2 Computer model of the LB6 building frame

live loads was 10 kN/m^2. The design check on ultimate limit state (ULS) considers the factored load i.e (Mu) and nominal strength of the section (Mn) multiplied by the strength reduction factor (Φ) according to ACI 318. Fig. 2 shows a 3D FE model of the LB6 building. Design verifications on serviceability limit state (SLS) of the structural elements according to the current ACI 318 [8] was then performed and it was found that all existing structural elements satisfied the design. However, there is no available information on the vibration checks and on maximum floor deflections, which suggests that these calculations may have not been performed. Therefore, vibration levels and deflections of the 2nd floor under normal operation were needed to reassess the structure.

4 Strengthening of Hollow Core Slabs for SLS

4.1 General Design Considerations

The cracks and concrete spalling in the 2nd floor are shown in Fig. 1. A conventional repair using epoxy injection and epoxy floor coating was initially applied after the problem was reported. After careful consideration and discussions with the stake-holders, it was decided to add steel beams under the hollow core slab as a span shortening technique. This aimed to reduce vibrations and deformations of the floor under normal operation.

4.2   Strengthening Steel Framing Plan

Hot-rolled steel beams of 200 × 200 mm, steel grade A36 (yield stress = 245
MPa) were used to reduce the floor span of the hollow core slab. The original span
of the hollow core was 6 m long and was reduced to 2 m after adding the steel
beams, as shown in Fig. 3a. The existing steel beams (300 × 200 mm) and newly-
added steel beams (200 × 200 mm) were then rechecked for their ultimate capacity
and serviceability requirements according to the current design guidelines [8–11].
Consequently, field measurement on deflection, concrete strain and vibration levels
before and after rehabilitation work were carried out to assess the performance of
the strengthening work.

Construction details and installation procedure were carried out according to the
local code of practice and law requirement [8]. Figure 3b shows an installation of steel
beams under the hollow core slabs with epoxy coating as a fire protection system.

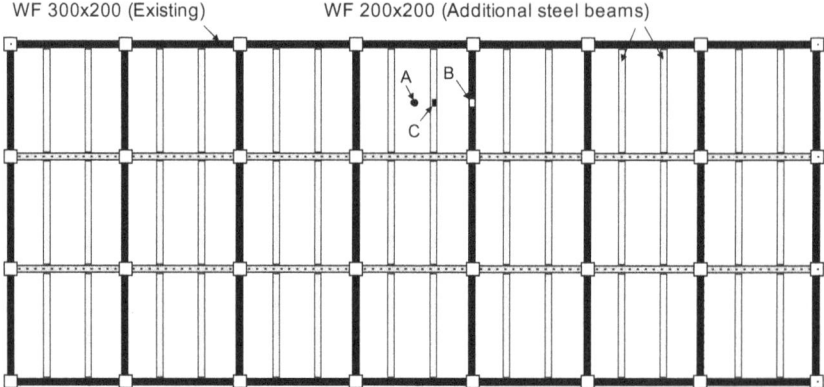

1st Floor after adding additional beams
(a) Key plan for additional steel beam installed under the 1st floor

(b) Overview of the 1st floor after installation of the steel beams

Fig. 3 Visual inspection of the hollow core concrete slabs

5 Field Vibration Measurement and Dynamic Properties

5.1 Vibration Field Measurement

Accelerometers, LVDTs and strain sensors were installed at the mid-span of the hollow core slab (point A in Fig. 3a), as shown in Fig. 1c. The uniaxial accelerometer model A2123 (Fig. 4b) can measure dynamic properties in tough field conditions. The accelerometer can measure the dynamic response ranges between ±2g (g = acceleration of gravity). A data acquisition system located in the 1st floor (Fig. 4a) continuously recorded the field data using a sampling rate of 10 Hz.

Vibrations of the structure were recorded for 10 hours for each channel location from about 18:00 to 16:00 hrs. The determined value of the acceleration level (a) is compared with values of human perception to vibration in different standards [9, 11], as summarized in Table 1. As seen in the table, there are 5 levels of sensitivity to vibration. Level one ("low probability of complaints") is comparable with a vibration perception threshold, whereas the highest level ("adverse comments probable") could be compared with a comfort limit.

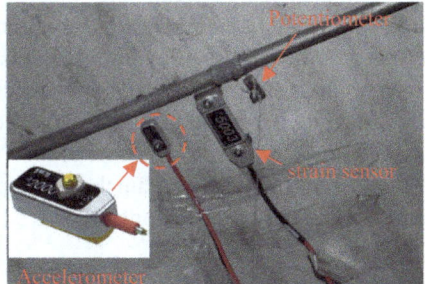

(a) Instrumentation and control point

(b) Accelerometer, LVDT and strain sensors

Fig. 4 Test setup and instrumentation

Table 1 Human perception to vibration

Human vibration effects	Acceleration range
Imperceptible	$a < 0.005\ g$
Perceptible	$0.005 < a < 0.015\ g$
Annoying	$0.015\ g < a < 0.05\ g$
Very annoying	$0.05\ g < a < 0.15\ g$
Intolerable	$0.15 < a$

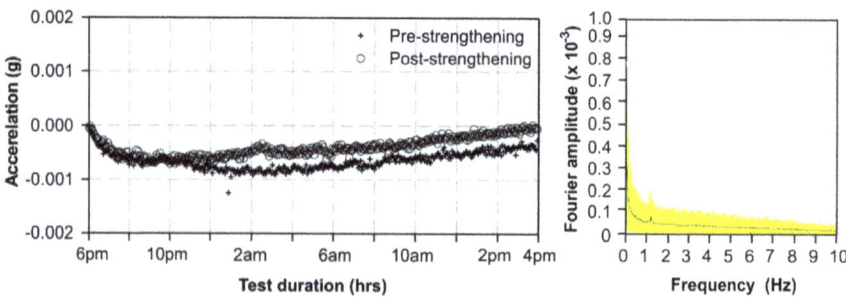

Fig. 5 Peak acceleration at top of concrete slab measured from accelerometer

5.2 Dynamic Response

Vibration records were divided into small segment (of 10 Hz) during the field measurement. Figure 5-left shows the root-mean-square of the acceleration in unit of g (9.81 m/s^2) measured at the mid-span of the hollow core slab before and after adding the steel strengthening beams. The analysis of the natural frequency of the dynamic response after the strengthening intervention is also presented in Fig. 5-right. As seen in Fig. 5-left, high levels of vibration were captured between 14:00–15:00 hrs as this was the peak operation period. It is also shown that the vibration level of the concrete floor (measured in the gravity direction) decreased by up to 35% after the strengthening intervention, and it remained within the imperceptible level described in Table 1 ($a < 0.005\ g$). The acceleration responses were analyzed by techniques in the frequency domain to obtain the Fourier spectrum. From the Fourier spectrum, the natural frequency was determined as 1.2 Hz for the strengthened hollow care slabs.

5.3 Deflections of the Hollow Core Slab

The deflection of the slab was monitored using LDVTs (precision = 0.001 mm) before and after the strengthening. As seen in Fig. 6, the maximum deflection of the slab before and after strengthening is approximately 1 mm during the peak operation period, which is well below the allowable value L/360 according to the current building code [8]. It is also shown that the strengthening reduced the deflection, although this reduction was marginal.

5.4 Strains in the Concrete Slabs

Figure 7 compares the strains measured at the mid-span of the hollow core slab. It is shown that the strains in the slab after strengthening reduced by up to 25% at the

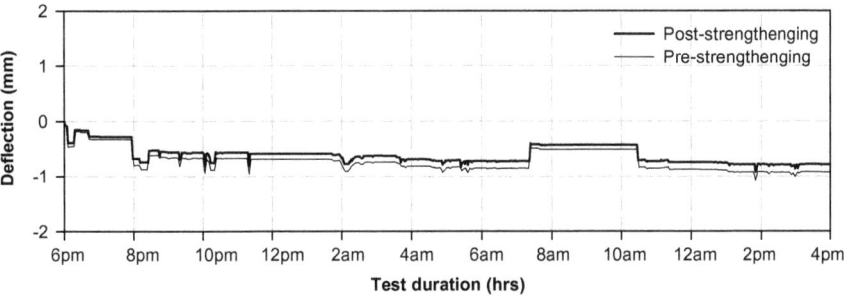

Fig. 6 Measured deflection of the concrete slab

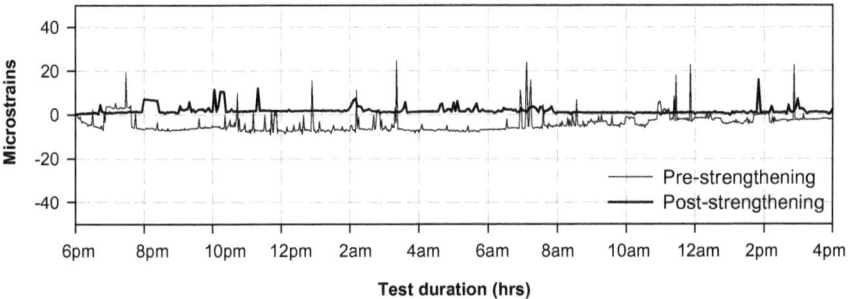

Fig. 7 Comparison of measured strains at mid-span of the slab

peak operation period when compared to the value measured before strengthening, thus confirming the effectiveness of the strengthening system. It was also noted that the excessive high "jumping" of the deflection readings over time was due to the presence of an unstable electric signal during the field measurement, which in turn resulted in sudden increases of the readings (particularly for the pre-strengthened slab).

5.5   Strains in Steel Beams

Figure 8 compares measured strains at the midspan of two adjacent steel beams: an existing steel beam (i.e. Point B in Fig. 3a for steel beam WF 300 × 200 mm), and an additional strengthening steel beam (Point B in Fig. 3a for steel beam WF 200 × 200 mm), as referred in Fig. 3a. The results confirm that the strain value of the additional steel beams is higher than those induced in the existing steel beam by up to 35% during the peak operation period, measured after strengthening work, thus confirming the good performance of the strengthening intervention.

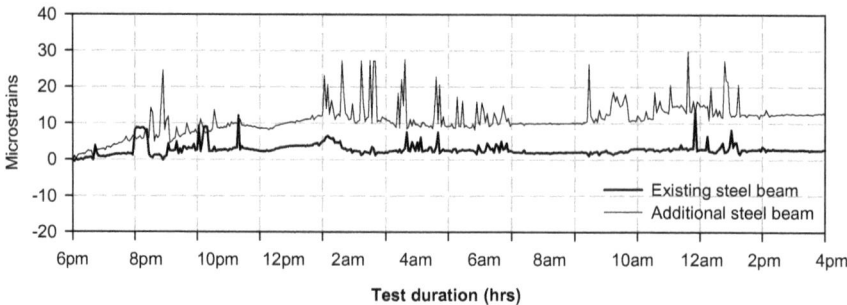

Fig. 8 Comparison of measured strains at mid-span of the existing and newly-added steel beams measured after post-strengthening works

6 Conclusions

Based on the field application discussed in this paper, the following remarks can be made:

- Design verifications based on the current ACI codes have shown that the capacity of the as-built RC beams was sufficient to resist original design live loads the 2nd floor.
- The span shortening technique (addition of steel beams) under the hollow core slab successfully reduced floor deflections and vibration levels.
- From the examination of vibration amplitudes, the maximum accelerations are less than 0.005 g and therefore this is classified to be an imperceptible level for humans.
- Analysis of the dynamic response showed that the first natural frequency of strengthened hollow concrete slab is 1.20 Hz
- The difference between pre and post-strengthening strain measurement was approximately 35%, thus confirming the effectiveness of the strengthening intervention.

Acknowledgement The authors acknowledge the financial supports provided by the Thailand Toray Science Foundation through the Science & Technology Research Grants (STRG 2019). The authors are also thankful to Agility Logistic Ltd, for providing access to the field measurement during the first phase of the project.

References

1. Lawson, R.M., Bode, H., Brekelmans, J.W.P.M., Wright, P.J., Mullett, D.L.: Slimflor' and 'Slimdek' construction: european developments. J. Struct. Eng. **77**(8), 22–30 (1999)
2. Hicks, S.: Current trend in modern floor construction. Magaz. British Constr. Steelwork Assoc. (BCSA) **11**(1), 32–33 (2003)

3. Mahmood, I., Konstantinos, A., Tsavdaridi, D.: The evolution of composite flooring systems: applications, testing, modelling and Eurocode design approaches. J. Constr. Steel Res. **155**, 286–300 (2019)
4. Baldassinoa, N., Roversoa, G., Ranzib, G., Zandoninia, R.: Service and ultimate behaviour of slim floor beams: an Experimental Study. Structures **17**, 74–86 (2019)
5. Ghindea, C.L., Cretu, D., Popescu, M., Cruciat, R., Tulei, E.: On-site experimental testing to study the vibration of composite floors. Key Eng. Mater. **601**, 231–234 (2014)
6. Douglas, I., Donato, A., Haritos, N.: Experimental observation of forklift-induced vibrations on a PSC floor in the Australian Earthquake Engineering Society 2014 Conference, Nov 21–23, Cumberland Lorne Resort, Vic (2014)
7. RILEM Technical Committee 104.: Damage classification of concrete structures. Mater. Struct. **24**(4), 253–259 (1991)
8. ACI 318.: Building code requirements for structural concrete and commentary. In: Johnson, B.M., Wilson, A.H., (eds.) Terminology of Building Conservation Industry, Division of Building Research, NRC Canada (2019)
9. BS 6472-1: Guide to Evaluation of Human Exposure to Vibration in Buildings, Part 1: Vibration Sources Other Than Blasting (2008)
10. ISO 2631-1: Mechanical Vibration and Shock–Evaluation of Human Exposure to Whole-Body Vibration–Part 1: General Requirements (1997)
11. AS 2670.2: Evaluation of Human Exposure to Whole-Body Vibration–Continuous and Shock-Induced Vibration in Buildings (1–80 Hz). Council of Standards Australia (1990)

3D Thermo-Hygro-Mechanical Simulation of a RC Slab Under Restrained Shrinkage and Applied Loads: Influence of the Reinforcement Ratio on Service Life Behaviour

José Gomes, Rui Carvalho, Carlos Sousa, José Granja, Rui Faria, Dirk Schlicke, and Miguel Azenha

Abstract The structural design of reinforced concrete (RC) structures requires the use of adequate methodologies for control of crack widths, in order to ensure proper service life behavior. When RC elements are subjected to the combined effect of restrained deformations and external loads, which is a relatively common situation in civil engineering structures, current design codes and recommendations generally fail to provide specific/clear instructions for measures to be taken for reinforcement design in view of crack width control. This paper intends to show the influence of different approaches for quantification of the necessary reinforcement on the service life behavior of these structures, through conduction of a parametric study. Such study is performed in order to understand how the variation of the longitudinal reinforcement affects the structural behavior of a solid one-way RC slab subjected to restrained shrinkage and external applied vertical loads (in quasi-permanent load combination). The structural behavior of the slab is simulated for 5 distinct quantities of longitudinal reinforcement, with a 3D thermo-higro-mechanical model, where the non-uniform distribution of stresses due to heat of hydration and drying shrinkage are considered. The attained results allowed conclusions to be withdrawn in regard to the expected behavior when reinforcement is underdesigned (by neglecting restraint) or overdesigned (by considering the cracking force of concrete when evaluating axial restraint). Intermediate solutions, matching some simplified approaches in the literature where found to be the most reasonable.

J. Gomes (✉) · R. Carvalho · J. Granja · M. Azenha
University of Minho, School of Engineering, Guimarães, Portugal

M. Azenha
e-mail: miguel.azenha@civil.uminho.pt

C. Sousa · R. Faria
Faculty of Engineering of the University of Porto, Porto, Portugal

D. Schlicke
Graz University of Technology, Graz, Austria

© The Author(s), under exclusive license to Springer Nature Switzerland AG 2022   321
J. Sena-Cruz et al. (eds.), *Proceedings of the 3rd RILEM Spring Convention and Conference (RSCC 2020)*, RILEM Bookseries 34,
https://doi.org/10.1007/978-3-030-76465-4_29

Keywords Service life behavior · Thermo-hygro-mechanical modelling ·
Restrained shrinkage · One-way slab · Structural design

1 Introduction

A proper service life behavior and durability of reinforced concrete (RC) structures
are known to be related, among other factors, to the capacity to control the maximum
crack width under a certain pre-defined limit. The calculation of crack width due to the
independent effects of external loads or imposed deformations have been subject of
study over the last decades [1–3]. Nonetheless, the recommended design practices [4–
6], which inherently need to be of simple application, end up being somewhat vague
and limited, in what concerns the estimation of crack width in elements subjected to
the combined effect of restrained deformations and external loads.

The potentially over-simplified and inadequate nature of current design provisions
can be explained with the following reasons: (i) the consideration of non-uniform
distribution of moisture in concrete and consequent non-uniform distribution of
stresses due to drying shrinkage, which plays a key role in the process of crack
development, requires the use of complex numerical models and material characterization that are not compatible with common design practice and; (ii) the complex
interaction between the effects of bending moments, viscoelasticity and imposed
deformations is usually avoided with empirical solutions. The former point is mainly
relevant during the crack formation stage, as observed in the work of Gomes et al.
[7], where non-linear FEM simulations of a restrained slab were performed with
uniform and non-uniform drying shrinkage, showing convergence of the structural
behavior at long term for both simulations. The latter point, however, has practical
implications at long term: the deployment of different simplified approaches based
on design code provisions will lead to differences in reinforcement for controlling
crack width as large as 50% [8].

In this paper, a 3D thermo-higro-mechanical simulation (THMS) framework [9–
11], in which the self-induced stresses caused by non-uniform temperature and moisture fields are considered, is deployed to assess the structural behavior of a solid
one-way restrained slab subjected to service loads. Five different scenarios of longitudinal reinforcement are considered, regarding different hypothetically possible
design approaches using the Model Code 2010 formulation for estimation of crack
width, for comparison of the service life behavior. A brief outline of the THMS
framework is presented on Sect. 2, the analyzed case study is described on Sect. 3
and the respective results are addressed on Sect. 4. Finally, in Sect. 4, the main
conclusions from the present work are drawn.

2 Thermo-Hygro-Mechanical Model

2.1 Thermal Model

The temperature fields are determined by considering the standard formulation for the energy balance [12]:

$$k\, div\,(grad\, T) + \dot{Q} = \rho c\, \partial T / \partial t \tag{1}$$

where T is the temperature (K), c is the specific heat of the material ($Jkg^{-1}K^{-1}$), ρ is the mass density (kgm^{-3}) and k is the thermal conductivity of the material ($Wm^{-1}K^{-1}$). $\dot{Q}$ is the material heat generation rate (kgm^{-3}), assessed experimentally by calorimetric testing, and expressed through an Arrhenius-type function based on the degree of heat development α_T [9, 13]:

$$\alpha_T = f(\alpha_T)\, A_T exp(E_a / RT) \tag{2}$$

where $f(\alpha_T)$ is the normalized heat generation function, A_T is the rate constant, R is the ideal gas constant (8.314 $Jmol^{-1}K^{-1}$) and E_a is the apparent activation energy ($Jmol^{-1}$).

Equation (3) expresses the boundary conditions as a function of a lumped convection-radiation coefficient h_{cr} [9, 14]:

$$q_h = h_{cr}(T - T_{env}) \tag{3}$$

where q_h is the heat flux through boundary at temperature T and T_{env} is the environment temperature.

2.2 Hygrometric Model

The humidity fields are modelled with a macroscale approach in which the moisture transfer is simulated through a diffusion-type equation, and the internal relative humidity is the considered to be the driving potential. The moisture diffusion in concrete is defined with a similar approach to the one presented in Model Code 2010 [6, 9]:

$$\partial H / \partial t = div\,(D_{H*} grad\, H) \tag{4}$$

where D_{H*} is a lumped diffusion coefficient that is determined with Eq. (5) [6]:

$$D_{H*} = D_1\big[(D_0/D_1) + (1 - (D_0/D_1))/\big(1 + [(1 - H)/(1 - H_c)]^n\big)\big] \qquad (5)$$

where D_1 is the value of D_{H*} for $H = 1$, D_0 is the value of D_{H*} for $H = 0$, H_c is the relative humidity for which $D_{H*} = 0.5D_1$ and n is a material property (constant).

Equation (6) expresses the boundary conditions as a function of a boundary transfer coefficient h_m:

$$q_m = h_m\big(H_{surf} - H_{env}\big) \qquad (6)$$

where q_m is the moisture flux between concrete pores and the environment and H_{surf} and H_{env} are the values of H on the surface and surrounding environment, respectively.

2.3 Mechanical Model and Framework Integration

The mechanical model includes the effect of the imposed strains generated by the thermal and moisture fields calculated with the models presented on the previous subsections. Whereas the thermal stains are determined considering a constant value of the concrete coefficient of thermal dilatation α_c, the drying shrinkage strains are calculated according the formulation proposed by [15]:

$$\varepsilon_{sh} = \varepsilon_{sh,ult}\big[0.97 - 1.895(H - 0.2)^3\big] \qquad (7)$$

where ε_{sh} is the drying shrinkage strain for a given variation of the humidity H and $\varepsilon_{sh,ult}$ is the ultimate shrinkage of concrete for $H = 0$.

The viscoelastic behavior of the concrete is simulated with a Kelvin chain [16], in which the spring stiffness of each Kelvin unit is determined with an algorithm based on the adaptation of the Dirichlet series approximation of the Double Power Law [17].

Concrete cracking is simulated with a smeared, multi-directional fixed crack model with strain decomposition [18]. After crack formation, the tension stiffening is with constitutive models for concrete and steel that simulate the stresses in each material averaged along a transfer length L_s, according to the modeling approach proposed by Sousa et al. [19].

3 Numerical Simulation of One-Way Restrained Slab

3.1 Overall Description of the Case Study

The RC slab (C30/37 concrete and S500C steel) studied in this paper and schematically presented in Fig. 1, is 0.15 m thick and 5 m wide, it develops over a length of 50 m and is supported at every 5 m by 0.3 × 0.5 m^2 transverse beams. These beams are supported by columns with a cross section of 0.3 × 0.3 m^2. The longitudinal deformation of the slab is restrained by two massive concrete blocks at its extremities.

The slab is subjected to an ideal environment of constant temperature (20 °C) and relative humidity (60%), while the thermal effects of daily and seasonal temperature variations are being disregarded.

The following loads for category A floor, according to Eurocode 1 [20], are considered: permanent $g_k = 2$ kN/m^2 (additionally to the self-weight of all structural elements) and live $q_k = 2$ kN/m^2 ($\psi_2 = 0.3$).

The slab is completely moisture-sealed during the first 7 days of age. Hence, during such period, it is solely subjected to the effect of restrained thermal deformations. At 7 days, propping and formwork are removed, activating the concrete drying and the self-weight of the structural elements. At 28 days, live and additional permanent loads are applied in correspondence to a quasi-permanent load combination, according to Eurocode 1 [20].

This case study is the same as the one analyzed by Gomes et al. [7] for assessment of the influence of non-uniform temperature and moisture fields in the structural behavior of the slab, mentioned in Sect. 1. A more detailed description regarding the case study and deployed THMS can be found on the aforementioned work.

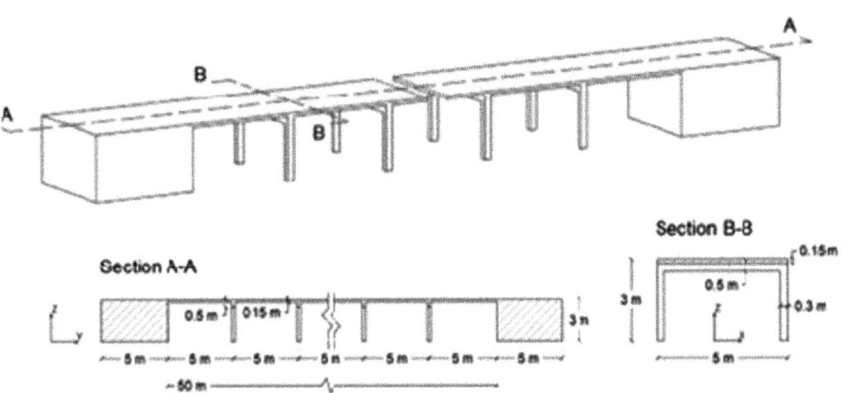

Fig. 1 Schematic representation of the studied slab

3.2 Design of Reinforcement

The upcoming paragraphs pertain to the disclosure of the 5 scenarios of reinforcement considered in the present work and presented on Table 1. Before definition of the longitudinal reinforcement for each scenario, a linear-elastic analysis was performed in order to determine the bending moments acting on the slab in the longitudinal direction over the columns and center of the slab. Because bending moments are higher near the free edges of the slab, it was decided to divide the slab into two reinforcement zones: (i) zone 1 corresponding to the 1m vicinity of the columns and (ii): zone 2 corresponding to the remaining 3m in the center of the slab.

Scenario 1 ($\rho_s = 0.8\%$) is the reference scenario and was designed according to standard recommendations for ultimate limit state (ULS), while the service limit state (SLS) was checked but did not affect the reinforcement solution. For both SLS and ULS, the effect of the restraint axial force was completely disregarded (which is allowed by Eurocode 2 in cases where joints are incorporated at every 30 m). In the design of the longitudinal reinforcement, the negative bending moments at ULS were reduced with a coefficient of redistribution of 0.8.

Scenario 2 ($\rho_s = 1.2\%$) was designed in order to meet the crack width requirement in SLS ($w_k < 0.3$ mm, for an exposure class of XC3). In this scenario, the Model Code 2010 [6] formulation for estimation of crack width was deployed for a situation of composed bending, considering the long-term restraint force determined in the THMS performed for Scenario 1. The resultant longitudinal restraint force (integral extended over the slab width) is constant over the entire structure development. At the slab's mid-span section, the transverse variation of the longitudinal force value is negligible. But the variation is not negligible over the transverse axis crossing the supporting column, varying between 247 kN/m (57% of N_{cr}) at the column position and 181 kN/m (42% of N_{cr}) at the middle point in-between columns. See further details in [7].

Scenario 3 ($\rho_s = 1.0\%$) is the reinforcement solution that corresponds to the average reinforcement area between Scenario 1 and Scenario 2. Scenario 4 ($\rho_s = 0.6\%$) and Scenario 5 ($\rho_s = 1.9\%$) are, respectively, the lower and higher boundaries of the parametric analysis, in which the design criteria is the same as for Scenario 2,

Table 1 Reinforcement solution for each scenario

Scenario	Zone 1 top $M_{qp} = 19.2$kNm	Zone 1 bottom $M_{qp} = 8.4$kNm	Zone 2 top $M_{qp} = 9.1$kNm	Zone 2 bottom $M_{qp} = 6.6$kNm
1	φ10//0.100	φ10//0.100	φ10//0.150	φ10//0.150
2	φ12//0.100	φ12//0.150	φ12//0.130	φ12//0.150
3	φ12//0.118	φ12//0.147	φ12//0.162	φ12//0.177
4	φ10//0.110	φ10//0.200	φ10//0.180	φ10//0.250
5	φ16//0.130	φ16//0.150	φ16//0.150	φ16//0.150

but considering the extreme situations where the restraint force acting on the slab is equal to zero ($N = 0$) and equal to the cracking force of the concrete ($N = N_{cr}$).

3.3 Thermo-Hygro-Mechanical Simulation (THMS)

Table 2 shows the material properties considered in the numerical simulations performed in this work. The average mechanical properties of concrete and steel are in consistency with the Model Code 2010 provisions for a concrete class C30/37 and steel S500, respectively. The humidity diffusion is calculated considering the same material properties adopted by Azenha et al. [21]. The thermal conductivity and specific heat of concrete were determined on the basis of a weighted average of the corresponding thermal properties of the mix constituents, which is composed by granite type aggregate, and the heat generation potential was determined based on previous experiments of the cement in an isothermal calorimeter [9].

The THMS begins at the setting time of concrete and it is performed over a period of 50 years (in which time-dependent tensile strength and viscoelasticity of concrete are considered) as to infer the entire service life behavior in the expected lifespan.

The thermo-hygrometric modeling is performed with a in-house developed FE software [9] for determination of the slab temperature and moisture fields, based on the governing equations presented on Sect. 2. The temperature and humidity flows are considered unidirectional along the slab thickness and the following moisture and thermal boundary conditions are considered: (i) during the first 7 days, before formwork removal and while the slab is completely sealed, $h_m = 0$ ms^{-1} for both top and bottom surfaces, $h_{cr} = 10$ Wm^{-2} °C^{-1} for the top surface in direct contact with the environment (duly calibrated by Azenha et al. [22]) and $h_{cr} = 5$ Wm^{-2} °C^{-1} for the bottom surface in contact with the 18 mm thick plywood formwork ([9]); (ii) after formwork removal, the boundary conditions are considered the same for both top and bottom surfaces: $h_{cr} = 5$ Wm^{-2} °C^{-1} and $h_m = 4.81 \times 10^{-8}$ ms^{-1}.

The temperature and drying shrinkage fields calculated in the thermo-hygrometric analysis are induced, respectively, in the nodes and in the different material layers of the mechanical model mesh, shown in Fig. 2. This mesh is composed by solid FE type CHX60 [16] with 20 nodes and an integration scheme of $3 \times 3 \times 3$ Gauss

Table 2 Material properties

Mechanical		Thermal		Hygrometric	
ρ (kgm^{-3})	2500	c (Jkg^{-1} °C^{-1})	960	D_0	2.98×10^{-11}
Poisson ratio ν	0.2	k (Wm^{-1} °C^{-1})	2.6	D_1	3.08×10^{-10}
E_{cm} (GPa)	33	α_c ($\mu\varepsilon$ °C^{-1})	10	H_c	0.8
f_{cm} (MPa)	38	α_s ($\mu\varepsilon$ °C^{-1})	12	n	2
f_{ctm} (MPa)	2.9	A_t	2.645×10^7	$\varepsilon_{sh,ult}$ ($\mu\varepsilon$)	625
G_f (Nm)	76	E_a (kJmol^{-1})	38.38	$\varepsilon_{sh,60\%}$ ($\mu\varepsilon$)	530

Fig. 2 FE mesh of the slab
and symmetry conditions

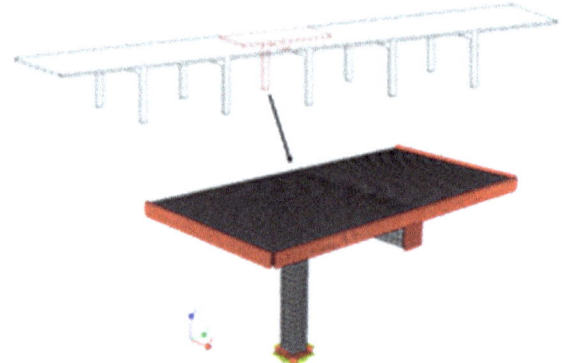

Points. The slab is divided along its thickness in 8 layers of $100 \times 100 \times 18.75$ mm elements whereas the beam and column are composed by cubic FE with 100 mm edge. The numerical analysis is optimized by taking into account the structural symmetry: only a quarter of two adjacent panels are modeled, along with half the beam and respective column.

4 Discussion of Results

4.1 Thermo-Hygrometric Results

Results from the thermo-hygrometric simulation (temperature and humidity fields and consequent imposed strains) are presented on Fig. 3, with reference to several key instants, labeled A to D.

During the first hours after setting of concrete, the temperature of the slab evolves from a uniform environment temperature of 20 °C to a maximum temperature of 29.1 °C, reached near the core of the slab, at 10h age (instant A). Due to the small

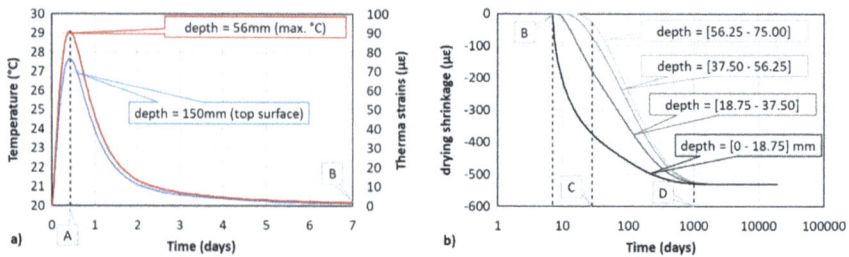

Fig. 3 Results from thermo-hygrometric analysis: **a** Temperature and thermal strains during the first 7 (linear scale of time); **b** drying shrinkage over 50 years (log scale of time)

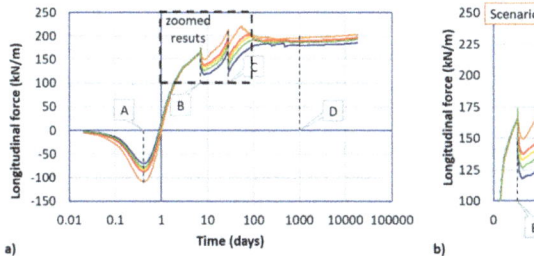 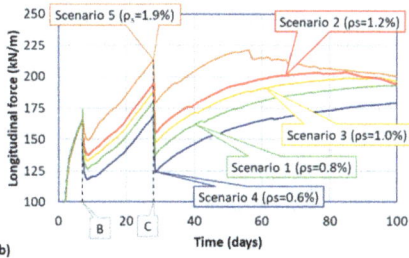

Fig. 4 Evolution of the longitudinal force for different scenarios: **a** over 50 years (log scale of time); **b** zoomed results for the first 100 days of analysis (linear scale of time)

thickness of the slab, the maximum gradient of temperature observed over the slab thickness, also reached at instant A, is only 1.4 °C.

At 7 days (instant B) the formwork is removed, and the slab is directly exposed to an environment with constant relative humidity of 60%, inducing drying of the concrete. In contrast to the temperature strains field, a significant gradient of drying shrinkage is observed between the core and the surface of the slab, motivated by a slower decrease of moisture in the interior layers. The equilibrium of drying shrinkage over the slab thickness is only reached at 1000 days (instant D) when all layers of the slab are at the same relative humidity of the surrounding environment.

4.2 Structural Behavior of the Slab

Figure 4 shows the evolution of the global longitudinal axial force, averaged along the slab width, for all considered scenarios of reinforcement.

During the first hours, the restraint of the thermal strains caused by heat of hydration induces an axial compression force in the slab that reaches its maximum value at 10 h (instant A). Due to the additional self-induced stresses caused by different thermal strains acting on concrete and steel reinforcement (coefficient of thermal dilation is slightly different for both materials, as shown in Table 2), the maximum axial force reached at instant A is higher for higher reinforcement ratios and varies between -70 kN/m for Scenario 4 ($\rho_s = 0.6\%$) and -109 kN/m for Scenario 5 ($\rho_s = 1.9\%$).

Before formwork removal at 7 days (instant B), a tensile force is induced by the cooling that occurs after the temperature peak at instant A. Even though the development of this tensile force depends on the early age viscoelastic behavior of the concrete at early ages, the differences regarding the peak compressive force induced in the slab at instant A do not influence significantly the value of the tensile restraint force observed immediately before instant B, which varies between 164.6 kN/m for Scenario 5 ($\rho_s = 1.9\%$) and 165.4 kN/m for Scenario 4 ($\rho_s = 0.6\%$).

After formwork removal and consequent activation of the self-weight and drying of concrete, cracking is induced near the support and at mid-span (where bending moments are higher), which is responsible for the stiffness decrease and consequent reduction of the restraint force. This is also observed at 28 days (instant C), when live and additional permanent loads are applied, resulting in an increase of cracking and consequent relief of the restraint load. These drops of restraint force and further evolution over time are conditioned by the degree of cracking and consequent loss of axial rigidity, therefore, the axial load acting on the slab during the crack formation stage is higher for higher reinforcement ratios.

As the concrete drying advances, the restraint force continues to increase up to an instant that varies between 56 days (for Scenario 5) and 240 days (for Scenario 4), slightly decreasing afterwards until 1000 days (instant D), when the drying shrinkage becomes uniform and constant. This decrease of restraint force can be justified by the dominance of creep effects in comparison to the imposed deformations and by the specific nature of concrete shrinkage (as opposed to a global imposed deformation).

At long term, the restraint force observed for all reinforcement scenarios tend to converge until instant D, being approximately constant afterwards, in correspondence to the stabilization of loads and drying, and the fact that creep effects are already very small at these late ages. The restraint force observed at the end of the analysis (50 years) varies between 186 kN/m (43% of $N_{cr} = A_c\ f_{ctm}$) for Scenario 4 and 203 kN/m (47% of N_{cr}) for Scenario 5. These values are at the same magnitude as the maximum restraint force, observed for Scenario 5 (with higher reinforcement ratio), which reached a value of 221 kN/m (51% of N_{cr}).

The evolution of the crack width, at reinforcement level, is presented on Fig. 5 for 4 points representative of the location of maximum crack strains observed on

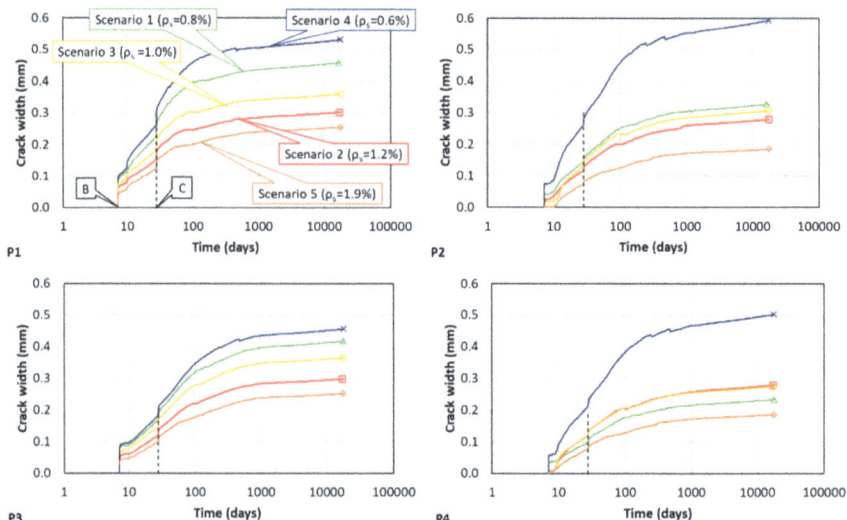

Fig. 5 Evolution of the crack width at the reinforcement level

top and bottom layers of each reinforcement zone: (i) two points P1 and P2 (near the support and at mid-span, respectively), located on the reinforcement zone Z2 at 1.25 m from the outer border of the slab and (ii) two points P3 and P4 (near the support and at mid-span, respectively) located on the reinforcement zone Z1 over the longitudinal axis of the supporting column. The crack width is calculated by integrating, for each point, the crack strains at reinforcement level (which is an output from the mechanical model) over a distance equal to twice the transfer length $l_{s,\max}$ (quantified according with Model Code 2010 [6] formulation, taking into account the local axial and bending forces at long term as well as the local reinforcement ratio).

It can be observed that cracking is induced at support (P1 and P3) and mid-span (P2 and P4), in both reinforcement zones, immediately after removal of formwork at instant B due to the combined effect of the slab's restrained shrinkage and self-weight. At instant C, when live and additional permanent loads are applied, the growth of bending moments causes a sudden increase of the crack width that is, as expected, more pronounced for Scenario 4 ($\rho_s = 0.6\%$) with lower reinforcement ratio.

At the end of the analysis, the maximum crack width observed in all representative points was below the limit crack width value (0.3 mm for exposure class XC3, according to Model Code 2010) only for Scenarios 2 ($\rho_s = 1.2\%$) and 5 ($\rho_s = 1.9\%$), in which the design was performed taking into account a restraint force caused be imposed deformation.

Whereas for Scenario 5 a very conservative approach was followed, in which a restraint force equal to the axial cracking force of the concrete ($N_{cr} = 435\text{kN/m}$) was considered, the reinforcement for Scenario 2 was designed taking into account the exact restraint force determined with the numerical THM simulation (considering the amount of reinforcement previously quantified for Scenario 1). The maximum crack width observed for Scenario 2 is close to the expected crack width according to Model Code formulation (the adopted restraint force is similar to the actual restraint force for Scenario 2, as observed in Sect. 4.2) and not much higher than the maximum crack width observed for Scenario 5 (differences of 0.05 mm at the support and 0.1 mm at mid-span).

It is worth comparing the results analyzed in this section with the value of restraint force estimated by simplified methods such as the ones proposed by Camara and Luís [23] and Fehling [24].

In the former method, the long-term restraint force is estimated as ξN_{cr}, where ξ is a tabulated reduction coefficient that depends on the reinforcement ratio (in correspondence to a realistic service load level) and the value and type of the imposed deformation. Considering an imposed drying shrinkage of 500 $\mu\varepsilon$ and a service load in correspondence to a reinforcement ratio of 0.8% (scenario that best fits the presented case study), a reduction coefficient of $\xi = 0.45$ is estimated, which is within the range of values observed for all studied scenarios.

The later method provides diagrams for estimation of the non-dimensional normal force, $v_{ct} = N / (A_{c \cdot} f_{ctm})$, which in turn gives the restraint force N. The estimated v_{ct} value depends on the bending moments due to loading, the expected imposed deformation, the target value for crack width limitation and the diameter of the reinforcement bars, as well as the effective area of concrete in tension and the influence of non-uniform self-equilibrating stresses. For each scenario, the reduction factor v_{ct} was calculated for points P1 and P2 (considering for each point the bending moments and maximum crack width determined in the THMS). The average v_{ct} determined for Scenarios 1 to 5 are, respectively: 0.88, 0.89, 0.89, 0.83 and 0.95. Even though these results indicate that this method is less accurate for very thin slabs (for development of the diagrams, the stresses on reinforcement were calculated considering fixed values for the effective depth d and internal lever arm z of h-0.083h and 0.83h, respectively), they are on the safe side.

5   Conclusions

The sensitivity analysis performed in this paper, where the structural behavior of a slab with different solutions of steel reinforcement are assessed with a non-linear thermo-hygro-mechanical simulation, allowed to draw the following conclusions:

Even though the gradient of temperature induced in the slab due to heat of hydration is negligible, on account of the small thickness of the slab, the restraining of thermal deformations induces a tensile stress on the slab at the time of formwork removal that decreases its capacity to resist further stresses over the period of analysis.

Different levels of reinforcement influence the restraint force during the crack formation stage, however, this influence tends to vanish at long term, as drying shrinkage develops.

The evolution of crack width over time, observed for all reinforcement scenarios in Sect. 4.2, shows the importance of considering the effect of restrained deformations for controlling the crack width under a pre-determined limit. However, it is also shown that conservative approaches where the cracking force of the concrete is considered as the restraint force will lead to an overdesign situation where the adopted reinforcement is significantly higher than the optimal solution.

Finally, it was also concluded that simplified procedures for determination of the stabilized crack width based on direct quantification of the restraint force will predictably yield, in general, safe results.

Acknowledgements Funding provided by the Portuguese Foundation for Science and Technology (FCT) to the Research Project IntegraCrete (PTDC/ECM-EST/1056/2014–POCI-01-0145-FEDER-016841) and to the PhD Grant SFRH/BD/148558/2019 is acknowledge. This work was also financially supported by UIDB/04029/2020–ISISE, and Base Funding–UIDB/04708/2020 of the CONSTRUCT–Instituto de I&D em Estruturas e Construções, both funded by national funds through the FCT/MCTES (PIDDAC).

References

1. Favre, R., Beeby, A.W., Falkner, H., Koprna, M., Schiessl, P.: CEB design manual on cracking and deformations. Lausanne, Switzerland: Ecole Polytechnique Fédérale de Lausanne (1985)
2. Nejadi, S., Gilbert, I.: Shrinkage cracking and crack control in restrained reinforced concrete members. ACI Struct. J. **101**(6), 840–845 (2004)
3. Schlicke, D., Tue, N.V.: Minimum reinforcement for crack width control in restrained concrete members considering the deformation compatibility (in English). Struct. Concr. **16**(2), 221–232 (2015)
4. EN 1992-1-1: Eurocode 2 Design of concrete structures Part 1-1 General rules and rules for buildings (2004)
5. EN 1992-3: Eurocode 2 Design of Concrete Structures Part 3: Liquid retaining and containment structures (2006)
6. FIB, fib Model code for concrete structures 2010. Berlin, Germany: Ernst & Sohn (2013)
7. Gomes, J., Carvalho, R., Sousa, C., Granja, J., Faria, R., Schlicke, D., Azenha, M.: 3D numerical simulation of the cracking behaviour of a RC one-way slab under the combined effect of thermal, shrinkage and external loads. Eng. Struct. **212** (2020)
8. Azenha, M., Granja, J.: Seminar: design of reinforcement for RC elements under the combined effect of applied loads and restrained shrinkage. E-book of presentations. Zenodo (2017). https://doi.org/10.5281/zenodo.800693
9. Azenha, M.: Numerical Simulation of the structural behaviour of concrete since its early ages Ph.D. Universidade do Porto, FEUP, Porto (2009)
10. Azenha, M., Sousa, C., Faria, R., Neves, A.: Thermo–hygro–mechanical modelling of self-induced stresses during the service life of RC structures. Eng. Struct. **33**(12), 3442–3453 (2011)
11. Azenha, M., Leitão, L., Granja, J., Sousa, C., Faria, R., Barros, J.A.O.: Experimental validation of a framework for hygro-mechanical simulation of self-induced stresses in concrete. Cement Concr. Compos. **80**, 41–54 (2017)
12. Incropera, F., DeWitt, D.: Fundamentals of mass transfer. Wiley, New York (1996)
13. Reinhardt, H., Blaauwendraad, J., Jongedijk, J.: Temperature development in concrete structures taking account of state dependent properties presented at the International Conference on concrete at early ages (1982)
14. Jonasson, J.-E.: Modelling of temperature, moisture and stresses in young concrete, 153 Doctoral thesis, comprehensive summary, Doktorsavhandling/Högskolan i Luleå, Luleå tekniska universitet, Luleå (1994)
15. Kwak, H.-G., Ha, S.-J., Kim, J.-K.: Non-structural cracking in RC walls: Part I. Finite Element Formul. Cement Concr. Res. **36**(4), 749–760 (2006). Accessed 04 Jan 2006
16. Manie, J.: DIANA 10.2 user's manual. Delft, Netherlands: TNO DIANA BV (2017)
17. Bažant, Z.P., Osman, E.: Double power law for basic creep of concrete Matériaux et Construction. J. **9**(1), 3–11 (1976). Accessed 01 Jan 1976
18. Rots, J.G., Blaauwendraad, J.: Crack models for concrete, discrete or smeared? Fixed, multi-directional or rotating? Heron **34**(1) (1989)
19. Sousa, C., Leitão, L., Faria, R., Azenha, M.: A formulation to reduce mesh dependency in FE analyses of RC structures under imposed deformations. Eng. Struct. **132**, 443–455 (2017)
20. EN 1991-1-1: Eurocode 1 Actions on structures-Part 1-1: General actions -Densities, self-weight, imposed loads for buildings (2002)
21. Azenha, M., Leitão, L., Granja, J.L., de Sousa, C., Faria, R., Barros, J.A.O.: Experimental validation of a framework for hygro-mechanical simulation of self-induced stresses in concrete. Cement Concr. Compos. **80**, 41–54 (2017)
22. Azenha, M., Faria, R., Ferreira, D.: Identification of early-age concrete temperatures and strains: monitoring and numerical simulation. Cement Concr. Compos. **31**(6), 369–378 (2009)

23. Camara, J., Luís, R.: Structural response and design criteria for imposed deformations superimposed to vertical loads presented at the FIB congress. Naples (2006)
24. Fehling, E., Lautbecher, T.: Beschränkung der Rißbreite bei kombinierter Beanspruchung aus Last und Zwang Beton-und Stahlbetonbau **98**(7), 377–388 (2003)

Morphological and Chemical Characterization of Self-Healing Products in MgO Concrete

Maria Amenta, Stamatoula Papaioannou, Vassilis Kilikoglou, and Ioannis Karatasios

Abstract This work presents the results of the measurements carried out by the Laboratory of Archaeological and Building Materials of NCSR "Demokritos" in the framework of the Interlaboratory Test (ILT) organized by SARCOS COST action CA15202. The overall aim of ILT was to assess the self-healing properties of concrete with the addition of a self-healing agent. Following the ILT tentative methodology, the concrete samples were initially subjected to controlled cracking after curing for 30 days. Their sealing efficiency—as part of the self-healing process—were assessed by water permeability tests carried out in disk specimens cured for 1, 3 and 6 months under water, as a healing regime. At the same time, crack width of all samples was measured each time, by examination at the stereomicroscope. Besides the measurements regarding the sealing efficiency, Scanning Electron Microscopy (SEM) was also performed, in order to understand the contribution of the healing agent on the sealing mechanism. In this context, samples were collected from the crack face and the secondary healing products inside the cracks were studied by SEM combined with X-ray spectroscopy (EDS) in samples with and without the healing agent. The interpretation of the results suggests that the main methodological parameter affecting the sealing efficiency is the initial crack width. This indicates that during the quantitative evaluation of the sealing results, the initial level of damage attributed to the specimens should always be considered. In addition, SEM examination revealed the composition and the morphology of the healing products that were identified in the fractured surfaces and thus, the relative contribution of both healing agent and cement hydration.

Keywords Self-healing concrete · MgO · Water permeability · Sealing efficiency · Healing products

M. Amenta (✉) · S. Papaioannou · V. Kilikoglou · I. Karatasios
Institute of Nanoscience and Nanotechnology, NCSR "Demokritos", Athens, Greece
e-mail: m.amenta@inn.demokritos.gr

© The Author(s), under exclusive license to Springer Nature Switzerland AG 2022   335
J. Sena-Cruz et al. (eds.), *Proceedings of the 3rd RILEM Spring Convention
and Conference (RSCC 2020)*, RILEM Bookseries 34,
https://doi.org/10.1007/978-3-030-76465-4_30

1   Introduction

Self-healing concrete is considered one of the most promising solutions for the enhancement of building materials' durability and sustainability. The successful application of self-healing concrete in residential, commercial and industrial constructions can have a significant impact both in the safety of structures and on their environmental footprint [1]. The study of self-healing in cementitious materials has gained a lot of attention in the last two decades, and several promising laboratory approaches have been proposed [2–5]. At the same time, researchers use a multitude of techniques for assessing the efficiency of the self-healing mechanism, although there is not a standardized methodology yet for the expression and quantification of self-healing efficiency [6].

In this context an Interlaboratory Test (ILT) was organized by SARCOS COST action CA15202. One of the objectives of the ILT was to assess the self-healing efficiency of concrete with the addition of a self-healing agent, by developing and applying a specific methodology for the measurement of the crack width and water permeability through the crack at preset time periods [7]. Although the goal of the project is the evaluation of the healing efficiency, the methodology developed and discussed in this paper concerns the sealing effect of MgO added in the mixtures.

This work presents the results of the measurements carried out in the Laboratory of Archaeological and Building Materials of NCSR "Demokritos". The self-healing agent selected in this case was a combination of three types of expansive minerals: magnesium oxide (MgO), bentonite clay, and quicklime, replacing equal part of the cement. These minerals can increase the self-healing efficiency of concrete as they react with water and produce expansive healing phases inside the crack [8, 9].

The aim of this work is to comment on the feasibility of the proposed methodology as well as, to discuss and interpret the experimental results derived. Moreover, this paper focuses on the morphological, chemical characterization of the healing products by electron microscopy combined with X-ray spectroscopy (SEM/EDS), in order to provide insights on the contribution of the healing agent on the sealing mechanism.

2   Materials and Methods

2.1   Materials

All specimens were produced in Riga TU following the ILT tentative methodology.

Two series of concrete cylindrical specimens (Ø100XH200 mm) were produced; one with the addition of expansive minerals (number of specimens = 4) as a self-healing agent (ADDS) and a reference mixture (number of specimens = 4) (REF). The mix designs of the reference and self-healing mortars are shown in Table 1.

Table 1 Mix design

kg	Reference (REF)	Self-healing (ADDS)
CEM I 42.5 N	360	315
Water	198	198
Natural sand 0/4	980	980
Crushed dolomite gravel 2/16	910	910
Steel fibers	40	40
Plasticizer	~3lt/m³	~3–3.2lt/m³
Hydrated lime (L)	–	18
MgO (M)	–	18
Bentonite (B)	–	9

Specimens were wrapped in plastic cling film and stored in humidity chamber at 20 °C, 95 ± 5% RH for 14 days until shipping.

2.2 Sample Preparation and Controlled Cracking

Three disks (Ø 100 mm × 50 mm height) were cut from each cylindrical specimen, discarding the ends of the cylinder (Fig. 1a). In all discs, a notch (3 mm depth, 1 mm width) was generated on both sides, by an electric hand saw (Fig. 1b), in order to ensure the controlled crack propagation, parallel to the loading direction, in the centre of the specimens.

Pre-cracking was performed on all disks through a splitting tensile test carried out in an Instron 100 kN machine under closed-loop displacement control, 30 days after their production. The crack opening was monitored by Crack Mouth Opening Displacement (CMOD) gauges attached to one side of the specimen during the test

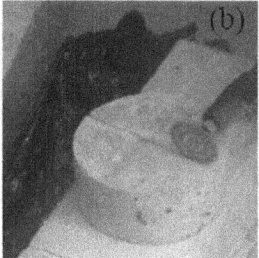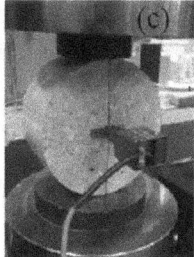

Fig. 1 Three discs were cut from the cylindrical specimens (**a**), notches were generated by an electric hand-saw on both sides of the specimens (**b**), CMOD gauges were used to control the crack width during cracking (**c**)

(Fig. 1c). According to the proposed methodology, the target opening of the crack was set at 300 μm, in the loaded state.

2.3   Crack Measurement

After controlled cracking, the generated crack width was measured with a Leica S6D optical microscope in three equidistant points of the length of the crack on each side of the specimen. Six points were measured in total and the average crack width was calculated for each specimen. This was repeated after each healing period (1, 3, 6 months) where the same points were measured and compared with the initial measurement. Crack closing was calculated according to Eq. (1):

$$CCE(t) = [(C\_unheal - C\_heal(t))/C\_unheal] x 100\% \qquad (1)$$

Where: $CCE(t)$ = crack closing efficiency at age t, C_unheal = CMOD at cracking age after unloading, C_healed (t)= healed crack width at test time (t)

2.4   Water Permeability/Sealing Efficiency

The cracked specimens were immediately stored submerged in deionized water, for the designated healing period (1, 3 and 6 months), in waterproof containers. Following the ILT tentative methodology, the specimens were removed from the water containers and dried for 24 h at 40 °C. PVC tubes of 200 mm height were glued on the top of the disks using one-component acrylic-based sealant. The sealant was allowed to dry for 24 hours in laboratory conditions. The tubes were filled with 1.5 liters of tap water. During the test, the water flow was determined by weighting the water passing through the crack and collected every 5 minutes. The final measurement was taken after 30 minutes. The output of the test was a water flow measure (litres / minute). The sealing efficiency was calculated by Eq. (2):

$$SE = [(W\_unheal - W\_heal (t))/W\_unheal] x 100\% \qquad (2)$$

Where: SE = sealing efficiency calculated efficiency at age t, W_unheal = amount of water that has passed through the specimen's unhealed crack at cracking age, W_heal (t) = amount of water that has passed through the specimen's healed crack at test time (t)

2.5 Characterization of Healing Products/Carbonation Depth

Besides the measurements regarding the sealing efficiency, further tests were also performed, in order to understand the contribution of the healing agent on the healing mechanism.

More specifically, following the water permeability measurement after the 6 months healing period, one representative specimen of each group was selected to be further studied regarding the characterization of the healing products that contributed to the healing mechanism. In this context, the specimens were severed in two parts following their initial crack so that the crack faces would be exposed. One part was used for carbonation test by spraying phenolphthalein solution (1% w/v in ethanol). Representative samples were collected from the crack face of the second part, and the secondary healing products inside the cracks were studied by Scanning Electron Microscopy (SEM) combined with X-ray spectroscopy (EDS).

3 Results and Discussion

3.1 Controlled Cracking

During the splitting tests of the specimens the CMOD was monitored and the test was stopped when the CMOD reading was 300 μm. During the test, it was evident that the mechanical properties of the two groups differ. Representative loading curves of reference (REF) and self-healing group (ADDS) are presented in Fig. 2. The values in brackets mentioned in the two graphs, correspond to the maximum crack-width measured in the stereo-microscope after unloading. Since the CMOD was measured in the centre of the notch length, in a relatively small area (i.e. 1 cm), the detection of

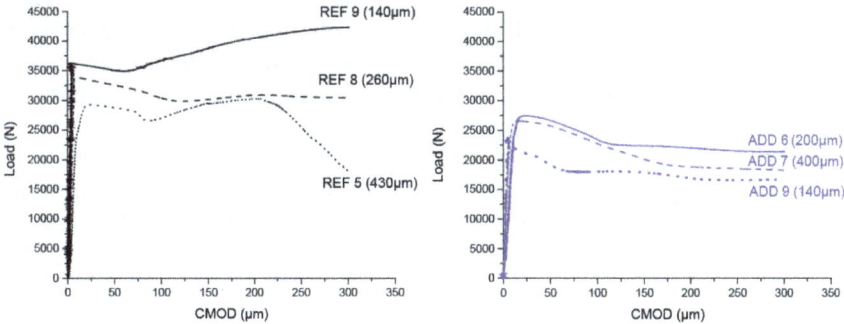

Fig. 2 Representative loading curves of reference and self-healing concrete specimens during splitting test. The values in brackets correspond to the maximum crack-width value measured at the stereomicroscope, after unloading

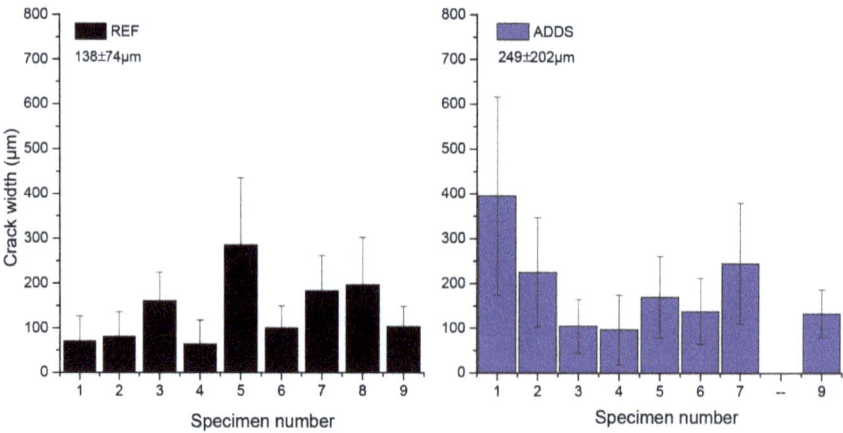

Fig. 3 Average crack width measurements of reference and self-healing concrete specimens

crack widths larger than the target CMOD values indicate the irregular development and propagation of the cracks that is uneven along the notch ligament. The latter is also affected by the random distribution and orientation of the steel fibres, that result in different stress bridging capability of the fibres intersecting the active crack surfaces.

The reduced load values of self-healing specimens (ADDS) can be attributed to the partial replacement of cement with magnesium oxide (MgO), bentonite clay, and quicklime, which do not take part in the hydration of the binder matrix.

Additionally, crack width was measured in 6 equidistant points in each specimen after cracking. Stereomicroscopic examination along the length of the notch revealed high standard deviation on the crack width. The average crack width measurements at the stereomicroscope along with their standard deviation are presented in Fig. 3.

The inhomogeneity of the results and the high standard deviation values indicate that the addition of steel fibers in the concrete mix may have influenced the cracking behavior of the specimens, introducing an elastic behavior after unloading that leads to a reduction of the final crack opening. The spatial orientation of the fibers, as well as their varying distribution between different specimens could also be a factor of the unevenness of crack width measured.

In Fig. 3 it can be seen that the average crack width of all measurements in the self-healing group (ADDS) is higher (249 μm) when compared with the reference specimens (138 μm). This could be also the direct effect of the reduced mechanical properties of this mixture.

3.2 Crack Healing

Self-healing of cracks was examined by stereomicroscopic examination of cracked specimens after curing in water for 1, 3 and 6 months. Crack width was systematically measured in the same 6 equidistant points in each specimen and it was compared to the previously measured crack width of the same point. The average crack width measurements showed a consistent reduction after 1, 3 and 6 months of healing under water (Fig. 4). Nevertheless, the data present very high standard deviation. That was expected, since the initial crack generated by the splitting test was characterized by high variation among the crack width measured in each specimen (Fig. 3).

Overall, the crack closing of each individual point after the healing process was significantly affected by the initial crack width (Fig. 5). In the scatter plot of the initial crack opening versus the crack closing ratio, the points of the highest opening that exhibit 100% closing could be considered as an indication of the maximum opening that could be healed.

In the above context, the interpretation of the plots presented in Fig. 5 indicates that in the reference group, crack openings up to 360 μm may exhibit 100% closing ratio (Fig. 6). In contrast, in the self-healing (ADDS) group this threshold decreases to cracks below 150 μm.

Moreover, it is shown that 67.9% of the points measured in reference specimens showed 100% closing ratio, while only 12.5% was counted in the self-healing specimens. This is indicated also by the crack closing ratio of the two groups after 1, 3 and 6 months as it is shown in Fig. 7. However, the great variability of closing ratio values for cracks in the range of 80–150 μm highlights the difficulty of the accurate quantitative modelling of the sealing phenomena, especially by studying only the crack surface.

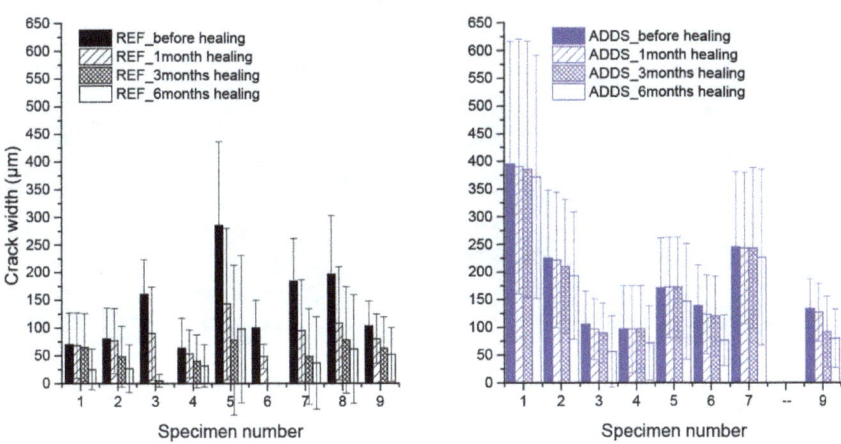

Fig. 4 Average crack width measurements of reference (REF) and self-healing (ADDS) concrete specimens after 0, 1, 3, 6 months of healing

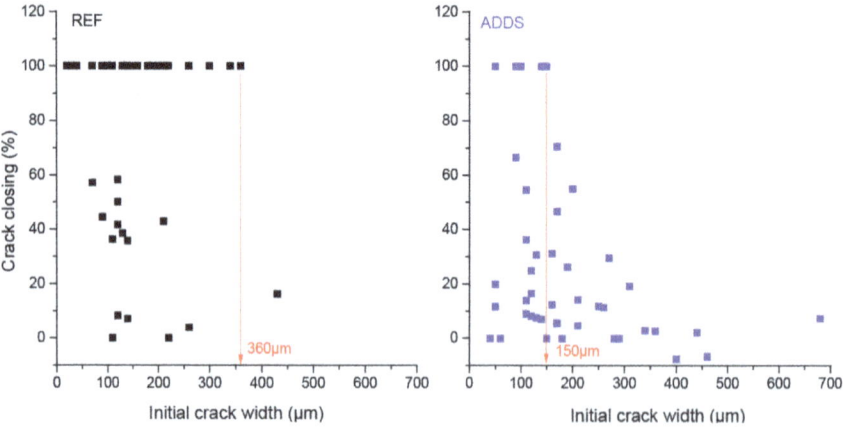

Fig. 5 Scatter plot of individual crack points in reference (REF) and self-healing (ADDS) concrete specimens, indicating their crack closing ratio after 6 months of healing as a function of the initial crack width

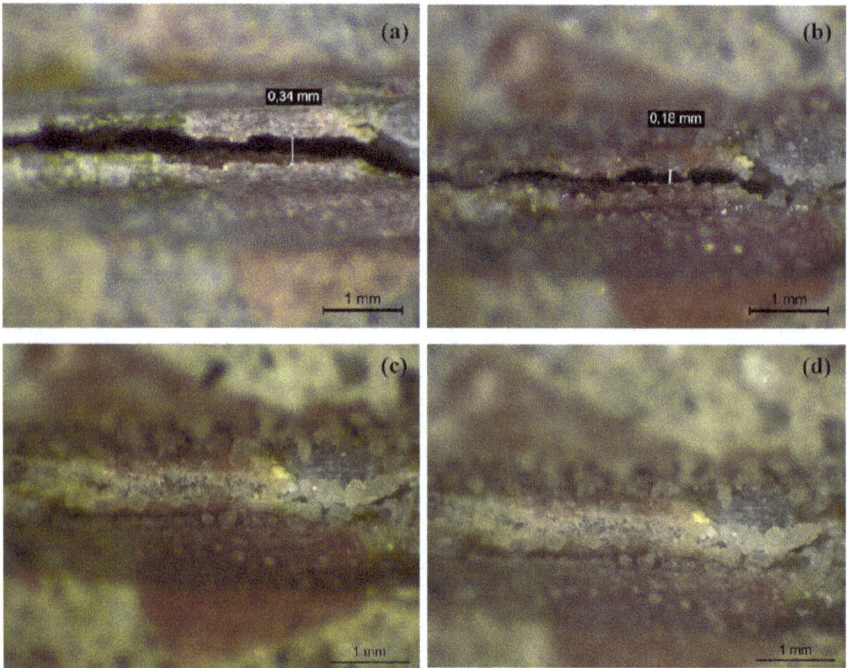

Fig. 6 Indicative example of the evolution of crack healing, in a reference specimen by monitoring of the same point after 0 (**a**), 1 (**b**), 3 (**c**) and 6 (**d**) months

Fig. 7 Crack closing (%) of reference (REF) and self-healing (ADDS) concrete after 1, 3, 6 months of healing

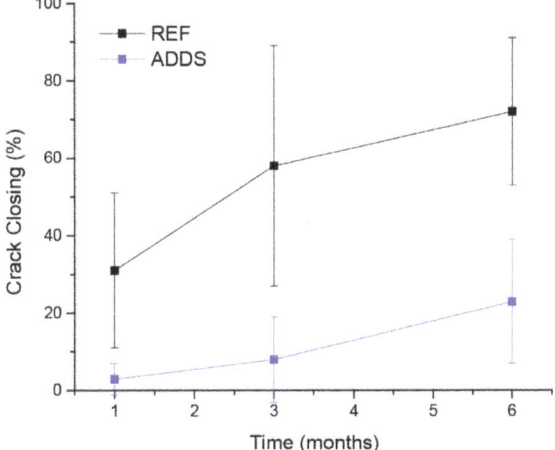

Crack closing ratio was improved with healing time for both mixtures. Nevertheless, the crack closing efficiency is much higher for the reference specimens, indicating that the wider average crack opening of the self-healing mixture (249 μm at ADDS versus 138 μm at REF) affected greatly their healing efficiency. This shows that the quantitative evaluation of the sealing results should be always consider the initial level of damage attributed to the specimens [10].

3.3 Water Permeability

The water permeability was measured in both groups of specimens, in four distinct time periods and was expressed as water flow (WF) values in Table 2. The first water permeability measurement was performed immediately after the specimens cracking (0 heal) and later, after 1, 3 and 6 months of healing under water. The specimens that exhibited no water flow after cracking (0 months) were excluded considering that no sufficient damage was induced.

The water flow values of the reference specimens (REF) follows a decreasing trend through the evolution of healing time (Table 2). Similarly, self-healing specimens (ADDS) present a decrease of water permeability only between the first and the sixth

Table 2 Water flow (WF) of reference (REF) and self-healing (ADDS) concrete after 0, 1, 3 and 6 months of healing

	0 months		1 month		3 months		6 months	
	WF (g/min)	std	WF (g/min)	std	WF (g/min)	std	WF (g/min)	std
REF	12.3	10.7	7.8	10.8	3.7	6.2	2.2	4.4
ADDS	21.6	20.2	25.0	20.5	14.9	16.6	15.5	16.0

month, exhibiting similar values between three and six months. The high standard deviation values are attributed to the wide variation of crack values between the different specimens.

It is noteworthy that in the ADDS group the measurement performed after one month of healing exhibited higher values, meaning that the amount of water passing through the crack of the specimens after one month of healing, was higher than the amount initialy passed (0 heal). Considering that no crack width increase was recorded in any specimen, this behaviour could be attributed to the methodology followed. In particular, the first measurement carried out in dry specimens while that after one month of healing took place in water saturated specimens dried for 24 hours at 40 °C. It is therefore assumed that the lower water flow rate in the initial measurement (0 heal) could be attributed to the partial water absorption through the crack, until saturation. In order to overcome this pottential error, the water flow measurements could take place twice and consider only the second for interpretation purposes.

3.4 Characterization of Healing Products

The phenolphthalein test performed after six months revealed the areas of the crack face were pH value is lower than 10 suggesting that carbonation has occurred. The distribution of the violet color areas showed that carbonation has occurred on the external parts of the crack faces (no color change) (Fig. 8). This indicates that carbonation is facilitated near the external surfaces of the crack, since these areas are easily accessed by the atmospheric carbon dioxide [11]. Consequently, calcium carbonate was one of the main healing products which were produced as part of the autogenic self-healing mechanism.

Fig. 8 Reference (REF) (left) and self-healing (ADDS) (right) concrete specimens after Phenolph-thalein test. Carbonated areas are not colored and are specified by the white line

Fig. 9 Typical healing products observed in the crack face of Reference specimens, Calcite (**a**), Ettringite and Portlandite (**b**) and Hydraulic phases (**c**)

Besides calcium carbonate, microscopic examination revealed the presence of several other secondary healing products on the crack faces. More specifically, ettringite, portlandite and a variety of hydraulic phases were observed in both reference and self-healing specimens. In both groups, calcite crystals were found near the surface of the crack (Figs. 9a and 10a), which agrees with the carbonation test performed on the same specimens.

Ettringite and portlandite crystals were formed mainly inside macropores (Figs. 9b and 10b). Finally, hydraulic phases were abundantly observed in both reference and self-healing specimens. More specifically, in reference specimens the hydraulic products present a typical low-density bee-hive formation (Fig. 9c). This morphology is often encountered in healed cracks as a product of the autogenic self-healing mechanism.

In self-healing specimens, a denser formation of gel-like hydraulic phases was observed (Fig. 10c). Moreover, the presence of dolomite crystals was observed, together with calcite (Fig. 10d).

The characterization of the observed healing products was based on their elemental analysis. EDS analysis results are presented in Tables 3 and 4.

4 Concluding Remarks

In this study the sealing efficiency of concrete with the addition of three types of expansive minerals (magnesium oxide (MgO), bentonite clay, and quicklime) as self-healing agents was examined. It was shown that the addition of the expansive minerals had a negative effect on the mechanical properties of the concrete. This consequently resulted in the formation of wider cracks on these specimens at lower loading, when compared to the reference ones.

Concerning the methodology developed for the controlled damage and the assessment of sealing capacity of cement-based materials, the following conclusions can be drawn. Although steel fibres were added in order to prevent the splitting of the specimens during loading, the examination in stereomicroscope indicated the irregular

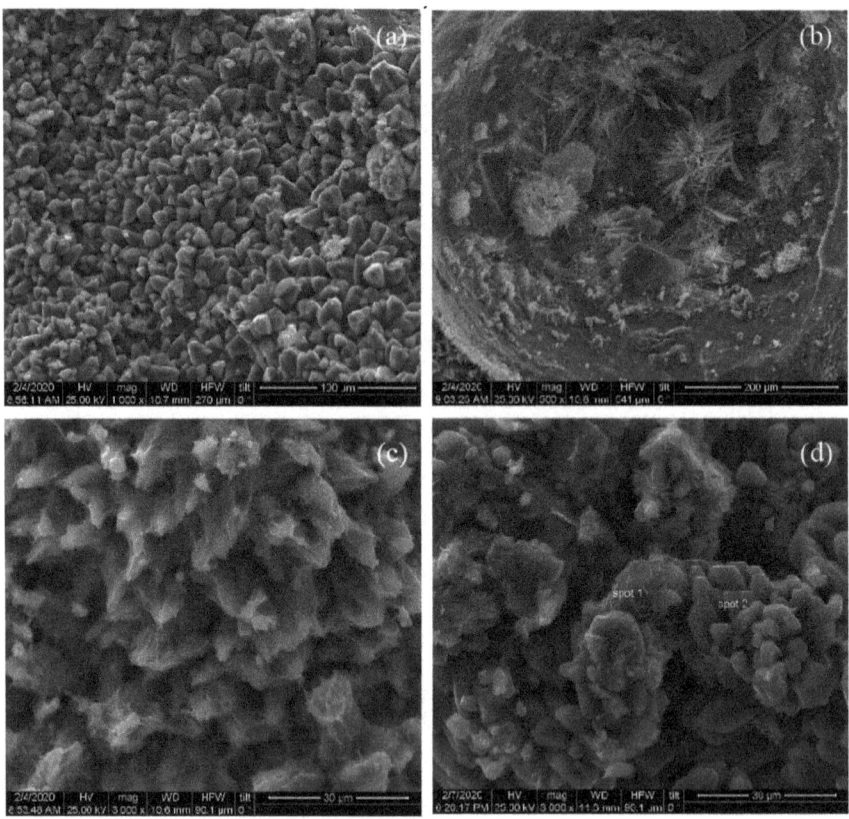

Fig. 10 Typical healing products observed in the crack face of self-healing specimens, Calcite (**a**), Ettringite and Portlandite (**b**) Hydraulic phases (**c**), Dolomite (sp. 1) and Calcite (sp 2) (**d**)

Table 3 EDS analysis of healing products in Reference sample

	(a)	(b)	(c)
MgO	n.d.	n.d.	n.d.
Al_2O_3	n.d.	12.6	2.3
SiO_2	1.9	5.3	14.9
SO_3	n.d.	27.3	n.d.
CaO	98.1	54.7	82.8
Total	100	100	100

development and propagation of the cracks that is uneven along the notch ligament. Therefore, in some cases the maximum crack width is larger than the CMOD target. The latter is also affected by the random distribution and orientation of the steel fibres, that resulted in different stress bridging capability of the fibres intersecting

Table 4 EDS analysis of healing products in Self-healing sample

	(a)	(b)	(c)	(d)
MgO	1.2	1.3	1.1	38.3
Al_2O_3	1.3	16.9	1.5	5.3
SiO_2	2.7	4.6	13.0	3.9
SO_3	n.d.	29.2	n.d.	n.d.
CaO	94.8	46.3	84.5	50.4
Total	100	100	100	100

the active crack surfaces. At the same time, the deformation behaviour of the specimens was also affected, resulting in an elastic behaviour of the active crack surfaces and thus, immediate crack closing during unloading. It is proposed that in case of steel fibre reinforced concrete, a cyclic loading splitting test might be preferable in order to achieve more consistent crack width between the examined specimens.

The microscopic observation of the crack faces in SEM revealed the presence of dolomitic healing products, which might have a positive effect on the self-healing properties of the concrete.

Regarding the water permeability experiment, in order to overcome the potential error of water absorption by the dry specimens, it is proposed that the water flow measurements could take place twice and consider only the second run for the interpretation purposes. Concerning the healing properties, the contribution of the expansive agents in the self-healing mechanism was not clear. Sealing efficiency and crack closing ratio were significantly affected by the inconsistency of the cracks' width. This highlights that the quantitative evaluation of the healing results should be always consider the initial level of damage attributed to the specimens.

Acknowledgements The authors would like to acknowledge the opportunity provided by EU COST Action CA 15202 "SARCOS" to participate in the ILT of WG2.

References

1. Ferrara, L.: Self-healing cement-based materials: an asset for sustainable construction industry. In: IOP Conference Series Materials Science and Engineering, vol. 1, p. 012007 (2018). https://doi.org/10.1088/1757-899x/442/1/012007
2. Souradeep, G., Kua, H.W.: Encapsulation technology and techniques in self-healing concrete. J. Mater. Civ. Eng. **28**(12), 04016165 (2016)
3. Xue, C., Li, W., Li, J., Tam, V.W.Y., Ye, G.: A review study on encapsulation-based self-healing for cementitious materials. Struct. Concr. **1**(20), 198–212 (2019)
4. De Belie, N., et al.: A review of self-healing concrete for damage management of structures. Adv. Mater. Interf. **5**, 1800074 (2018)
5. Amenta, M., Karatasios, I., Maravelaki, P., Kilikoglou, V.: Monitoring of self-healing phenomena towards enhanced sustainability of historic mortars. Appl. Phys. A **5**(122), 554 (2016)

6. Ferrara, L., et al.: Experimental characterization of the self-healing capacity of cement based materials and its effects on the material performance: a state of the art report by COST action SARCOS WG2. Constr. Build. Mater. **167**, 115–142 (2018)

7. Litina, C., Bumanis, G., Anglani, G., Dudek, M., Maddalena, R., Amenta, M., Al-tabbaa, A., et al.: Evaluation of methodologies for assessing self-healing performance of concrete with mineral expansive agents: An interlaboratory study. Materials **14**(8) (2021). https://doi.org/10.3390/ma14082024

8. Qureshi, T., Kanellopoulos, A., Al-Tabbaa, A.: Autogenous self-healing of cement with expansive minerals-I: impact in early age crack healing. Constr. Build. Mater. **192**, 768–784 (2018)

9. Qureshi, T., Al-Tabbaa, A., Self-healing of drying shrinkage cracks in cement-based materials incorporating reactive MgO. Smart Mater. Struct. **25**, 8 (2016)

10. Amenta, M., Metaxa, Z.S., Papaioannou, S., et al.: Quantitative evaluation of self-healingcapacity in cementitious materials. Mat. Des. Proc. Comm. **152** (2020). https://doi.org/10.1002/mdp2.152

11. Huang, H., Ye, G., Qian, C., Schlangen, E.: Self-healing in cementitious materials: materials, methods and service conditions. Mater. Des. **92**, 499–511 (2016)

Activated CFRP NSMR Ductile Strengthening System

Jacob Wittrup Schmidt, Christian Overgaard Christensen, Per Goltermann, and José Sena-Cruz

Abstract This paper presents some of the initial results from an ongoing research project which concerns a new NSMR strengthening method, expected to provide high ductility and strengthening effect as well as increased response consistency. A prestress of 1100 MPa activation was applied to the ductile NSMR system and compared to a reference beam. Ductility in this system is provided by a response controlling anchorage system, which interacts with the adhesively bonded NSMR. The proposed adhesive has a low E-modulus which is expected to reduce stress concentration issues of adhesively bonded joints when compared to stiffer epoxy adhesives are used, where IC-debonding seems to be the dominating failure mode. A beam deflection increase of approximately 105% has been observed when using the low E-modulus adhesive compared to the the epoxy adhesive. Change in the failure modes were furthermore observed, where the brittle de-bonding failure modes were mitigated thus extending the yielding regime with approximately 150% .

Keywords Strengthening · NSMR · CFRP · Activation · Ductility · Flexible adhesive · Ductility mechanism · Anchorage

1 Introduction

Near surface mounted reinforcement (NSMR) strengthening using pull-truded Carbon Fiber Reinforced Polymers (CFRP) has been used for upgrading of concrete structures for several decades [1–5]. It is generally seen, that such strengthening applications provide a significant strengthening effect to the concrete structure. However,

J. W. Schmidt (✉)
Department of Building Environment, University of Aalborg, 9220 Aalborg, Denmark
e-mail: jws@build.aau.dk

C. O. Christensen · P. Goltermann
Technical University of Denmark, 2800 Kongens, Lyngby, Denmark

J. Sena-Cruz
ISISE/IB-S, University of Minho, 4800-058 Guimarães, Portugal

© The Author(s), under exclusive license to Springer Nature Switzerland AG 2022   349
J. Sena-Cruz et al. (eds.), *Proceedings of the 3rd RILEM Spring Convention
and Conference (RSCC 2020)*, RILEM Bookseries 34,
https://doi.org/10.1007/978-3-030-76465-4_31

brittle failure modes and lack of collapse warning seems to cause reluctance towards the use of these systems. Intermediate crack debonding (IC-debonding), [6–8], end-peeling [9], and concrete cover separation [10] are some of the most common failure modes. However, several additional failure modes can occur when anchoring the NSMR CFRP material [11, 12]. Concrete delamination failure modes (IC-debonding, end peeling and concrete cover separation) leads to a failure state, where the CFRP material is not fully utilized. This seems undesirable, since significant expenses are related to the material cost and because more controlled failure mechanisms could reduce safety requirements (safety factors, durability factors, installation, etc.) related to such systems.

Anchorage of CFRP systems is recognized as a method, which provide increased utilization of the CFRP [13, 14]. These studies indicate that failure mechanisms may be controllable with regard to both ductility and strength. However, mechanical anchoring of the CFRP material often pose a great challenge due to the weak transverse properties and material brittleness. Crushing of the CFRP, soft slip, power slip, cutting of the fibers, bending of fibers, frontal overload, and fiber failure [15] are some of the mechanical anchor failure modes, which can result in a premature failure. Only fiber failure and power slip occur as high magnitude failure modes, which may meet the guideline requirements [16, 17].

Anchoring and activation of a NSMR CFRP strengthening system is deemed to be a method, which provide high utilization of the CFRP material, if activated before adhesively bonding to the structure. Such a study is presented in [3]. A good control of the failure development observed, where analytical and experimental results had good correlation. Fiber failure was reached for beams exposed to 70% activation indicating that the CFRP material was fully utilized. It is worth noticing that CFRP rupture was achieved at a strain of approximately 0.02 mm/mm and 3300 MPa. A high and stabile strengthening effect was seen from this study, but it still seems to be an open question if the strain magnitude of the CFRP is sufficient for providing an acceptable failure warning. Nevertheless, the obtained strengthening effect and ductility magnitude seems to be higher than seen from conventional freely anchored strengthening system. These results provided some of the first milestones in an ongoing research project at the Danish Technical University (DTU). The findings enabled the next research step, which is presented in this paper. The main research objective is to develop a generic NSMR CFRP system, which provides a high strengthening and ductility magnitude, similar or higher than seen from steel reinforcement.

This paper shows some of the initial results related to this research, where a novel ductile anchored NSMR CFRP strengthening system is used to strengthen beams identical to the beams tested in the prior study [3]. One of the main research questions in this study is: Can a ductile anchor zone combined with low E-modulus adhesive provide ductility and strengthening throughout the full yielding regime of the internal reinforcement or until concrete crushing occurs?

2 Activated Strengthening

2.1 RC Test Beam Geometry, Applied Loading and Material Properties

Figure 1 shows the T-section reinforced concrete (RC) beams with a length of 6.4 m on which the strengthening system was mounted. The beams were tested until failure, in a four-point-bending configuration with a deformation controlled loading of 2 mm/min. The beam had a support distance of 5 m and the loading was applied in two loading points, which were 0.7 m from the beam centerline. Figure 2 depicts the reinforcement configuration and cross section geometry.

It is seen that the longitudinal tensile reinforcement consisted of 2Ø25 deformed steel reinforcement bars, whereas the top reinforcement consists of longitudinal Ø10 bars and two reinforcement grids. The Ø6 stirrups were spaced by 100 mm with a double spacing after 1.4 m until the beam middle. From the material characterization the following values were obtained: steel reinforcement—average yield strength of 565 MPa and an E-modulus of approximately 200 GPa; mean compressive concrete strength of approximately 62 MPa (at the age of the tested beams). The single Ø8 CFRP rod, mounted in the ductile activated NSMR CFRP system, was positioned in a pre-cut longitudinal groove of 15 mm width and 27 mm depth, Fig. 2.

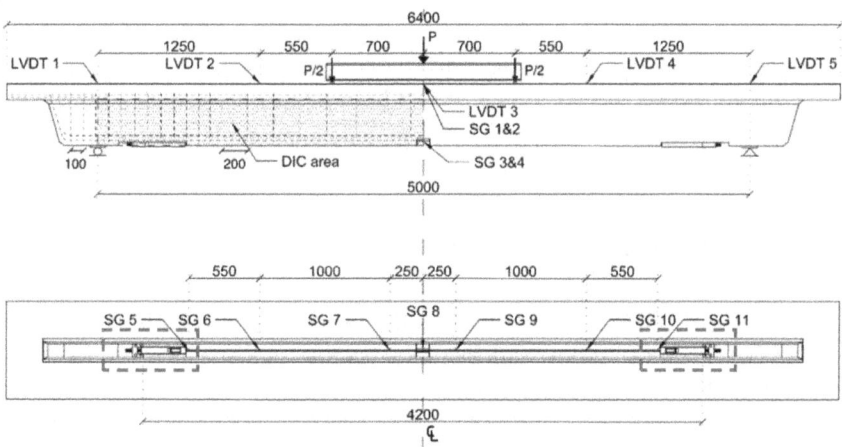

Fig. 1 Beam dimensions along with load setup and monitoring plan. Red retangular marked areas indicates the anchor points. *Note* units in [mm]

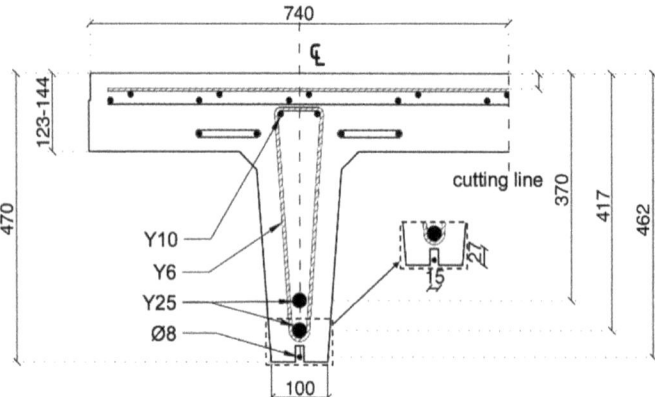

Fig. 2 Cross section and reinforcement configuration of the ANSMR RC beams. *Note* units in [mm]

3 Ductile Strengthening System

Two main mechanisms were hypothesized to have an effect on the ductility and strengthening magnitude: (i) a ductile anchorage mechanism and (ii) a low stiffness adhesive used in the CFRP concrete interface. A combination of these is hypothesized to prevent the brittle debonding failure mechanism seen from NSMR CFRP strengthening using an epoxy adhesive. Consequently, a novel ductile anchor was developed and tested in conjunction with different flexible adhesives. However, only one of these flexible adhesives is presented in this paper as example of the study presented.

3.1 Ductile Anchorage System

The novel ductile CFRP anchor system is depicted in Fig. 3 [18]. It is deemed to enable a fully tailored response with increased ductility in the anchor zone as well as high utilization of the CFRP. The assembled system consists of three main elements: (i) the anchor wedge, which anchor the CFRP rod, (ii) the anchor barrel, in which the pre-set wedge compresses and anchor the CFRP and, (iii) the ductile mechanism including a response control pin. Specifically for the tensile tests, a circular base plate provides anchor support.

Initially the anchor system was tested without the ductile mechanism system in order to provide a necessary capacity reference value to which the ductile mechanism could be designed. Verification of the full system response was performed by using the anchor system in one end of the ductile anchor system (CFRP anchor + ductile mechanism) applied to the other end. Figure 4 shows the un-deformed and deformed ductile mechanism (right) together with load-deformation test curve (left).

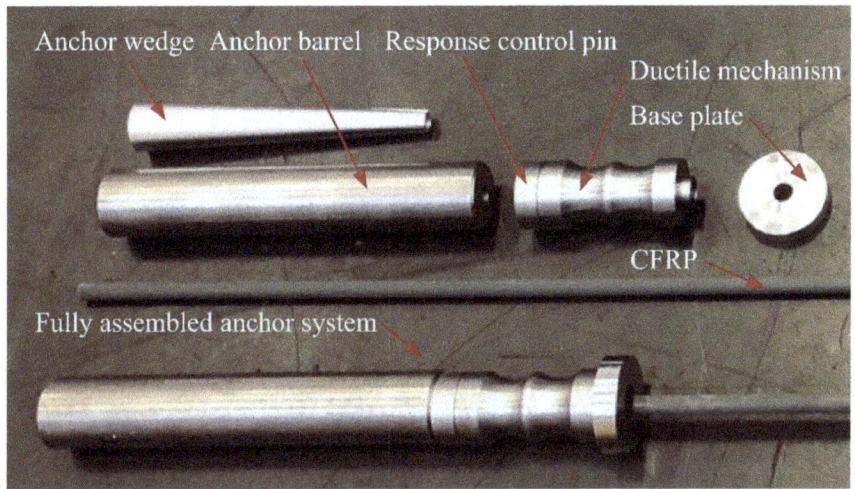

Fig. 3 Anchor parts and fully assembled ductile anchor [18]

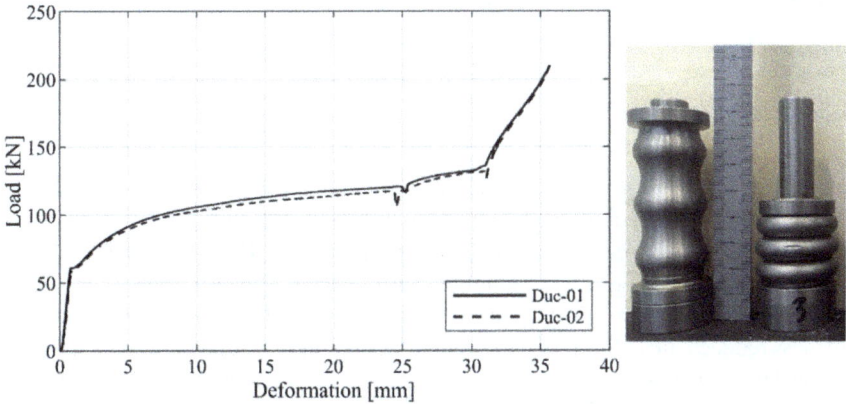

Fig. 4 Ductile mechanism load-deformation test curves (left) and ductile mechanism before and after testing (right)

The aim is a ductile mechanism, which can be tailored to different responses depending on the application. It is seen that the residual anchor capacity of a given anchor system (if any) is utilized when the ductile mechanism is fully collapsed. The high consistency between the tests should be noted since it affects the anchor response directly. The mechanism is designed to fit the CFRP anchor system. It is in these tests seen that yielding initiates at 60 kN where the ultimate capacity is reached at approximately 130 kN where a deformation of 30 mm is obtained. A strength increase throughout the yielding regime is observed, which is beneficial as

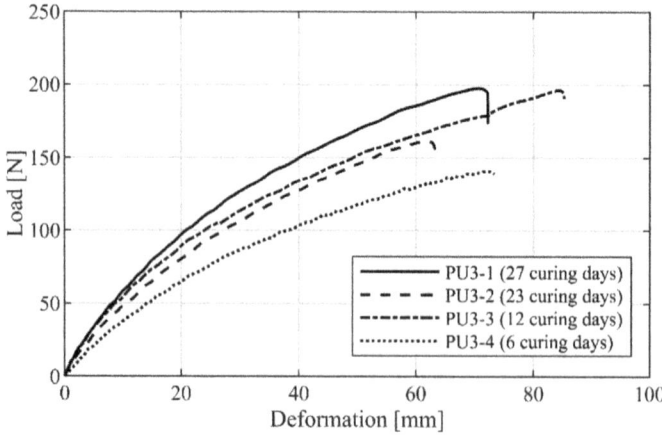

Fig. 5 Load/deformation response of flexible adhesives (left)

it allows redistribution of forces between the two anchors and thus more equalized deformations in the mechanisms when mounted on the beam.

3.2   Flexible Adhesive

In order to evaluate material properties and curing time sensitivity of the flexible low E-modulus adhesives, dogbone specimens were cast and tested in uniaxial tension.

Load-deformation curves related to one of the low E-modulus adhesive are presented in Fig. 5 along with the number of curing days. The tested adhesive appears to have a short initial linear response followed by a parabolic development, where the stiffness reduces. In addition, there seems to be a distinct relation between number of curing days and the strength/stiffness properties. A low E-modulus flexible adhesive, PU3 (Montage Ekstra 292), was chosen for this paper since it seemed to be a good reference when comparing with the other test results.

3.3   Mounting of the Ductile Anchored CFRP NSMR Strengthening System

Activation of the anchored CFRP NSMR system was done by the use of a novel anchor block system which consist of four main elements: (i) the anchor block, (ii) tensioning tray, (iii) ductile mechanism and (iv) the EW (Enclosed Wedge) CFRP anchor, Fig. 6.

The lower web surface of the T-section beam was prepared before mounting of the anchor block system. This preparation included cutting of the longitudinal

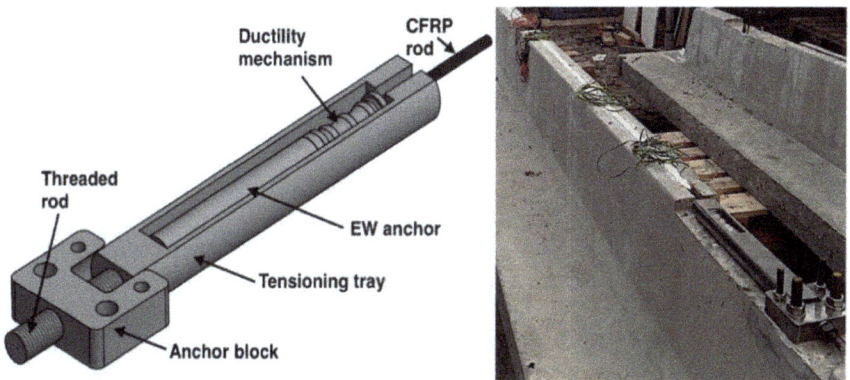

Fig. 6 Anchor block system components and installation

groove for the installation of the NSMR CFRP system, a cut-out for the anchor block system and drilling four holes for adhesively bonded threaded rods. Once prepared, the ductile strengthening system could be mounted through six steps: (i) the anchor block mounting on the beam web through the four threaded bars, (ii) installation of the EW anchor at each end of the CFRP rod with a pre-determined pre-setting force (approximately 130 kN), (iii) adhesive is injected into the groove bottom, (iv) ductile mechanism and EW anchor are positioned in the tensioning tray and inserted into the anchor block, (v) activation of the strengthening system using the threaded rod, and (vi) finalization of the adhesive groove injection layer.

4   Test Configurations and Results

Table 1 show some of the test configurations in the research program [19]. All activations of the strengthening system was done by using 50% of the guaranteed tensile capacity given by the manufacturer (Activation magnitude + beam dead load = approximately 1100 MPa). Two reference configurations were tested: REF-00, which is the non-activated reference beam without any strengthening system applied, whereas PCF-50 is 50% activation of the CFRP NSMR system using an epoxy

Table 1 Test configurations of T-beams

Beam-ID	Adhesive	Description
REF-00	–	Unstrengthened
PCF-50	Epoxy	Strengthened with 50% ANSMR
PU3-50	Montage Ekstra 292 (PU3)	Strengthened with 50% ANSMR

adhesive. PU3-50 is the beam strengthened with 50% activation in combination with the low E-modulus adhesive PU3 and the ductile anchor system.

4.1   Results

Figure 7 depict the load–deflection curves related to beam configuration REF-00, PCF-50 and PU3-50. It is seen from PCF-50, that a higher stiffness and crack initiation level is achieved in the linear elastic regime. A short yielding regime and high strengthening magnitude is furthermore obtained. Debonding of the CFRP NSMR occur at ultimate capacity. At this stage, the beam capacity is reduced to the level of REF-00.

However, when evaluating the failure mode from PU3-50, significantly larger ductility is achieved. This configuration shows increase of stiffness in the linear

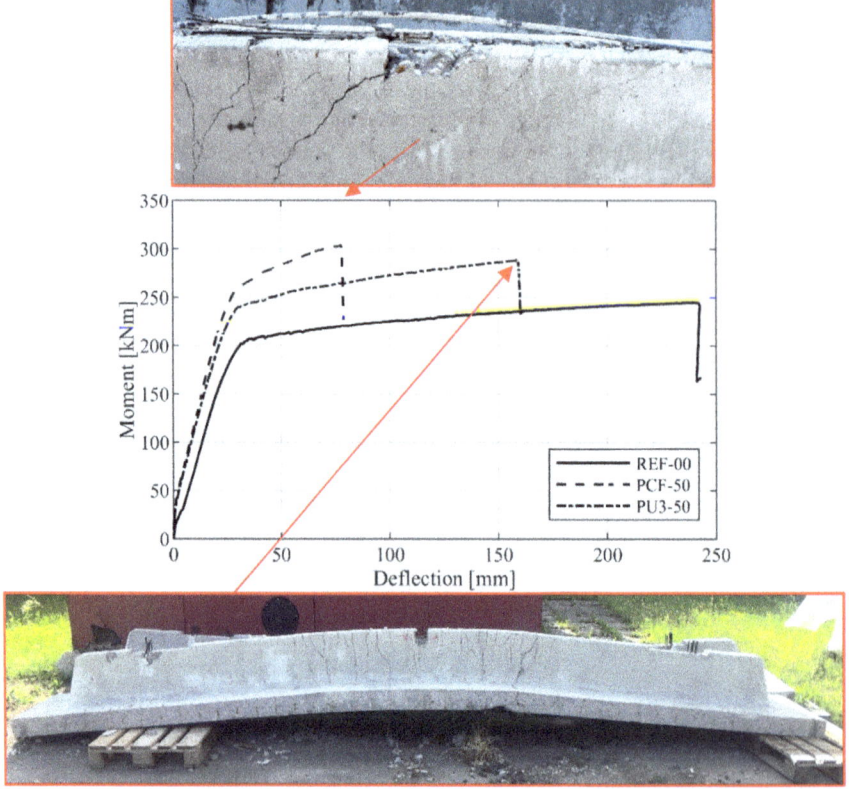

Fig. 7   Load–deflection curves for the presented beam (middle) along with *failure mode assessment of ANSMR beams with low E-modulus adhesive (bottom) and epoxy (top)*

elastic regime and a higher yielding initiation level compared to REF-00. Yielding initiates at a lower level than seen from PCF-50, which is due to the activation of the ductile mechanisms. It is seen, that the stiffness changes occur around 30–40 and 240 kN·m, due to concrete cracking and steel yielding, respectively. Significant yielding and a large deflection with termination in a concrete compression failure was achieved when testing this configuration.

PU3-50 (with the ductile anchor system and flexible adhesive) showed a significant increase in ductility compared to the epoxy configuration PCF-50. More than double deflection and approx. 150% extension of the yielding regime was achieved. The ductility increase was dedicated to the activation of the ductile mechanisms as well as a more even distribution of stresses in the flexible adhesive between CFRP and concrete. The results indicate that the brittle failure modes, usually obtained when applying CFRP strengthening, can be reduced or even prevented when using the ductile CFRP NSMR strengthening system. Finally, the failure threshold related to PU3-50 seems to be concrete crushing.

Figure 8 depicts moment-stress in the CFRP relationship, measured by representative strain gauges. For both configurations a high CFRP utilization is seen, which indicate tensile capacities, which meet the recommended values given by the manufacturer. The stiffness changes, seen from the moment-deflection curves, can also be identified on the CFRP stress development curves. Stress changes occur when the steel reinforcement yielding initiates as well as when transition from the un-cracked to a cracked regime initiates.

The optimal solution related to the ductile strengthening system seem to be located between two boundaries, fully bonded and fully un-bonded. This means that a balance between the ductile anchor response and adhesive stiffness must be considered in order to provide sufficient stress transfer in the adhesive bond between

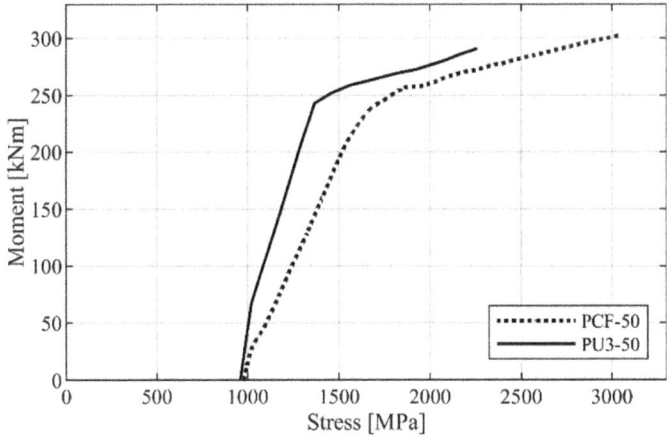

Fig. 8 Moment-stress in the CFRP relationship for representative beams

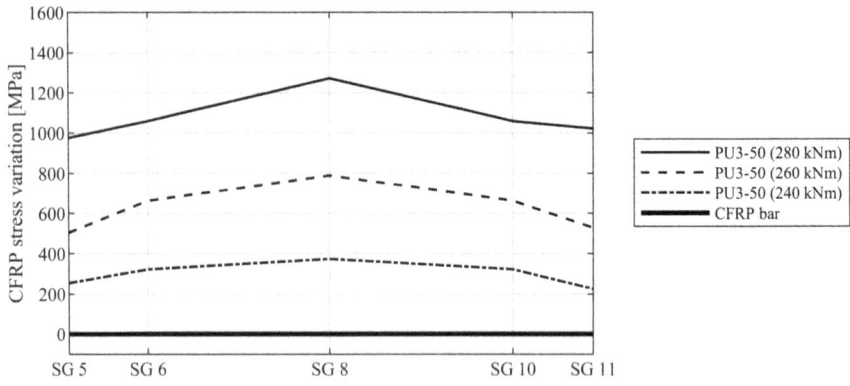

Fig. 9 Stress variation along CFRP bar for the different adhesives

CFRP and concrete. Simultaneously, sufficient activation of the ductile anchor should be addressed in order to control the strengthening system response sufficiently.

To further evaluate the stress variation along the CFRP rod, stress measurements at mid-span and at the anchor location are presented in Fig. 9 at different load levels for the PU3-50 beam. The activation level is subtracted from the measurements, thus the figure presents measured stress development along the CFRP NSMR reinforcement during testing. Three different load levels are considered: (i) Initiation of steel yielding (240 kN·m) and (ii) just before failure (280 kN·m) and (iii) in-between these values (260 kN·m). The mid-span CFRP stresses are clearly increased between the load levels, 240–280 kN·m compared to the stress levels measure at the anchor location. The PU3-50 provide stress transfer in the bond but ensures significant activation of the ductile anchors. A slight peak stress at the beam midspan is seen at 260–280 kN·m, whereas a more constant stress along the CFRP rod is seen at 240 kN·m. The measured stress difference between anchor points and midspan for PU3-50 is approximately 250 MPa at 260–280 kN·m.

Table 2 summarizes the test results from the three configurations. It is seen, that the strengthening configurations provide a strengthening magnitude of approximately 20%. However, the more brittle IC-debonding failure mode observed for the epoxy

Table 2 Test results and comparison

Configuration	Yielding [kNm]	Ultimate [kNm]	Capacity increase [%][a]	Deflection [mm]	Deflection increase [%][b]
REF-00	188.3	247.1	–	231.0	–
PCF-50	233.7	303.3	23	76.2	–
PU3-50	223.1	292.2	18	156.1	105

[a]Compared to the reference beam (REF-00)
[b]Compared to the epoxy strengthening configuration (PCF-50)

configuration (PCF-50) result in a deflection, which is approximately half of the deflection of the flexible adhesive configuration (PU3-50).

Consequently, the flexible adhesive bond provided significantly increased beam deflection and thus failure warning compared to the epoxy configuration. In addition, the activation magnitude did not seem to affect the ductility magnitude due to the ductile anchor system, which provide the desired failure mode control. Finally, the failure threshold related to PU3-50 seems to be concrete crushing when mounted on the given T-section concrete beams.

5 Conclusion

The indication research, presented in this paper concerns strengthening evaluation related to an Ø8 CFRP rod adhesively bonded in an activated NSMR system setup. Flexible adhesive and ductile anchor systems are deemed to be the main mechanisms to provide desired strength and ductility to such a strengthening system. The research addresses whether a ductile anchor zone combined with low E-modulus adhesive can provide ductility and strengthening throughout the full yielding regime of the internal reinforcement or until concrete crushing occurs. The current study consisted of a beam configuration with three beams: (i) REF-00, a non-activated reference beam without any strengthening system applied, (ii) a PCF-50, reference beam with 50% activation of the CFRP NSMR system and using an epoxy adhesive, and (iii) PU3-50, a beam strengthened with 50% activation in combination with a novel ductile anchor and low E-modulus adhesive PU3. Following conclusions can be drawn from the indication results:

- Significant strengthening (approx. 20%) of the linear elastic regime and the ultimate capacity was obtained when applying the PU3-50 test configuration;
- The low E-modulus adhesive enabled increased anchor activation and thereby activation of the ductility mechanism;
- PU3-50 showed increased beam ductility and failure warning where a 150% extension of the yielding regime was achieved compared to PCF-50;
- A beam deflection increase of more than 100% was reached compared to the PCF-50 thus insuring significant failure warning;
- Concrete compression failure was the governing failure mode of PU3-50, whereas IC-debonding was experienced for PCF-50;
- The PU3-50 configuration reached approximately 70% of the REF-00 deflection, where concrete crushing was experienced.

The results showed that ductility can be introduced into an otherwise linear elastic and brittle system by the use of the presented ductile strengthening system. Based on the design and geometry of the ductility mechanism, the initiation and progression of the yielding regime seems to be tailorable. More tests are however needed to fully verify the system response in connection with different applications. Furthermore, durability investigations and implementation into real life applications are some of the main objectives of the ongoing and future research.

Acknowledgements A sincere gratitude is addressed to S&P Denmark for supporting the ongoing research. Also a great thank you to Jorcks foundation for their acknowledgement and funding. Thank you to Perstrup Beton industri A/S. Furthermore the contributions from former students Frederik Alexander Meinzer Vind, Frederik Jensen, Nathalie Lückstädt Nielsen, Frits Eið, Frederik Munck, Matias Brix Mikkelsen, Klavs Foged Skovby and Nicolaj Jacob Birkebæk Thomsen is greatly appreciated. The last author acknowledge the grant SFRH/BSAB/150266/2019 provided by FCT financed by European Social Fund and by national funds through the FCT/MCTES, supporting his stay at DTU.

References

1. Rashid, R., Oehlers, D.J., Seracino, R.I.C.: Debonding of FRP NSM and EB retrofitted concrete: plate and cover interaction tests. J. Comp. for Const. **12**(2), 160–167 (2008)
2. El-Hacha, R., Rizkalla, S.H.: Near-surface-mounted fiber-reinforced polymer reinforcement for flexural strengthening of concrete structures. ACI Struct. J. **101**(5), 717–726 (2004)
3. Schmidt, J.W., Christensen, C.O., Goltermann, P., Hertz, K.D.: Shared CFRP activation anchoring method applied to NSMR strengthening of RC beams. Compos. Struct. **230**, 111487 (2019). https://doi.org/10.1016/j.compstruct.2019.111487
4. Sabau, C., Popescu, C., Sas, G., Schmidt, J.W., Blanksvärd, T., Täljsten, B.: Strengthening of RC beams using bottom and side NSM reinforcement. Comp. Part B **149**, 82–91 (2018)
5. Nordin, H., Täljsten, B.: Concrete beams strengthened with pre-stressed near surface mounted CFRP. J. Comp. Const. **10**(1), 60–68 (2006)
6. Smith, S.T.: Modeling debonding failure in FRP flexurally strengthened RC members using a local deformation model. J. Comp. Const. **11**(2), 184–191 (2007)
7. Teng, J.G.: Intermediate crack-induced debonding in RC beams and slabs. J. Constr. Build. Mater. **17**(6), 447–462 (2003)
8. Said, H., Wu, Z.: Evaluating and proposing models of predicting IC debonding failure. J. Comp. Const. **12**(3), 284–299 (2008)
9. Täljsten, B.: Strengthening of beams by plate bonding. J. Mater. Civil Eng. **9**(4), 206–212 (1997)
10. Corden, G., Ibell, T., Darby, A.: Concrete cover separation failure in near-surface mounted CFRP strengthened concrete structures. Struct. Eng. **86**(4), 19–21 (2008)
11. Cuntze, R.G., Freund, A.: The predictive capability of failure mode concept-based strength criteria for multidirectional laminates. J. Comp. Scien. Techn. **64**, 343–377 (2004)
12. Schmidt, J.W. Bennitz, A., Goltermann, P., Ravn, D.L.: External post-tensioning of CFRP tendons using integrated sleeve wedge anchorage. In: Proceedings of 6th International Conference on FRP Composites In Civil Engineering Rome, pp. 13–15 (2012)
13. Smith, S.T., Hua, S., Kima, S.J., Seracino, R.: FRP-strengthened RC slabs anchored with FRP anchors. Eng. Struct. **33**(4), 1075–1087 (2011)
14. Kalfat, R., Al-Mahaidi, R., Smith. S.T.: Anchorage devices used to improve the performance of reinforced concrete beams retrofitted with FRP composites: state-of-the-art review. J. Comp. Const. **17**(1), 14–33 (2013)
15. Schmidt J.W.: External strengthening of building structures with prestressed CFRP. Ph.D. report. Kgs. Lyngby, Denmark: Technical University of Denmark (DTU) (2011)
16. PTI, Acceptance Standards for Post-Tensioning Systems, Post-Tensioning Institute (1998)
17. EOTA, Post-Tensioning Kits for Prestressing of Structures - Guideline for European technical approval, EOTA (2002)

18. Schmidt, J.W., Christensen, C.O., Goltermann, P.: Ductile response controlled EW CFRP anchor system. Compos. Part B Eng. **201**, 108371 (2020)
19. Schmidt, J.W., Christensen, C.O., Goltermann, P., Sena-Cruz, J.: Activated ductile cfrp nsmr strengthening. Materials **14**, 2821 (2021). https://doi.org/10.3390/ma14112821

Lightning Source UK Ltd.
Milton Keynes UK
UKHW020606180722
406005UK00002B/18

9 783030 764678